Alexandrov Geometry

Foundations

GRADUATE STUDIES IN MATHEMATICS 236

Alexandrov Geometry

Foundations

Stephanie Alexander
Vitali Kapovitch
Anton Petrunin

AMERICAN MATHEMATICAL SOCIETY
Providence, Rhode Island

2020 *Mathematics Subject Classification.* Primary 53C23, 53C45, 30L99.

For additional information and updates on this book, visit
www.ams.org/bookpages/gsm-236

Library of Congress Cataloging-in-Publication Data

Names: Alexander, Stephanie, 1941– author. | Kapovitch, Vitali, 1969– author. | Petrunin, Anton, 1968– author.
Title: Alexandrov geometry : foundations / Stephanie Alexander, Vitali Kapovitch, Anton Petrunin.
Description: Providence, Rhode Island : American Mathematical Society, [2023] | Series: Graduate studies in mathematics, 1065-7339 ; Volume 236 | Includes bibliographical references and index.
Identifiers: LCCN 2023028258 | ISBN 9781470473020 (hardcover) | ISBN 9781470475369 (paperback) | ISBN 9781470475352 (ebook)
Subjects: LCSH: Metric spaces. | Global differential geometry. | Analytic spaces. | AMS: Differential geometry – Global differential geometry – Global geometric and topological methods (à la Gromov); differential geometric analysis on metric spaces. | Differential geometry – Global differential geometry – Global surface theory (convex surfaces & à la A. D. Aleksandrov). | Functions of a complex variable – Analysis on metric spaces – None of the above, but in this section.
Classification: LCC QA611.28 .A35 2023 | DDC 514/.32–dc23/eng/20230927
LC record available at https://lccn.loc.gov/2023028258

♾ The paper used in this book is acid-free and falls within the guidelines
established to ensure permanence and durability.
Visit the AMS home page at `https://www.ams.org/`

10 9 8 7 6 5 4 3 2 1 29 28 27 26 25 24

In loving memory of
Stephanie Alexander

Contents

Preface

Alexandrov spaces are defined via axioms similar to those given by Euclid. The Alexandrov axioms replace certain equalities with inequalities. Depending on the signs of the inequalities, we obtain Alexandrov spaces with *curvature bounded above* (CBA) and *curvature bounded below* (CBB). The definitions of the two classes of spaces are similar, but their properties and known applications are quite different.

The goal of this book is to give a comprehensive exposition of the structure theory of Alexandrov spaces with curvature bounded above and below. It includes all the basic material as well as selected topics inspired by considering the two contexts simultaneously. We only consider the intrinsic theory, leaving applications aside. Our presentation is linear, with a few exceptions where topics are deferred to later chapters to streamline the exposition. This book includes material *up to the definition of dimension*. Another volume still in preparation will cover further topics.

Brief history

The first synthetic description of curvature is due to Abraham Wald [**159**]; it was given in a lone publication on a "coordinateless description of Gauss surfaces" published in 1936. In 1941, similar definitions were rediscovered by Alexandr Alexandrov [**16**].

In Alexandrov's work the first fruitful applications of this approach were given. Mainly, *Alexandrov's embedding theorem* [**11, 12**], which describes closed convex surfaces in Euclidean 3-space, and the *gluing theorem* [**13**], which gave a flexible tool to modify nonnegatively curved metrics on a sphere. These two results together gave an intuitive geometric tool to study embeddings and bending of surfaces in Euclidean space and changed the subject dramatically. They formed the foundation of the branch of geometry now called *Alexandrov geometry*.

Curvature bounded below. The theory grew out of studying intrinsic and extrinsic geometry of convex surfaces without the smoothness condition. It was developed by Alexandr Alexandrov and his school. Here is a very incomplete list of contributors to the subject: Yuriy Borisov, Yuriy Burago, Boris Dekster, Iosif Liberman, Sergey Olovyanishnikov, Aleksey Pogorelov, Yuriy Reshetnyak, Yuriy Volkov, Viktor Zalgaller.

The first result in higher dimensional Alexandrov spaces was the splitting theorem. It was proved by Anatoliy Milka [**118**] and appeared in 1967. Milka used a global definition similar to the one used in this book.

In the 1980s the interest in convergence of Riemannian manifolds spurred by *Gromov's compactness theorem* [**71**] turned attention toward the singular spaces that can occur as limits of Riemannian manifolds. Immediately it was recognized that if the manifolds have a uniform lower sectional curvature bound, then the limit spaces have a lower curvature bound in the sense of Alexandrov. There followed during the 1990s an explosion of work on intrinsic theory of Alexandrov spaces starting with papers of Yuriy Burago, Grigory Perelman, and Michael Gromov [**44, 127**]. Similar ideas were developed independently by Karsten Grove and Peter Petersen, whose work was not converted into a publication, and also by Conrad Plaut [**137**].

Around the same time an implicit application of higher-dimensional Alexandrov geometry was given by Michael Gromov in his bound on Betti numbers [**75**]. Another implicit application, which essentially used Alexandrov geometry before it was was actually introduced, given later by Wu-Yi Hsiang and Bruce Kleiner in their paper on nonnegatively curved 4-manifolds with infinite symmetry groups [**83**]. The work of Hsiang and Kleiner and its extension by Karsten Grove and Burkhard Wilking [**78**] are some of the most beautiful applications of this branch of Alexandrov geometry.

The above activity was very much related to so-called *comparison geometry*, a branch of differential geometry that compares Riemannian manifolds to spaces of constant curvature. In addition to the already mentioned Gromov's compactness theorem, the following results had a big influence on the development of Alexandrov geometry: *Toponogov comparison theorem* [**156**], which is a generalization of the theorem of Alexandrov [**14**]; *Toponogov splitting theorem* [**156**], which is a generalization of Cohn-Vossen's theorem [**55**]; *Finiteness theorems* of Cheeger and Grove–Petersen [**52, 77**]; Gromov's bound on the number of generators of the fundamental group [**73**, 1.5]; and the *Yamaguchi fibration theorem* [**162**].

Let us give a list of available introductory texts on Alexandrov spaces with curvature bounded below:

- The first introduction to Alexandrov geometry is given in the original paper of Yuriy Burago, Michael Gromov, and Grigory Perelman [**44**] and its extension [**127**] written by Perelman.
- A brief and reader-friendly introduction was written by Katsuhiro Shiohama [**150**, Sections 1–8].
- Another reader-friendly introduction, written by Dmiti Burago, Yuriy Burago, and Sergei Ivanov, is given in [**37**, Chapter 10].
- The book of Athanase Papadopoulos [**124**] gives a comprehensive introduction to nonpositive curvature in the sense of Busemann.

In addition, let us mention two surveys, one by Conrad Plaut [**139**] and the other by the third author [**132**].

Curvature bounded above. The study of spaces with curvature bounded above started later, inspired by analogy with the theory of curvature bounded below. The first paper on the subject was written by Alexandrov [**18**], appearing in 1951. An analogous weaker definition was considered earlier by Herbert Busemann [**45**].

Contributions to the subject were made by Valerii Berestovskii, Arne Beurling, Igor Nikolaev, Dmitry Sokolov, Yuriy Reshetnyak, Samuel Shefel; this list is not complete as well. The most fundamental results were obtained by Yuriy Reshetnyak. They include his *majorization theorem* and *gluing theorem*. The gluing theorem states that if two nonpositively curved spaces have isometric convex sets, then the space obtained by gluing these sets along an isometry is also nonpositively curved.

The development of Alexandrov geometry was greatly influenced by the *Hadamard–Cartan theorem*. Its original formulation states that the exponential map at any point of a complete Riemannian manifold with nonpositive sectional curvature is a covering. In particular, it implies that the universal cover is diffeomorphic to Euclidean space of the same dimension. See further discussion below 9.65.

An influential implicit application of Alexandrov spaces with curvature bounded above can be seen in *Euclidean buildings*, introduced by Jacques Tits as a means to study algebraic groups.

Here is a list of available texts covering the basics of Alexandrov spaces with curvature bounded above:

- The book of Martin Bridson and André Haefliger [**34**] gives the most comprehensive introduction available today.
- The lecture notes of Werner Ballmann [**21**, **22**] include a brief and clear introduction.
- The volume [**37**, Chapter 9] gives another reader-friendly introduction by Yuriy Burago, Dmitry Burago, and Sergei Ivanov.
- The book [**10**] by the three authors of the present volume gives an introduction aiming at reaching interesting applications and theorems with a minimum of preparation.
- The book of Jürgen Jost [**90**] gives a more analytic viewpoint to the subject.

One of the most striking applications of CAT(0) spaces was given by Dmitry Burago, Sergei Ferleger, and Alexey Kononenko [**38**], who used them to study *billiards*; this idea was developed further in [**39–43**]. Another beautiful application is the construction of *exotic aspherical manifolds* by Michael Davis [**59**]; related results are surveyed in [**50**, **60**]. Both of these topics are discussed in [**10**]. The study of group actions on CAT(0) spaces and CAT(0) cube complexes played a key role in the proof of the *virtually fibered conjecture* that a finite cover of every closed hyperbolic 3-manifold fibers over the circle.

Satellites and successors

Surfaces with *bounded integral curvature* were studied by Alexandrov's school. An excellent book on the subject was written by Alexandr Alexandrov and Viktor Zalgaller [**15**]; see also a more up-to-date survey by Yuriy Reshetnyak [**141**].

Spaces with *two-sided bounded curvature* is another subject already studied by Alexandrov's school; a good survey is written by Valerij Berestovskij and Igor Nikolaev [**24**].

A spin-off of the idea of synthetically defining upper curvature bounds was given by Michael Gromov [**76**]. He defined so-called *δ-hyperbolic spaces*, which satisfy a coarse version of the negative curvature condition, applying in particular to discrete metric spaces. This notion and its various generalizations such as semihyperbolicity (a coarse version of nonpositive curvature) and relative hyperbolicity have led to the emergence of the subject of *geometric group theory*, which relates geometric properties of groups to their algebraic ones. This is a well-developed subject with a large number of subfields and applications, such as the theory of small cancellation groups, automatic groups, mapping class groups, automorphisms of free groups, isoperimetric inequalities on groups, actions on $\mathbb{R}$-trees, and Gromov's boundaries of groups.

The so-called *curvature dimension condition* introduced by John Lott, Cédric Villani, and Karl-Theodor Sturm gives a synthetic description of Ricci curvature bounded below; see the book of Villani [**158**] and references therein. A striking application of this theory to geodesic flow in CBB spaces was found recently by Elia Bruè, Andrea Mondino, and Daniele Semola [**35**].

Alexandrov geometry influenced the development of *analysis on metric spaces*. An excellent book on the subject was written by Juha Heinonen, Pekka Koskela, Nageswari Shanmugalingam, and Jeremy Tyson [**82**].

Acknowledgments

We thank Semyon Alesker, Valerii Berestovskii, I. David Berg, Richard Bishop, Yuriy Burago, Alexander Christie, Nicola Gigli, Sergei Ivanov, Bernd Kirchheim, Bruce Kleiner, Rostislav Matveyev, John Lott, Greg Kuperberg, Nikolai Kosovsky, Nina Lebedeva, Wilderich Tuschmann, and Sergio Zamora Barrera. A special thanks to Alexander Lytchak, who contributed deeply to this book during the long process of writing, but refused to be one of us.

We thank the mathematical institutions where we worked on this book: BIRS, MFO, Henri Poincaré Institute, IHES, University of Cologne, Max Planck Institute for Mathematics, EIMI.

During the long writing of this book, we were partially supported by the following grants: Stephanie Alexander—Simons Foundation grant 209053; Vitali Kapovitch—NSF grant DMS-0204187, NSERC Discovery grants, and Simons Foundation grant 390117; Anton Petrunin—NSF grants DMS-0406482, DMS-0905138, DMS-1309340, DMS-2005279, Simons Foundation grants 245094 and 584781, and Minobrnauki, grant 075-15-2022-289.

Part 1

Preliminaries

Chapter 1

Model plane

A. Trigonometry

Given a real number κ, the *model κ-plane* will be a complete simply connected two-dimensional Riemannian manifold of constant curvature κ.

The model κ-plane will be denoted by $\mathbb{M}^2(\kappa)$.

- If $\kappa > 0$, $\mathbb{M}^2(\kappa)$ is isometric to a sphere of radius $\frac{1}{\sqrt{\kappa}}$; the unit sphere $\mathbb{M}^2(1)$ will be also denoted by $\mathbb{S}^2$.
- If $\kappa = 0$, $\mathbb{M}^2(\kappa)$ is the Euclidean plane, which is also denoted by $\mathbb{E}^2$.
- If $\kappa < 0$, $\mathbb{M}^2(\kappa)$ is the Lobachevsky plane with curvature κ.

Let $\varpi^\kappa = \operatorname{diam} \mathbb{M}^2(\kappa)$, so $\varpi^\kappa = \infty$ if $\kappa \leqslant 0$ and $\varpi^\kappa = \pi/\sqrt{\kappa}$ if $\kappa > 0$; ϖ is just a cursive form of π.

The distance between points $x, y \in \mathbb{M}^2(\kappa)$ will be denoted by $|x - y|$, and $[xy]$ will denote the geodesic segment connecting x and y. The segment $[xy]$ is uniquely defined for $\kappa \leqslant 0$ and for $\kappa > 0$ it is defined uniquely if $|x - y| < \varpi^\kappa = \pi/\sqrt{\kappa}$.

A triangle in $\mathbb{M}^2(\kappa)$ with vertices x, y, z will be denoted by $[xyz]$. Formally, a triangle is an ordered set of its sides, so $[xyz]$ is just a short notation for the triple $([yz], [zx], [xy])$.

The angle of $[xyz]$ at x will be denoted by $\measuredangle\left[x\,{}^{y}_{z}\right]$.

By $\tilde{\triangle}^\kappa\{a, b, c\}$ we denote a triangle in $\mathbb{M}^2(\kappa)$ with side lengths a, b, c, so $[xyz] = \tilde{\triangle}^\kappa\{a, b, c\}$ means that $x, y, z \in \mathbb{M}^2(\kappa)$ are such that

$$|x - y| = c, \quad |y - z| = a, \quad |z - x| = b.$$

For $\tilde\triangle^\kappa\{a,b,c\}$ to be defined, the sides a, b, c must satisfy the triangle inequality. If $\kappa > 0$, we require in addition that $a+b+c < 2\cdot\varpi^\kappa$; otherwise $\tilde\triangle^\kappa\{a,b,c\}$ is considered to be undefined.

Trigonometric functions. We will need three *trigonometric functions* in $\mathbb{M}^2(\kappa)$: *cosine*, *sine*, and *modified distance*, denoted by cs^κ, sn^κ, and md^κ, respectively.

They are defined as the solutions of the following initial value problems respectively:

$$\begin{cases} x''+\kappa\cdot x=0,\\ x(0)=1,\\ x'(0)=0. \end{cases} \quad \begin{cases} y''+\kappa\cdot y=0,\\ y(0)=0,\\ y'(0)=1. \end{cases} \quad \begin{cases} z''+\kappa\cdot z=1,\\ z(0)=0,\\ z'(0)=0. \end{cases}$$

Namely, we set $\operatorname{cs}^\kappa(t)=x(t)$, $\operatorname{sn}^\kappa(t)=y(t)$, and

$$\operatorname{md}^\kappa(t)=\begin{cases} z(t) & \text{if } 0\leqslant t\leqslant\varpi^\kappa,\\ \frac{2}{\kappa} & \text{if } t>\varpi^\kappa. \end{cases}$$

Here are the tables which relate our trigonometric functions to the standard ones, where we take $\kappa > 0$:

$$\operatorname{sn}^{\pm\kappa}=\tfrac{1}{\sqrt{\kappa}}\cdot\operatorname{sn}^{\pm1}(x\cdot\sqrt{\kappa});\quad \operatorname{cs}^{\pm\kappa}=\operatorname{cs}^{\pm1}(x\cdot\sqrt{\kappa});\quad \operatorname{md}^{\pm\kappa}=\tfrac{1}{\kappa}\cdot\operatorname{md}^{\pm1}(x\cdot\sqrt{\kappa});$$

$$\operatorname{sn}^{-1}x=\sinh x;\quad \operatorname{cs}^{-1}x=\cosh x;\quad \operatorname{md}^{-1}x=\cosh x-1;$$

$$\operatorname{sn}^{0}x=x;\quad \operatorname{cs}^{0}x=1;\quad \operatorname{md}^{0}x=\tfrac{1}{2}\cdot x^2;$$

$$\operatorname{sn}^{1}x=\sin x;\quad \operatorname{cs}^{1}x=\cos x;\quad \operatorname{md}^{1}x=\begin{cases} 1-\cos x & \text{for } x\leqslant\pi,\\ 2 & \text{for } x>\pi. \end{cases}$$

Note that

$$\operatorname{md}^\kappa(x)=\int\limits_0^x \operatorname{sn}^\kappa(t)\cdot \boldsymbol{d}t \qquad \text{for} \qquad x\leqslant\varpi^\kappa.$$

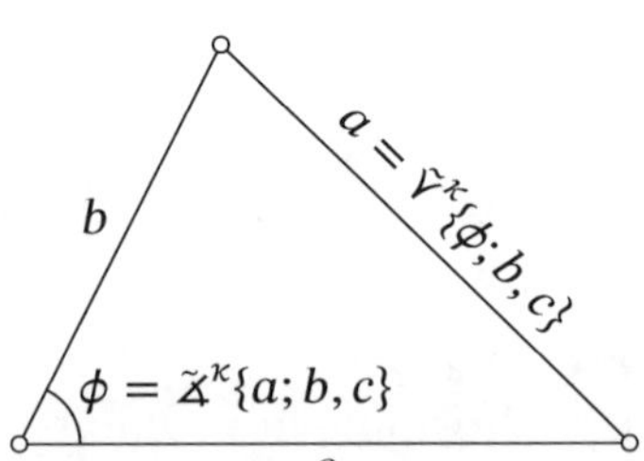

Let ϕ be the angle of $\tilde\triangle^\kappa\{a,b,c\}$ opposite to a. In this case, we will write

$$a=\tilde{\curlyvee}^\kappa\{\phi;b,c\} \quad\text{or}\quad \phi=\tilde\measuredangle^\kappa\{a;b,c\}.$$

The functions $\tilde{\curlyvee}^\kappa$ and $\tilde\measuredangle^\kappa$ will be called the *model side* and the *model angle*, respectively. Let

$$\tilde{\curlyvee}^\kappa\{\phi;b,-c\}=\tilde{\curlyvee}^\kappa\{\phi;-b,c\}:=\tilde{\curlyvee}^\kappa\{\pi-\phi;b,c\};$$

in this way we define $\tilde{\curlyvee}^\kappa\{\phi;b,c\}$ when one of the numbers b and c is negative.

1.1. Properties of standard functions.

(a) *For fixed a and ϕ, the function* $y(t) = \operatorname{md}^\kappa\left(\tilde{\gamma}^\kappa\{\phi; a, t\}\right)$ *satisfies the differential equation*

$$y'' + \kappa\cdot y = 1.$$

(b) *Let* $\alpha\colon [a, b] \to \mathbb{M}^2(\kappa)$ *be a unit-speed geodesic, and let A be the image of a complete geodesic. If $f(t)$ is the distance from $\alpha(t)$ to A, the function* $y(t) = \operatorname{sn}^\kappa(f(t))$ *satisfies the differential equation*

$$y'' + \kappa\cdot y = 0$$

for $y \neq 0$.

(c) *For fixed κ, b, and c, the function*

$$a \mapsto \tilde{\measuredangle}^\kappa\{a; b, c\}$$

is increasing and defined on a real interval. Equivalently, the function

$$\phi \mapsto \tilde{\gamma}^\kappa\{\phi; b, c\}$$

is increasing and defined if $b, c < \varpi^\kappa$, *and* $\phi \in [0, \pi]$. *(Formally speaking, if* $\kappa > 0$ *and* $b + c \geqslant \varpi^\kappa$, *it is defined only for* $\phi \in [0, \pi)$, *but* $\tilde{\gamma}^\kappa\{\phi; b, c\}$ *can be extended to* $[0, \pi]$ *as a continuous function.)*

(d) *For fixed ϕ, a, b, c, the function*

$$\kappa \mapsto \tilde{\measuredangle}^\kappa\{a; b, c\} \quad \textit{and} \quad \kappa \mapsto \tilde{\gamma}^\kappa\{\phi; b, c\}$$

are nondecreasing (in fact, increasing, if $|b - c| < a < b + c$*) and nonincreasing (in fact, increasing, if* $0 < \phi < \pi$*), respectively.*

(e) Alexandrov's lemma. *Assume that for real numbers a, b, a', b', x, and κ, the following two expressions are defined:*

$$\tilde{\measuredangle}^\kappa\{a; b, x\} + \tilde{\measuredangle}^\kappa\{a'; b', x\} - \pi, \quad \tilde{\measuredangle}^\kappa\{a'; b + b', a\} - \tilde{\measuredangle}^\kappa\{x; a, b\},$$

Then they have the same sign.

All the properties except (e), Alexandrov's lemma, can be shown by direct calculation. Alexandrov's lemma is reformulated in 6.3 and is proved there.

Cosine law. The formulas $a = \tilde{\gamma}^\kappa\{\phi; b, c\}$ and $\phi = \tilde{\measuredangle}^\kappa\{a; b, c\}$ can be rewritten using the cosine law in $\mathbb{M}^2(\kappa)$:

$$\cos\phi = \begin{cases} \dfrac{b^2 + c^2 - a^2}{2\cdot b\cdot c} & \text{if } \kappa = 0, \\[2ex] \dfrac{\operatorname{cs}^\kappa a - \operatorname{cs}^\kappa b\cdot \operatorname{cs}^\kappa c}{\kappa\cdot \operatorname{sn}^\kappa b\cdot \operatorname{sn}^\kappa c} & \text{if } \kappa \neq 0. \end{cases}$$

However, rather than using these explicit formulas, we mainly will use the properties of $\tilde{\measuredangle}^\kappa$ and $\tilde{\gamma}^\kappa$ listed in 1.1.

B. Hemisphere lemma

1.2. Hemisphere lemma. *For $\kappa > 0$, any closed path of length less than $2\cdot\varpi^\kappa$ (respectively, at most $2\cdot\varpi^\kappa$) in $\mathbb{M}^2(\kappa)$ lies in an open (respectively, closed) hemisphere.*

Proof. Applying rescaling, we may assume that $\kappa = 1$, and thus $\varpi^\kappa = \pi$ and $\mathbb{M}^2(\kappa) = \mathbb{S}^2$. Let α be a closed curve in $\mathbb{S}^2$ of length $2\cdot\ell$.

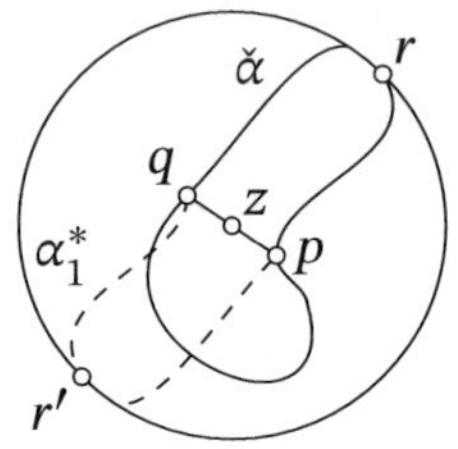

Assume $\ell < \pi$. Let $\check\alpha$ be a subarc of α of length ℓ, with endpoints p and q. Since $|p-q| \leqslant \ell < \pi$, there is a unique geodesic $[pq]$ in $\mathbb{S}^2$. Let z be the midpoint of $[pq]$. We claim that α lies in the open hemisphere centered at z. If not, α intersects the boundary great circle of this hemisphere; let r be a point in the intersection. Without loss of generality, we may assume that $r \in \check\alpha$. The arc $\check\alpha$ together with its reflection in z form a closed curve of length $2\cdot\ell$ that contains r and its antipodal point r'. Thus

$$\ell = \operatorname{length}\check\alpha \geqslant |r - r'| = \pi,$$

a contradiction.

If $\ell = \pi$, then either α is a local geodesic and, hence, a great circle, or α may be strictly shortened by substituting a geodesic arc for a subarc of α whose endpoints p^1, p^2 are arbitrarily close to a point p on α. In both cases α lies in a closed hemisphere; the former case is trivial, and in the latter case, α lies in a closed hemisphere obtained as a limit of closures of open hemispheres containing the shortened curves as p^1, p^2 approach p. □

1.3. Exercise. Give a proof of 1.2 based on Crofton's formula.

Chapter 2

Metric spaces

In this chapter we fix conventions and notations. We are assuming that the reader is familiar with basic notions in metric geometry.

A. Metrics and their relatives

Definitions. Let $\mathbb{I}$ be a subinterval of $[0,\infty]$. A function ρ defined on $\mathcal{X}\times\mathcal{X}$ is called an $\mathbb{I}$-valued metric if the following conditions hold:

- $\rho(x,x)=0$ for any x;
- $\rho(x,y)\in\mathbb{I}$ for any pair $x\neq y$;
- $\rho(x,y)+\rho(x,z)\geqslant\rho(y,z)$ for any triple of points x, y, z.

The value $\rho(x,y)$ is also called the *distance* between x and y.

The above definition will be used for four choices of the interval $\mathbb{I}$: $(0,\infty)$, $(0,\infty]$, $[0,\infty)$, and $[0,\infty]$. Any $\mathbb{I}$-valued metric can be referred to briefly as a metric; the interval should be apparent from context but by default, a metric is $(0,\infty)$-valued. If we need to be more specific, we may also use the following names:

- a $(0,\infty)$-valued metric may be called a *genuine metric.*
- a $(0,\infty]$-valued metric may be called an ∞*-metric.*
- a $[0,\infty)$-valued metric may be called a *genuine pseudometric.*
- a $[0,\infty]$-valued metric may be called a *pseudometric* or an ∞*-pseudometric.*

A metric space is a set equipped with a metric. The distance between points x and y in a metric space $\mathcal{X}$ will usually be denoted by

$$|x-y| \quad \text{or} \quad |x-y|_{\mathcal{X}};$$

the latter will be used if we need to emphasize that we are working in the space $\mathcal{X}$.

The function $\operatorname{dist}_x : \mathcal{X} \to \mathbb{R}$ defined as

$$\operatorname{dist}_x : y \mapsto |x-y|$$

will be called the *distance function* from x.

Any subset A in a metric space $\mathcal{X}$ will be also considered as a *subspace*; that is, a metric space with the metric defined by restricting the metric of $\mathcal{X}$ to $A \times A \subset \mathcal{X} \times \mathcal{X}$.

The *direct product* $\mathcal{X} \times \mathcal{Y}$ of two metric spaces $\mathcal{X}$ and $\mathcal{Y}$ is defined as the metric space carrying the metric

$$|(p,\phi)-(q,\psi)| = \sqrt{|p-q|^2 + |\phi-\psi|^2}$$

for $p, q \in \mathcal{X}$ and $\phi, \psi \in \mathcal{Y}$.

A map between two metric spaces is called an *isometry* if it is a bijection and preserves distances between points.

Zero and infinity. Genuine metric spaces are the main objects of study in this book. However, the generalizations above are useful in various definitions and constructions. For example, the construction of length metric (see Section 2.C) uses infinite distances. The following definition gives another example.

2.1. Definition. Assume $\{\mathcal{X}_\alpha\}_{\alpha\in\mathcal{A}}$ is a collection of ∞-metric spaces. The disjoint union

$$\boldsymbol{X} = \bigsqcup_{\alpha\in\mathcal{A}} \mathcal{X}_\alpha$$

has a natural ∞-metric on it defined as follows: given two points $x \in \mathcal{X}_\alpha$ and $y \in \mathcal{X}_\beta$, let

$$\begin{aligned} |x-y|_{\boldsymbol{X}} &= \infty & &\text{if } \alpha \neq \beta, \\ |x-y|_{\boldsymbol{X}} &= |x-y|_{\mathcal{X}_\alpha} & &\text{if } \alpha = \beta. \end{aligned}$$

The resulting ∞-metric space $\boldsymbol{X}$ will be called the *disjoint union* of $\{\mathcal{X}_\alpha\}_{\alpha\in\mathcal{A}}$, denoted by

$$\bigsqcup_{\alpha\in\mathcal{A}} \mathcal{X}_\alpha.$$

Now let us give examples showing that vanishing and infinite distance between distinct points can appear naturally and useful in constructions.

Suppose a set $\mathcal{X}$ comes with a set of metrics $|*-*|_\alpha$ for $\alpha \in \mathcal{A}$. Then

$$|x-y| = \sup\{\,|x-y|_\alpha : \alpha \in \mathcal{A}\,\}$$

is in general only an ∞-metric; that is, even if the metrics $|*-*|_\alpha$ are genuine, then $|*-*|$ might be $(0,\infty]$-valued.

Let $\mathcal{X}$ be a set, let $\mathcal{Y}$ be a metric space, and let $\Phi : \mathcal{X} \to \mathcal{Y}$ be a map. If Φ is not injective, then the *pullback*

$$|x-y|_{\mathcal{X}} = |\Phi(x) - \Phi(y)|_{\mathcal{Y}}$$

defines only a pseudometric on $\mathcal{X}$.

Corresponding metric space and metric component. The following two observations show that nearly any question about metric spaces can be reduced to a question about genuine metric spaces.

Assume $\mathcal{X}$ is a pseudometric space. Set $x \sim y$ if $|x-y| = 0$. Note that if $x \sim x'$, then $|y-x| = |y-x'|$ for any $y \in \mathcal{X}$. Thus, $|*-*|$ defines a metric on the quotient set $\mathcal{X}/\sim$. This way we obtain a metric space $\mathcal{X}'$. The space $\mathcal{X}'$ is called the *corresponding metric space* for the pseudometric space $\mathcal{X}$. Often we do not distinguish between $\mathcal{X}'$ and $\mathcal{X}$.

Set $x \approx y$ if and only if $|x-y| < \infty$; this is another equivalence relation on $\mathcal{X}$. The equivalence class of a point $x \in \mathcal{X}$ will be called the *metric component* of x; it will be denoted by $\mathcal{X}_x$. One could think of $\mathcal{X}_x$ as $\mathrm{B}(x,\infty)_{\mathcal{X}}$, the open ball centered at x and radius ∞ in $\mathcal{X}$; see definition below.

It follows that any ∞-metric space is a *disjoint union* of genuine metric spaces, the metric components of the original ∞-metric space; see Definition 2.1.

To summarize this discussion: given a $[0,\infty]$-valued metric space $\mathcal{X}$, we may pass to the corresponding $(0,\infty]$-valued metric space $\mathcal{X}'$ and break the latter into a disjoint union of metric components, each of which is a genuine metric space.

B. Notations

Balls. Given $R \in [0,\infty]$ and a point x in a metric space $\mathcal{X}$, the sets

$$\mathrm{B}(x,R) = \{\, y \in \mathcal{X} : |x-y| < R \,\},$$
$$\overline{\mathrm{B}}[x,R] = \{\, y \in \mathcal{X} : |x-y| \leqslant R \,\}$$

are called the *open* and the *closed balls* of radius R with center x, respectively.

If we need to emphasize that these balls are taken in the space $\mathcal{X}$, we write

$$\mathrm{B}(x,R)_{\mathcal{X}} \quad \text{and} \quad \overline{\mathrm{B}}[x,R]_{\mathcal{X}},$$

respectively.

Since in the model space $\mathbb{M}^m(\kappa)$ all balls of the same radius are isometric, often we will not need to specify the center of the ball, and may write

$$\mathrm{B}(R)_{\mathbb{M}^m(\kappa)} \quad \text{and} \quad \overline{\mathrm{B}}[R]_{\mathbb{M}^m(\kappa)},$$

respectively.

A set $A \subset \mathcal{X}$ is called *bounded* if $A \subset \mathrm{B}(x, R)$ for some $x \in \mathcal{X}$ and $R < \infty$.

Distances to sets. For subset $A \subset \mathcal{X}$, let us denote the distance from A to a point x in $\mathcal{X}$ by $\operatorname{dist}_A x$; that is,

$$\operatorname{dist}_A x := \inf\{\, |a - x| \,:\, a \in A \,\}.$$

For any subset $A \subset \mathcal{X}$, the sets

$$\mathrm{B}(A, R) = \{\, y \in \mathcal{X} \,:\, \operatorname{dist}_A y < R \,\},$$
$$\overline{\mathrm{B}}[A, R] = \{\, y \in \mathcal{X} \,:\, \operatorname{dist}_A y \leqslant R \,\}$$

are called the *open* and *closed R-neighborhoods* of A, respectively.

Diameter, radius, and packing. Let $\mathcal{X}$ be a metric space. Then the *diameter* of $\mathcal{X}$ is defined as

$$\operatorname{diam} \mathcal{X} = \sup\{\, |x - y| \,:\, x, y \in \mathcal{X} \,\}.$$

The *radius* of $\mathcal{X}$ is defined as

$$\operatorname{rad} \mathcal{X} = \inf\{\, R > 0 \,:\, \mathrm{B}(x, R) = \mathcal{X} \text{ for some } x \in \mathcal{X} \,\}.$$

The *ε-pack* of $\mathcal{X}$ (or *packing number*) is the maximal number (possibly infinite) of points in $\mathcal{X}$ at distance $> \varepsilon$ from each other; it is denoted by $\operatorname{pack}_\varepsilon \mathcal{X}$. If $m = \operatorname{pack}_\varepsilon \mathcal{X} < \infty$, then a set $\{x^1, x^2, \dots, x^m\}$ in $\mathcal{X}$ such that $|x^i - x^j| > \varepsilon$ is called a *maximal ε-packing* in $\mathcal{X}$.

G-delta sets. Recall that an arbitrary union of open balls in a metric space is called an *open set*. A subset of a metric space is called a *G-delta set* if it can be presented as an intersection of a countable number of open subsets.

2.2. Baire's theorem. *Let $\mathcal{X}$ be a complete metric space, and let $\{\Omega_n\}$, $n \in \mathbb{N}$, be a collection of open dense subsets of $\mathcal{X}$. Then $\bigcap_{n\in\mathbb{N}} \Omega_n$ is dense in $\mathcal{X}$.*

Proper spaces. A metric space $\mathcal{X}$ is called *proper* if all closed bounded sets in $\mathcal{X}$ are compact. This condition is equivalent to each of the following statements:

(1) For some (and therefore any) point $p \in \mathcal{X}$ and any $R < \infty$, the closed ball $\overline{\mathrm{B}}[p, R] \subset \mathcal{X}$ is compact.

(2) The function $\operatorname{dist}_p : \mathcal{X} \to \mathbb{R}$ is proper for some (and therefore any) point $p \in \mathcal{X}$.

We will also often use the following two classical statements:

2.3. Proposition. *Proper metric spaces are separable and second countable.*

2.4. Proposition. *Let $\mathcal{X}$ be a metric space. Then the following are equivalent*

(a) *$\mathcal{X}$ is compact;*

(b) *$\mathcal{X}$ is* sequentially compact; *that is, any sequence of points in $\mathcal{X}$ contains a convergent subsequence;*

(c) *$\mathcal{X}$ is complete and for any $\varepsilon > 0$ there is a* finite ε-net *in $\mathcal{X}$; that is, there is a finite collection of points $p_1, \dots, p_N$ such that $\bigcup_i \mathrm{B}(p_i, \varepsilon) = \mathcal{X}$.*

(d) *$\mathcal{X}$ is complete and for any $\varepsilon > 0$ there is a* compact ε-net *in $\mathcal{X}$; that is, $\mathrm{B}(K, \varepsilon) = \mathcal{X}$ for a compact set $K \subset \mathcal{X}$.*

C. Length spaces

A *curve* in a metric space $\mathcal{X}$ is a continuous map $\alpha\colon \mathbb{I} \to \mathcal{X}$, where $\mathbb{I}$ is a *real interval* (that is, an arbitrary convex subset of $\mathbb{R}$).

2.5. Definition. Let $\mathcal{X}$ be a metric space. Given a curve $\alpha\colon \mathbb{I} \to \mathcal{X}$, we define its *length* as

$$\operatorname{length}\alpha := \sup\left\{\sum_{i\geqslant 1} |\alpha(t_i) - \alpha(t_{i-1})| \ :\ t_0, \dots, t_n \in \mathbb{I}, t_0 \leqslant \dots \leqslant t_n\right\}.$$

The following lemma is an easy exercise.

2.6. Lower semicontinuity of length. *Assume $\alpha_n\colon \mathbb{I} \to \mathcal{X}$ is a sequence of curves that converges pointwise to a curve $\alpha_\infty\colon \mathbb{I} \to \mathcal{X}$. Then*

$$\operatorname{length}\alpha_\infty \leqslant \varliminf_{n\to\infty} \operatorname{length}\alpha_n.$$

Given two points x and y in a metric space $\mathcal{X}$, consider the value

$$\|x - y\| = \inf_{\alpha}\{\operatorname{length}\alpha\},$$

where infimum is taken for all paths α from x to y.

It is easy to see that $\|{*} - {*}\|$ defines a $(0, \infty]$-valued metric on $\mathcal{X}$; it will be called the *induced length metric* on $\mathcal{X}$. Clearly

$$\|x - y\| \geqslant |x - y|$$

for any $x, y \in \mathcal{X}$.

It easily follows from the definition that the length of a curve α with respect to $\|{*}-{*}\|$ is equal to the length of α with respect to $|{*}-{*}|$. In particular, iterating the construction produces the same metric $\|{*} - {*}\|$.

2.7. Definition. If $\|x-y\| = |x-y|$ for any pair of points x, y in a metric space $\mathcal{X}$, then $|{*}-{*}|$ is called *length metric*, and $\mathcal{X}$ is called a *length space*.

In other words, a metric space $\mathcal{X}$ is a *length space* if for any $\varepsilon > 0$ and any two points $x, y \in \mathcal{X}$ with $|x-y| < \infty$ there is a path $\alpha\colon [0,1] \to \mathcal{X}$ connecting x to y such that

$$\operatorname{length}\alpha < |x-y| + \varepsilon.$$

In this book, most of the time we consider length spaces. If $\mathcal{X}$ is a length space and $A \subset \mathcal{X}$, then the set A comes with the inherited metric from $\mathcal{X}$, which might be not a length metric. The corresponding length metric on A will be denoted by $|{*}-{*}|_A$.

Variations of the definition. We will need the following variations of Definition 2.7:

- Assume $R > 0$. If $\|x-y\| = |x-y|$ for any pair $|x-y| < R$, then $\mathcal{X}$ is called an *R-length space*.
- If any point in $\mathcal{X}$ admits a neighborhood Ω such that $\|x-y\| = |x-y|$ for any pair of points $x, y \in \Omega$, then $\mathcal{X}$ is called a *locally length space*.
- A metric space is called *geodesic* if for any two points x, y with $|x-y| < < \infty$ there is a geodesic $[xy]$ in $\mathcal{X}$.
- Assume $R > 0$. A metric space is called *R-geodesic* if for any two points x, y such that $|x-y| < R$ there is a geodesic $[xy]$ in $\mathcal{X}$.

Note that the notions of ∞-length spaces and length spaces are the same. Clearly, any geodesic space is a length space and any R-geodesic space is R-length.

2.8. Example. Consider a metric space $\mathcal{X}$ obtained by gluing a countable collection of disjoint intervals $\mathbb{I}_n$ of length $1+\frac{1}{n}$ where for each $\mathbb{I}_n$ one end is glued to p and the other to q. Then $\mathcal{X}$ carries a natural complete length metric such that $|p-q| = 1$, but there is no geodesic connecting p to q.

2.9. Exercise. Let $\mathcal{X}$ be a metric space, and let $\|{*}-{*}\|$ be the length metric on it. Show the following:

(a) If $\mathcal{X}$ is complete, then $(\mathcal{X}, \|{*}-{*}\|)$ is complete.

(b) If $\mathcal{X}$ is compact, then $(\mathcal{X}, \|{*}-{*}\|)$ is geodesic.

2.10. Exercise. Give an example of a complete length space such that no pair of distinct points can be joined by a geodesic.

2.11. Exercise. Let $\mathcal{X}$ be a complete length space. Show that for any compact subset K in $\mathcal{X}$ there is a compact path-connected subset K' that contains K.

2.12. Definition. Consider two points x and y in a metric space $\mathcal{X}$.

(i) A point $z \in \mathcal{X}$ is called a *midpoint* between x and y if

$$|x - z| = |y - z| = \tfrac{1}{2} \cdot |x - y|.$$

(ii) Assume $\varepsilon \geqslant 0$. A point $z \in \mathcal{X}$ is called an *ε-midpoint* between x and y if

$$|x - z| \leqslant \tfrac{1}{2} \cdot |x - y| + \varepsilon \quad \text{and} \quad |y - z| \leqslant \tfrac{1}{2} \cdot |x - y| + \varepsilon.$$

Note that a 0-midpoint is the same as a midpoint.

The following lemma was essentially proved by Karl Menger [**116**, Section 6].

2.13. Lemma. *Let $\mathcal{X}$ be a complete metric space.*

(a) *Assume that for any pair of points $x, y \in \mathcal{X}$, and any $\varepsilon > 0$, there is an ε-midpoint z. Then $\mathcal{X}$ is a length space.*

(b) *Assume that for any pair of points $x, y \in \mathcal{X}$, there is a midpoint z. Then $\mathcal{X}$ is a geodesic space.*

(c) *If for some $R > 0$, the assumptions* (a) *or* (b) *hold only for pairs of points $x, y \in \mathcal{X}$ such that $|x - y| < R$, then $\mathcal{X}$ is an R-length or an R-geodesic space, respectively.*

Proof. Fix a pair of points $x, y \in \mathcal{X}$. Let $\varepsilon_n = \frac{\varepsilon}{2^{2 \cdot n}}$.

Set $\alpha(0) = x$, and $\alpha(1) = y$. Let $\alpha(\frac{1}{2})$ be an ε_1-midpoint between $\alpha(0)$ and $\alpha(1)$. Further, let $\alpha(\frac{1}{4})$ and $\alpha(\frac{3}{4})$ be ε_2-midpoints between the pairs $(\alpha(0), \alpha(\frac{1}{2}))$ and $(\alpha(\frac{1}{2}), \alpha(1))$, respectively. Applying the above procedure recursively, on the nth step we define $\alpha(\frac{k}{2^n})$, for every odd integer k such that $0 < \frac{k}{2^n} < 1$, to be an ε_n-midpoint between the already defined $\alpha(\frac{k-1}{2^n})$ and $\alpha(\frac{k+1}{2^n})$.

This way we define $\alpha(t)$ for all dyadic rationals t in $[0, 1]$. If $t \in [0, 1]$ is not a dyadic rational, consider a sequence of dyadic rationals $t_n \to t$ as $n \to \infty$. By completeness of $\mathcal{X}$, the sequence $\alpha(t_n)$ converges; let $\alpha(t)$ be its limit. It is easy to see that $\alpha(t)$ does not depend on the choice of the sequence t_n, and $\alpha \colon [0, 1] \to \mathcal{X}$ is a path from x to y. Moreover,

$$\begin{aligned} \operatorname{length} \alpha &\leqslant |x - y| + \sum_{n=1}^{\infty} 2^{n-1} \cdot \varepsilon_n \leqslant \\ &\leqslant |x - y| + \tfrac{\varepsilon}{2}. \end{aligned} \tag{1}$$

Since $\varepsilon > 0$ is arbitrary, we have (a).

To prove (b), one should repeat the same argument taking midpoints instead of ε_n-midpoints. In this case, (1) holds for $\varepsilon_n = \varepsilon = 0$.

The proof of (c) is obtained by a straightforward modification of the proofs above. □

2.14. Exercise. Suppose that for any two distinct points x and y in a complete metric space $\mathcal{X}$ there is yet another point z such that

$$|x-z|+|y-z|=|x-y|.$$

Show that $\mathcal{X}$ is geodesic.

Since in a compact set a sequence of $\frac{1}{n}$-midpoints (z_n) contains a convergent subsequence, Lemma 2.13 implies the following.

2.15. Corollary. *A proper length space is geodesic.*

2.16. Hopf–Rinow theorem. *Any complete, locally compact length space is proper.*

Proof. Let $\mathcal{X}$ be a locally compact length space. Given $x\in\mathcal{X}$, denote by $\rho(x)$ the supremum of all $R>0$ such that the closed ball $\overline{\mathrm{B}}[x,R]$ is compact. Since $\mathcal{X}$ is locally compact,

(2) $$\rho(x)>0 \quad \text{for any } x\in\mathcal{X}.$$

It is sufficient to show that $\rho(x)=\infty$ for some (and therefore any) point $x\in\mathcal{X}$.

Assume the contrary; that is, $\rho(x)<\infty$.

(3) $B=\overline{\mathrm{B}}[x,\rho(x)]$ *is compact for any* x.

Indeed, $\mathcal{X}$ is a length space; therefore for any $\varepsilon>0$, the set $\overline{\mathrm{B}}[x,\rho(x)-\varepsilon]$ is a compact ε-net in B. Since B is closed and hence complete, it is compact by Proposition 2.4. △

(4) $|\rho(x)-\rho(y)|\leqslant|x-y|_{\mathcal{X}}$ *for any* $x,y\in\mathcal{X}$*; in particular* $\rho\colon\mathcal{X}\to\mathbb{R}$ *is a continuous function.*

Indeed, assume the contrary; that is, $\rho(x)+|x-y|<\rho(y)$ for $x,y\in\mathcal{X}$. Then $\overline{\mathrm{B}}[x,\rho(x)+\varepsilon]$ is a closed subset of $\overline{\mathrm{B}}[y,\rho(y)]$ for $\varepsilon>0$. Then compactness of $\overline{\mathrm{B}}[y,\rho(y)]$ implies compactness of $\overline{\mathrm{B}}[x,\rho(x)+\varepsilon]$, a contradiction. △

Set $\varepsilon=\min_{y\in B}\{\rho(y)\}$; the minimum is defined since B is compact. From (2), we have $\varepsilon>0$.

Choose a finite $\frac{\varepsilon}{10}$-net $\{a_1,a_2,\dots,a_n\}$ in B. The union W of the closed balls $\overline{\mathrm{B}}[a_i,\varepsilon]$ is compact. Clearly $\overline{\mathrm{B}}[x,\rho(x)+\frac{\varepsilon}{10}]\subset W$. Therefore $\overline{\mathrm{B}}[x,\rho(x)+\frac{\varepsilon}{10}]$ is compact, a contradiction. □

2.17. Exercise. Construct a geodesic space that is locally compact, but whose completion is neither geodesic nor locally compact.

D. Convex sets

2.18. Definition. Let $\mathcal{X}$ be a geodesic space and let $A \subset \mathcal{X}$.

A is *convex* if for every two points $p, q \in A$ any geodesic $[pq]$ lies in A.

A is *weakly convex* if for every two points $p, q \in A$ there is a geodesic $[pq]$ that lies in A.

We say that A is *totally convex* if for every two points $p, q \in A$, every local geodesic from p to q lies in A.

If for some $R \in (0, \infty]$ these definitions are applied only for pairs of points such that $|p - q| < R$ and only for the geodesics of length $< R$, then A is called *R-convex*, *weakly R-convex*, or *totally R-convex*, respectively.

A set $A \subset \mathcal{X}$ is called *locally convex* if every point $a \in A$ admits an open neighborhood $\Omega \ni a$ such that for every two points $p, q \in A \cap \Omega$ every geodesic $[pq] \subset \Omega$ lies in A. Similarly one defines locally weakly convex and locally totally convex sets.

Remarks. Let us state a few observations that easily follow from the definition.

- The notion of (*weakly*) *convex set* is the same as (*weakly*) ∞-*convex set*.
- The inherited metric on a weakly convex set coincides with its length metric.
- Any open set is locally convex by definition.

The following proposition states that weak convexity survives under ultralimits. An analogous statement about convexity does not hold; for example, there is a sequence of convex discs in $\mathbb{S}^2$ that converges to a hemisphere, which is not convex.

2.19. Proposition. *Let $\mathcal{X}_n$, be a sequence of geodesic spaces. Let ω be an ultrafilter on $\mathbb{N}$ (see Definition 4.1). Assume that $A_n \subset \mathcal{X}_n$ is a sequence of weakly convex sets, $\mathcal{X}_n \to \mathcal{X}_\omega$, and $A_n \to A_\omega \subset \mathcal{X}_\omega$ as $n \to \omega$. Then A_ω is a weakly convex set of $\mathcal{X}_\omega$.*

Proof. Fix $x_\omega, y_\omega \in A_\omega$. Consider sequences $x_n, y_n \in A_n$ such that $x_n \to x_\omega$ and $y_n \to y_\omega$ as $n \to \omega$.

Denote by α_n a geodesic path from x_n to y_n that lies in A_n. Let

$$\alpha_\omega(t) = \lim_{n \to \omega} \alpha_n(t).$$

It remains to observe that α_ω is a geodesic path that lies in A_ω. □

E. Quotient spaces

Quotient spaces. Assume $\mathcal{X}$ is a metric space with an equivalence relation $\sim$. Note that given a family of pseudometrics ρ_α on $\mathcal{X}/\sim$, their least upper bound

$$\rho(x,y) = \sup_\alpha\{\rho_\alpha(x,y)\}$$

is also a pseudometric. If the projections $\mathcal{X} \to (\mathcal{X}/\sim, \rho_\alpha)$ are *short* (that is, *distance nonincreasing*), then so is $\mathcal{X} \to (\mathcal{X}/\sim, \rho)$.

It follows that the quotient space $\mathcal{X}/\sim$ admits a natural quotient pseudometric; this is the maximal pseudometric on $\mathcal{X}/\sim$ that makes the quotient map $\mathcal{X} \to \mathcal{X}/\sim$ short. The corresponding metric space will be also denoted as $\mathcal{X}/\sim$ and will be called the *quotient space* of $\mathcal{X}$ by the equivalence relation $\sim$.

In general, the points of the metric space $\mathcal{X}/\sim$ are formed by equivalence classes in $\mathcal{X}$ for a wider equivalence relation. However, in most of the cases we will consider, the set of equivalence classes will coincide with the set of points in the metric space $\mathcal{X}/\sim$.

2.20. Proposition. *Let $\mathcal{X}$ be a length space, and let $\sim$ be an equivalence relation on $\mathcal{X}$. Then $\mathcal{X}/\sim$ is a length space.*

Proof. Let $\mathcal{Y}$ be an arbitrary metric space. Since $\mathcal{X}$ is a length space, a map $f\colon \mathcal{X} \to \mathcal{Y}$ is short if and only if

$$\operatorname{length}(f \circ \alpha) \leqslant \operatorname{length} \alpha$$

for any curve $\alpha\colon \mathbb{I} \to \mathcal{X}$. Denote by $\|* - *\|$ the length metric on $\mathcal{Y}$. It follows that if $f\colon \mathcal{X} \to \mathcal{Y}$ is short, then so is $f\colon \mathcal{X} \to (\mathcal{Y}, \|* - *\|)$.

Consider the quotient map $f\colon \mathcal{X} \to \mathcal{X}/\sim$. Recall that the space $\mathcal{X}/\sim$ is defined by the maximal pseudometric that makes f short.

Denoting by $\|* - *\|$ the length metric on $\mathcal{X}/\sim$, it follows that

$$f\colon \mathcal{X} \to (\mathcal{X}/\sim, \|* - *\|)$$

is also short.

Note that

$$\|x - y\| \geqslant |x - y|_{\mathcal{X}/\sim}$$

for any $x, y \in \mathcal{X}/\sim$. From maximality of $|* - *|_{\mathcal{X}/\sim}$, we get

$$\|x - y\| = |x - y|_{\mathcal{X}/\sim}$$

for any $x, y \in \mathcal{X}/\sim$; that is, $\mathcal{X}/\sim$ is a length space. □

Group actions. Assume a group G acts on a metric space $\mathcal{X}$. Consider a relation $\sim$ on $\mathcal{X}$ defined by $x \sim y$ if there is $g \in G$ such that $x = g\cdot y$. Note that $\sim$ is an equivalence relation.

In this case, the quotient space $\mathcal{X}/\sim$ will also be denoted by $\mathcal{X}/G$, and can be regarded as the space of G-orbits in $\mathcal{X}$.

Assume that a group G acts on $\mathcal{X}$ by isometries. Then the distance between orbits $G\cdot x$ and $G\cdot y$ in $\mathcal{X}/G$ can be defined directly:

$$|G\cdot x - G\cdot y|_{\mathcal{X}/G} = \inf\{\, |x - g\cdot y|_{\mathcal{X}} = |g^{-1}\cdot x - y|_{\mathcal{X}} : g \in G\,\}.$$

If the G-orbits are closed, then $|G\cdot x - G\cdot y|_{\mathcal{X}/G} = 0$ if and only if $G\cdot x = G\cdot y$. In this case, the quotient space $\mathcal{X}/G$ is a genuine metric space.

The following proposition follows from the definition of a quotient space:

2.21. Proposition. *Assume $\mathcal{X}$ is a metric space and a group G acts on $\mathcal{X}$ by isometries. Then the projection $\pi\colon \mathcal{X} \to \mathcal{X}/G$ is a* submetry*; that is, $\pi(\mathrm{B}(p,r)) = \mathrm{B}(\pi(p),r)$ for any $p \in \mathcal{X}, r > 0$ (see Definition* 3.7*).*

F. Gluing and doubling

Gluing. Recall that the disjoint union of metric spaces can be also considered as a metric space; see Definition 2.1. Therefore the quotient space construction works as well for an equivalence relation on the disjoint union of metric spaces.

Consider two metric spaces $\mathcal{X}_1$ and $\mathcal{X}_2$ with subsets $A_1 \subset \mathcal{X}_1$ and $A_2 \subset \mathcal{X}_2$, and a bijection $\phi\colon A_1 \to A_2$. Consider the minimal equivalence relation on $\mathcal{X}_1 \sqcup \mathcal{X}_2$ such that $a \sim \phi(a)$ for any $a \in A_1$. In this case, the corresponding quotient space $(\mathcal{X}_1 \sqcup \mathcal{X}_2)/\sim$ will be called the *gluing of $\mathcal{X}$ and $\mathcal{Y}$ along ϕ* and is denoted by

$$\mathcal{X}_1 \sqcup_\phi \mathcal{X}_2.$$

Note that if the map $\phi\colon A_1 \to A_2$ is distance-preserving, then the inclusions $\iota_i\colon \mathcal{X}_i \to \mathcal{X}_1 \sqcup_\phi \mathcal{X}_2$ are also distance-preserving, and

$$|\iota_1(x_1) - \iota_2(x_2)|_{\mathcal{X}_1 \sqcup_\phi \mathcal{X}_2} = \inf_{a_2=\phi(a_1)} \{\, |x_1 - a_1|_{\mathcal{X}_1} + |x_2 - a_2|_{\mathcal{X}_2}\,\}$$

for any $x_1 \in \mathcal{X}_1$ and $x_2 \in \mathcal{X}_2$.

Doubling. Let $\mathcal{V}$ be a metric space, and let $A \subset \mathcal{V}$ be a closed subset. A metric space $\mathcal{W}$ glued from two copies of $\mathcal{V}$ along A is called the *doubling of $\mathcal{V}$ in A*.

The space $\mathcal{W}$ is completely described by the following properties:

- The space $\mathcal{W}$ contains $\mathcal{V}$ as a subspace; in particular the set A can be treated as a subset of $\mathcal{W}$.
- There is an isometric involution of $\mathcal{W}$ which is called a *reflection in A*; it will be denoted by $x \mapsto x'$.

- For any $x \in \mathcal{W}$ we have $x \in \mathcal{V}$ or $x' \in \mathcal{V}$, and
$$|x' - y|_{\mathcal{W}} = |x - y'|_{\mathcal{W}} = \inf_{a \in A}\{|x - a|_{\mathcal{V}} + |a - y|_{\mathcal{V}}\}$$
for any $x, y \in \mathcal{V}$.

The image of $\mathcal{V}$ under reflection in A will be denoted by $\mathcal{V}'$. The subspace $\mathcal{V}'$ is an isometric copy of $\mathcal{V}$. Clearly $\mathcal{V} \cup \mathcal{V}' = \mathcal{W}$ and $\mathcal{V} \cap \mathcal{V}' = A$. Moreover, $a = a' \iff a \in A$.

The following proposition follows directly from the definitions.

2.22. Proposition. *Assume $\mathcal{W}$ is the doubling of the metric space $\mathcal{V}$ in its closed subset A. Then:*

(a) *If $\mathcal{V}$ is a complete length space, then so is $\mathcal{W}$.*

(b) *If $\mathcal{V}$ is proper, then so is $\mathcal{W}$. In this case, for any $x, y \in \mathcal{V}$ there is $a \in A$ such that*
$$|x - a|_{\mathcal{V}} + |a - y|_{\mathcal{V}} = |x - y'|_{\mathcal{W}}.$$

(c) *Given $x \in \mathcal{W}$, let $\bar{x} = x$ if $x \in \mathcal{V}$, and $\bar{x} = x'$ otherwise. The map $\mathcal{W} \to \mathcal{V}$ defined by $x \mapsto \bar{x}$ is short and length-preserving. In particular, if γ is a geodesic in $\mathcal{W}$ with ends in $\mathcal{V}$, then $\bar{\gamma}$ is a geodesic in $\mathcal{V}$ with the same ends.*

G. Kuratowsky embedding

Given a metric space $\mathcal{X}$, let us denote by $\operatorname{Bnd}(\mathcal{X}, \mathbb{R})$ the space of all bounded functions on $\mathcal{X}$ equipped with the sup-norm
$$\|f\| = \sup_{x \in \mathcal{X}}\{|f(x)|\}.$$

Kuratowski embedding. Given a point $p \in \mathcal{X}$, consider the map $\operatorname{kur}_p : \mathcal{X} \to \operatorname{Bnd}(\mathcal{X}, \mathbb{R})$ defined by $\operatorname{kur}_p x = \operatorname{dist}_x - \operatorname{dist}_p$. The map kur_p will be called the *Kuratowski map at p*.

From the triangle inequality, we have
$$\|\operatorname{kur}_p x - \operatorname{kur}_p y\| = \sup_{z \in \mathcal{X}}\{||x - z| - |y - z||\} = |x - y|.$$
Therefore, for any $p \in \mathcal{X}$, the Kuratowski map gives a distance-preserving map $\operatorname{kur}_p : \mathcal{X} \hookrightarrow \operatorname{Bnd}(\mathcal{X}, \mathbb{R})$. Thus we can (and often will) consider the space $\mathcal{X}$ as a subset of $\operatorname{Bnd}(\mathcal{X}, \mathbb{R})$.

2.23. Exercise. Show that any compact metric space is isometric to a subspace in a compact length space.

Chapter 3

Maps and functions

Here we introduce several classes of maps between metric spaces and develop a language to describe various notions of convexity/concavity of real-valued functions on general metric spaces.

A. Submaps

We will often need maps and functions defined on subsets of a metric space. We call them *submaps* and *subfunctions*. Thus, given metric spaces $\mathcal{X}$ and $\mathcal{Y}$, a submap $\Phi\colon \mathcal{X} \multimap \mathcal{Y}$ is a map defined on a subset $\operatorname{Dom}\Phi \subset \mathcal{X}$.

A submap $\Phi\colon \mathcal{X} \multimap \mathcal{Y}$ is *continuous* if the inverse image of any open set is open. Since $\operatorname{Dom}\Phi = \Phi^{-1}(\mathcal{Y})$, the domain $\operatorname{Dom}\Phi$ of a continuous submap is open. The same holds for upper and lower semicontinuous functions $f\colon \mathcal{X} \multimap \mathbb{R}$ since they are continuous functions for a special topology on $\mathbb{R}$.

(Continuous partially defined maps could be defined via closed sets; namely, one could require that inverse images of closed sets are closed. While this condition is equivalent to continuity for functions defined on the whole space, it is different for partially defined functions. In particular, with this definition the domain of a continuous submap would have to be closed.)

B. Lipschitz conditions

3.1. Lipschitz maps. *Suppose $\mathcal{X}$ and $\mathcal{Y}$ are metric spaces, $\Phi\colon \mathcal{X} \multimap \mathcal{Y}$ is a continuous submap, and $\ell \in \mathbb{R}$.*

(a) *The submap Φ is called ℓ-*Lipschitz *if*

$$|\Phi(x) - \Phi(y)|_{\mathcal{Y}} \leqslant \ell \cdot |x - y|_{\mathcal{X}}$$

for any two points $x, y \in \operatorname{Dom}\Phi$.

- 1-*Lipschitz maps will be also called* short.

(b) *We say that* Φ *is* Lipschitz *if it is* ℓ-*Lipschitz for a constant* ℓ. *The minimal such constant is denoted by* $\operatorname{lip}\Phi$.

(c) *We say that* Φ *is* locally Lipschitz *if any point* $x \in \operatorname{Dom}\Phi$ *admits a neighborhood* $\Omega \subset \operatorname{Dom}\Phi$ *such that the restriction* $\Phi|_\Omega$ *is Lipschitz.*

(d) *Given* $p \in \operatorname{Dom}\Phi$, *we denote by* $\operatorname{lip}_p \Phi$ *the infimum of the real values* ℓ *such that* p *admits a neighborhood* $\Omega \subset \operatorname{Dom}\Phi$ *such that the restriction* $\Phi|_\Omega$ *is* ℓ-*Lipschitz.*

Note that $\Phi : \mathcal{X} \to \mathcal{Y}$ is ℓ-Lipschitz if and only if

$$\Phi(\mathrm{B}(x, R)_{\mathcal{X}}) \subset \mathrm{B}(\Phi(x), \ell \cdot R)_{\mathcal{Y}}$$

for any $R \geqslant 0$ and $x \in \mathcal{X}$. A dual version of this property is considered in the following definition.

3.2. Definitions. *Let* $\Phi : \mathcal{X} \to \mathcal{Y}$ *be a map between metric spaces, and* $\ell \in \mathbb{R}$.

(a) *The map* Φ *is called* ℓ-co-Lipshitz *if*

$$\Phi(\mathrm{B}(x, \ell \cdot R)_{\mathcal{X}}) \supset \mathrm{B}(\Phi(x), R)_{\mathcal{Y}}$$

for any $x \in \mathcal{X}$ *and* $R > 0$.

(b) *The map* Φ *is called* co-Lipschitz *if it is* ℓ-*co-Lipschitz for some constant* ℓ. *The minimal such constant is denoted by* $\operatorname{colip}\Phi$.

From the definition of co-Lipschitz maps we get the following:

3.3. Proposition. *Any co-Lipschitz map is open and surjective.*

In other words, ℓ-co-Lipschitz maps can be considered as a quantitative version of open maps. For that reason they are also called ℓ-open [**44**]. Also, be aware that some authors refer to our ℓ-co-Lipschitz maps as $\frac{1}{\ell}$-co-Lipschitz.

3.4. Proposition. *Let* $\mathcal{X}$ *and* $\mathcal{Y}$ *be metric spaces such that* $\mathcal{X}$ *is complete, and let* $\Phi : \mathcal{X} \to \mathcal{Y}$ *be a continuous co-Lipschitz map. Then* $\mathcal{Y}$ *is complete.*

Proof. Choose a Cauchy sequence y_n in $\mathcal{Y}$. Passing to a subsequence if necessary, we may assume that $|y_n - y_{n+1}|_{\mathcal{Y}} < \frac{1}{2^n}$ for each n. It is sufficient to show that y_n converges in $\mathcal{Y}$.

Denote by ℓ a co-Lipschitz constant for Φ. Note that there is a sequence x_n in $\mathcal{X}$ such that

$$\Phi(x_n) = y_n \quad \text{and} \quad |x_n - x_{n+1}|_{\mathcal{X}} < \frac{\ell}{2^n} \tag{1}$$

for each n. Indeed, such a sequence can be constructed recursively. Assuming that the points $x_1, \dots, x_{n-1}$ are already constructed, the existence of a sequence x_n satisfying (1) follows since Φ is ℓ-co-Lipschitz.

Notice that the sequence x_n is Cauchy. Since $\mathcal{X}$ is complete, x_n converges in $\mathcal{X}$; denote its limit by x_∞ and set $y_\infty = \Phi(x_\infty)$. Since Φ is continuous, $y_n \to y_\infty$ as $n \to \infty$. Hence the result. □

3.5. Lemma. *Let $\mathcal{X}$ be a metric space, and let $f : \mathcal{X} \to \mathbb{R}$ be a continuous function. Then for any $\varepsilon > 0$, there is a locally Lipschitz function $f_\varepsilon : \mathcal{X} \to \mathbb{R}$ such that $|f(x) - f_\varepsilon(x)| < \varepsilon$ for any $x \in \mathcal{X}$.*

Proof. Assume that $f \geqslant 1$. Construct a continuous positive function $\rho : \mathcal{X} \to \to \mathbb{R}_{>0}$ such that

$$|x - y| < \rho(x) \quad \Rightarrow \quad |f(x) - f(y)| < \varepsilon.$$

Consider the function

$$f_\varepsilon(y) = \sup\left\{ f(x)\cdot\left(1 - \tfrac{|x-y|}{\rho(x)}\right) : x \in \mathcal{X} \right\}.$$

It is straightforward to check that each f_ε is locally Lipschitz and $0 \leqslant f_\varepsilon - f < \varepsilon$.

Since any continuous function can be presented as the difference of two continuous functions bounded below by 1, the result follows. □

C. Isometries and submetries

3.6. Isometry. *Let $\mathcal{X}$ and $\mathcal{Y}$ be metric spaces, and let $\Phi : \mathcal{X} \to \mathcal{Y}$ be a map.*

(a) *The map Φ is* distance-preserving *if*

$$|\Phi(x) - \Phi(x')|_{\mathcal{Y}} = |x - x'|_{\mathcal{X}}$$

for any $x, x' \in \mathcal{X}$.

(b) *A distance-preserving bijection Φ is called an* isometry.

(c) *The spaces $\mathcal{X}$ and $\mathcal{Y}$ are called* isometric *(briefly $\mathcal{X} \stackrel{\text{iso}}{=\!=} \mathcal{Y}$) if there is an isometry $\Phi : \mathcal{X} \to \mathcal{Y}$.*

3.7. Submetry. *A map $\sigma : \mathcal{X} \to \mathcal{Y}$ between the metric spaces $\mathcal{X}$ and $\mathcal{Y}$ is called a* submetry *if*

$$\sigma(\mathrm{B}(p, r)_{\mathcal{X}}) = \mathrm{B}(\sigma(p), r)_{\mathcal{Y}}$$

for any $p \in \mathcal{X}$ and $r \geqslant 0$.

Note $\sigma : \mathcal{X} \to \mathcal{Y}$ is a submetry if it is 1-Lipschitz and 1-co-Lipschitz.

Note also that any submetry is an onto map.

The main source of examples of submetries comes from isometric group actions. Namely, assume $\mathcal{X}$ is a metric space and G is a subgroup of isometries of $\mathcal{X}$. Denote by $[x] = G\cdot x$ the G-orbit of $x \in \mathcal{X}$ and $\mathcal{X}/G$ the set of all G-orbits; let us equip it with the pseudometric defined by

$$|[x] - [y]|_{\mathcal{X}/G} = \inf\{\, |g\cdot x - h\cdot y|_{\mathcal{X}} : g, h \in G \,\}.$$

Note that if all the G-orbits form closed sets in $\mathcal{X}$, then $\mathcal{X}/G$ is a genuine metric space.

3.8. Proposition. *Let $\mathcal{X}$ be a metric space. Assume that a group G acts on $\mathcal{X}$ by isometries and in such a way that every G-orbit is closed. Then the projection map $\mathcal{X} \to \mathcal{X}/G$ is a submetry.*

Proof. We need to show that the map $x \mapsto [x] = G \cdot x$ is 1-Lipschitz and 1-co-Lipschitz. The co-Lipschitz part follows directly from the definitions of Hausdorff distance and co-Lipschitz maps.

Assume $|x - y|_{\mathcal{X}} < r$; equivalently $\mathrm{B}(x, r)_{\mathcal{X}} \ni y$. Since the action $G \curvearrowright \mathcal{X}$ is isometric, $\mathrm{B}(g \cdot x, r)_{\mathcal{X}} \ni g \cdot y$ for any $g \in G$.

In particular, the orbit $G \cdot y$ lies in the open r-neighborhood of the orbit $G \cdot x$. In the same way we can prove that the orbit $G \cdot x$ lies in the open r-neighborhood of the orbit $G \cdot y$. That is, the Hausdorff distance between the orbits $G \cdot x$ and $G \cdot y$ is less than r or, equivalently, $|[x] - [y]|_{\mathcal{X}/G} < r$. Since x and y are arbitrary, the map $x \mapsto [x]$ is 1-Lipschitz. □

3.9. Proposition. *Let $\mathcal{X}$ be a length space, and let $\sigma \colon \mathcal{X} \to \mathcal{Y}$ be a submetry. Then $\mathcal{Y}$ is a length space.*

Proof. Fix $\varepsilon > 0$ and a pair of points $x, y \in \mathcal{Y}$.

Since σ is 1-co-Lipschitz, there are points $\hat{x}, \hat{y} \in \mathcal{X}$ such that $\sigma(\hat{x}) = x$, $\sigma(\hat{y}) = y$, and $|\hat{x} - \hat{y}|_{\mathcal{X}} < |x - y|_{\mathcal{Y}} + \varepsilon$.

Since $\mathcal{X}$ is a length space, there is a curve γ joining $\hat{x}$ to $\hat{y}$ in $\mathcal{X}$ such that

$$\operatorname{length} \gamma \leqslant |x - y|_{\mathcal{Y}} + \varepsilon.$$

The curve $\sigma \circ \gamma$ joins x to y. Since σ is 1-Lipschitz and by the above,

$$\begin{aligned} \operatorname{length} \sigma \circ \gamma &\leqslant \operatorname{length} \gamma \\ &\leqslant |x - y|_{\mathcal{Y}} + \varepsilon. \end{aligned}$$

Since $\varepsilon > 0$ is arbitrary, $\mathcal{Y}$ is a length space. □

D. Speed of curves

Let $\mathcal{X}$ be a metric space. Recall that a *curve* in $\mathcal{X}$ is a continuous map $\alpha \colon \mathbb{I} \to \mathcal{X}$, where $\mathbb{I}$ is a real interval. A curve is called *Lipschitz* or *locally Lipschitz* if α is Lipschitz or locally Lipschitz, respectively. Length of curves is defined in Definition 2.5.

The following theorem follows from [**37**, 2.7].

3.10. Theorem. *Let $\mathcal{X}$ be a metric space, and let $\alpha\colon \mathbb{I} \to \mathcal{X}$ be a locally Lipschitz curve. Then the speed function*

$$\operatorname{speed}_{t_0} \alpha = \lim_{\substack{t \to t_0+ \\ s \to t_0-}} \frac{|\alpha(t) - \alpha(s)|}{|t-s|}$$

is defined for almost all $t_0 \in \mathbb{I}$, and

$$\operatorname{length} \alpha = \int\limits_{\mathbb{I}} \operatorname{speed}_t \alpha \cdot \mathbf{d}t,$$

where $\int$ denotes the Lebesgue integral.

A curve $\alpha\colon \mathbb{I} \to \mathcal{X}$ is *unit-speed* if

$$b - a = \operatorname{length}(\alpha|_{[a,b]})$$

for any subinterval $[a, b] \subset \mathbb{I}$. If α is Lipschitz, then, according to Theorem 3.10, this is equivalent to

$$\operatorname{speed} \alpha \overset{\text{a.e.}}{=} 1.$$

The following generalization of the standard Rademacher theorem on differentiability almost everywhere (a.e.) of Lipschitz maps between smooth manifolds [**37**, 5.5.2] was proved by Bernd Kirchheim [**95**].

The conclusion of the standard Rademacher theorem does not make sense for maps to a metric space since the target might have no linear structure. But the theorem does not hold even if we assume that the target is a Banach space. For example, consider the map $[0, 1] \to L^1[0, 1]$, defined by $x \mapsto \chi_{[0,x]}$, where χ_A denotes the characteristic function of A. This map is distance-preserving and in particular Lipschitz, but its differential is undefined at any point.

3.11. Theorem. *Let $\mathcal{X}$ be a metric space, and let $f\colon \mathbb{R}^n \multimap \mathcal{X}$ be 1-Lipschitz. Then for almost all $x \in \operatorname{Dom} f$ there is a pseudonorm $\|{*}\|_x$ on $\mathbb{R}^n$ such that*

$$|f(y) - f(z)|_{\mathcal{X}} = \|z - y\|_x + o(|y - x| + |z - x|).$$

Given f, the (pseudo)norm $\|{*}\|_x$ in the above theorem will be called its *differential of the induced metric* at x, or *metric differential* at x.

E. Convex real-to-real functions

In this section we will discuss generalized solutions of the differential inequalities

$$(1) \qquad y'' + \kappa \cdot y \geqslant \lambda \quad \text{and, respectively,} \quad y'' + \kappa \cdot y \leqslant \lambda$$

for fixed $\kappa, \lambda \in \mathbb{R}$. The solutions $y\colon \mathbb{R} \multimap \mathbb{R}$ are only assumed to be upper (respectively, lower) semicontinuous subfunctions.

Inequalities (1) are understood in the sense of distributions. That is, for any smooth function ϕ with compact support $\operatorname{Supp}\phi \subset \operatorname{Dom} y$, the following inequality should be satisfied:

$$\int\limits_{\operatorname{Dom} y} [y(t)\cdot\phi''(t) + \kappa\cdot y(t)\cdot\phi(t) - \lambda]\cdot \mathbf{d}t \geqslant 0, \tag{2}$$

$$\text{respectively} \quad \leqslant 0.$$

The integral is understood in the sense of Lebesgue; in particular inequality (2) makes sense for any Borel-measurable subfunction y. The proofs of the following propositions are straightforward.

3.12. Proposition. *Let $\mathbb{I} \subset \mathbb{R}$ be an open interval, and let $y_n : \mathbb{I} \to \mathbb{R}$ be a sequence of solutions of one of the inequalities in* (1). *Assume $y_n(t) \to y_\infty(t)$ as $n \to \infty$ for any $t \in \mathbb{I}$. Then y_∞ is a solution of the same inequality in* (1).

Assume y is a solution of one of the inequalities in (1). For $t_0 \in \operatorname{Dom} y$, let us define the *right* (*respectively, left*) *derivative* $y^+(t_0)$ ($y^-(t_0)$) at t_0 by

$$y^\pm(t_0) = \lim_{t\to t_0\pm} \frac{y(t) - y(t_0)}{|t - t_0|}.$$

Note that our sign convention for y^- is not standard—for $y(t) = t$ we have $y^+(t) = 1$ and $y^-(t) = -1$.

3.13. Proposition. *Let $\mathbb{I} \subset \mathbb{R}$ be an open interval, and let $y : \mathbb{I} \to \mathbb{R}$ be a solution of an inequality in* (1). *Then y is locally Lipschitz; its right and left derivatives $y^+(t_0)$ and $y^-(t_0)$ are defined for any $t_0 \in \mathbb{I}$. Moreover*

$$y^+(t_0) + y^-(t_0) \geqslant 0 \quad \textit{or, respectively,} \quad y^+(t_0) + y^-(t_0) \leqslant 0.$$

The next theorem gives a number of equivalent ways to define such generalized solutions.

3.14. Theorem. *Let $\mathbb{I}$ be an open real interval, and let $y : \mathbb{I} \to \mathbb{R}$ be a locally Lipschitz function. Then the following conditions are equivalent:*

(a) $y'' \geqslant \lambda - \kappa\cdot y$ (*respectively* $y'' \leqslant \lambda - \kappa\cdot y$).

(b) Barrier inequality. *For any $t_0 \in \mathbb{I}$, there is a solution $\bar{y}$ of the ordinary differential equation $\bar{y}'' = \lambda - \kappa\cdot\bar{y}$ with $\bar{y}(t_0) = y(t_0)$ such that $\bar{y} \geqslant y$ (respectively, $\bar{y} \leqslant y$) for all $t \in [t_0 - \varpi^\kappa, t_0 + \varpi^\kappa] \cap \mathbb{I}$. The function $\bar{y}$ is called a* lower (*respectively,* upper) barrier *of y at t_0.*

(c) Jensen's inequality. *For any pair of values $t_1 < t_2$ in $\mathbb{I}$ such that $|t_2 - t_1| < \varpi^\kappa$, the unique solution $z(t)$ of*

$$z'' = \lambda - \kappa\cdot z$$

such that

$$z(t_1) = y(t_1), \quad z(t_2) = y(t_2)$$

satisfies $y(t) \leqslant z(t)$ (respectively, $y(t) \geqslant z(t)$) for all $t \in [t_1, t_2]$.

Further, the following property holds:

(d) *Suppose $y'' \leqslant \lambda - \kappa \cdot y$. Let $t_0 \in \mathbb{I}$, and let $\bar{y}$ be a solution of the ordinary differential equation $\bar{y}'' = \lambda - \kappa \cdot \bar{y}$ such that $\bar{y}(t_0) = y(t_0)$ and $y^+(t_0) \leqslant \leqslant \bar{y}^+(t_0) \leqslant -y^-(t_0)$. (Note that such a $\bar{y}$ is unique if y is differentiable at t_0.)*

Then $\bar{y} \geqslant y$ for all $t \in [t_0 - \varpi^\kappa, t_0 + \varpi^\kappa] \cap \mathbb{I}$; that is, $\bar{y}$ is a barrier of y at t_0. (Similarly, by reversing inequalities, for $y'' \geqslant \lambda - \kappa \cdot y$.)

Note that Theorem 3.14 implies that y satisfies $y'' \geqslant \lambda$ ($y'' \leqslant \lambda$) on an interval $\mathbb{I} \subset \mathbb{R}$ if and only if $y(t) - \frac{\lambda}{2} \cdot t^2$ is convex (respectively, concave) on $\mathbb{I}$.

We will often need the following fact about convergence of derivatives of convex functions:

3.15. Two-shoulder lemma. *Let $\mathbb{I}$ be an open interval, and let $f_n : \mathbb{I} \to \mathbb{R}$ be a sequence of convex functions. Assume the functions f_n pointwise converge to a function $f_\infty : \mathbb{I} \to \mathbb{R}$. Then for any $t_0 \in \mathbb{I}$,*

$$f_\infty^\pm(t_0) \leqslant \varliminf_{n\to\infty} f_n^\pm(t_0).$$

Proof. Since the f_n are convex, we have $f_n^+(t_0) + f_n^-(t_0) \geqslant 0$, and for any t,

$$f_n(t) \geqslant f_n(t_0) \pm f^\pm(t_0) \cdot (t - t_0).$$

Passing to the limit, we get

$$f_\infty(t) \geqslant f_\infty(t_0) + \left[\varlimsup_{n\to\infty} f_n^+(t_0)\right] \cdot (t - t_0)$$

for $t \geqslant t_0$, and

$$f_\infty(t) \geqslant f_\infty(t_0) - \left[\varlimsup_{n\to\infty} f_n^-(t_0)\right] \cdot (t - t_0)$$

for $t \leqslant t_0$. Hence the result. □

3.16. Corollary. *Let $\mathbb{I}$ be an open interval, and let $f_n : \mathbb{I} \to \mathbb{R}$ be a sequence of functions such that $f_n'' \leqslant \lambda$ that converge pointwise to a function $f_\infty : \mathbb{I} \to \mathbb{R}$. Then:*

(a) *If f_∞ is differentiable at $t_0 \in \mathbb{I}$, then*

$$f_\infty'(t_0) = \pm \lim_{n\to\infty} f_n^\pm(t_0).$$

(b) *If all f_n and f_∞ are differentiable at $t_0 \in \mathbb{I}$, then*

$$f_\infty'(t_0) = \lim_{n\to\infty} f_n'(t_0).$$

Proof. Set $\hat{f}_n(t) = f_n(t) - \frac{\lambda}{2} \cdot t^2$ and $\hat{f}_\infty(t) = f_\infty(t) - \frac{\lambda}{2} \cdot t^2$. Note that the $\hat{f}_n$ are concave and $\hat{f}_n \to \hat{f}_\infty$ pointwise. Thus the theorem follows from 3.15, the Two-shoulder lemma. □

F. Convex functions on a metric space

In this section we define different types of convexity/concavity in the context of metric spaces; it will be mostly used for geodesic spaces. The notation refers to the corresponding second-order ordinary differential inequality.

3.17. Definition. Let $\mathcal{X}$ be a metric space. We say that an upper semicontinuous subfunction $f\colon \mathcal{X} \multimap (-\infty, \infty]$ satisfies the inequality

$$f'' + \kappa\cdot f \geqslant \lambda$$

if for *any* unit-speed geodesic γ in $\operatorname{Dom} f$, the real-to-real function $y(t) = f\circ\gamma(t)$ satisfies

$$y'' + \kappa\cdot y \geqslant \lambda$$

in the domain $\{\, t : y(t) < \infty \,\}$; see the definition in Section 3.E.

We say that a lower semicontinuous subfunction $f\colon \mathcal{X} \multimap [-\infty, \infty)$ satisfies the inequality

$$f'' + \kappa\cdot f \leqslant \lambda$$

if the subfunction $h = -f$ satisfies

$$h'' - \kappa\cdot h \geqslant -\lambda.$$

Functions satisfying the inequalities

$$f'' \geqslant \lambda \quad \text{and} \quad f'' \leqslant \lambda$$

are called *λ-convex* and *λ-concave*, respectively.

0-convex and 0-concave subfunctions will also be called *convex* and *concave*, respectively.

If f is λ-convex for $\lambda > 0$, then f will be called *strongly convex*; correspondingly, if f is λ-concave for $\lambda < 0$, then f will be called *strongly concave*.

If for any point $p \in \operatorname{Dom} f$ there is a neighborhood $\Omega \ni p$ and a real number λ such that the restriction $f|_\Omega$ is λ-convex (or λ-concave), then f is called *semiconvex* (respectively, *semiconcave*).

Various authors define the class of λ-convex (λ-concave) functions differently. Their definitions may correspond to $\pm\lambda$-convex ($\pm\lambda$-concave) or $\pm\frac{\lambda}{2}$-convex ($\pm\frac{\lambda}{2}$-concave) functions in our definitions.

3.18. Proposition. *Let $\mathcal{X}$ be a metric space. Assume that $f\colon \mathcal{X} \multimap \mathbb{R}$ is a semiconvex subfunction and $\phi\colon \mathbb{R} \to \mathbb{R}$ is a nondecreasing semiconvex function. Then the composition $\phi\circ f$ is a semiconvex subfunction.*

The proof is straightforward.

Chapter 4

Ultralimits

Here we introduce ultralimits of sequences of points, metric spaces, and functions. Our presentation is based on [**97**].

Ultralimits are closely related to Gromov–Hausdorff limits. We use them only as a canonical way to pass to convergent subsequences. We could avoid using them at the cost of saying "pass to a convergent subsequence" too many times. Doing this might be cumbersome and it obscures ideas of the proof; see for example the proof of the globalization theorem for general CBB spaces in Section 8.F. Also the use of ultralimits is convenient when dealing with CAT spaces due to the lack of compactness results.

A. Ultrafilters

We will need the existence of a selective ultrafilter ω that will be fixed once and for all. The existence follows from the axiom of choice and the continuum hypothesis.

Measure-theoretic definition. Recall that $\mathbb{N}$ denotes the set of natural numbers, $\mathbb{N} = \{1, 2, \dots\}$.

4.1. Definition. A finitely additive measure ω on $\mathbb{N}$ is called an *ultrafilter* if it satisfies

(a) $\omega(S) = 0$ or 1 for any subset $S \subset \mathbb{N}$. An ultrafilter ω is called *nonprincipal* if in addition

(b) $\omega(F) = 0$ for any finite subset $F \subset \mathbb{N}$. A nonprincipal ultrafilter ω is called *selective* if in addition

(c) for any partition of $\mathbb{N}$ into sets $\{C_\alpha\}_{\alpha\in\mathcal{A}}$ such that $\omega(C_\alpha) = 0$ for each α, there is a set $S \subset \mathbb{N}$ such that $\omega(S) = 1$ and $S \cap C_\alpha$ is a one-point set for each $\alpha \in \mathcal{A}$.

If $\omega(S) = 0$ for some subset $S \subset \mathbb{N}$, we say that S is *ω-small*. If $\omega(S) = 1$, we say that S contains *ω-almost all* elements of $\mathbb{N}$.

4.2. Advanced exercise. Let ω be an ultrafilter, and let $f : \mathbb{N} \to \mathbb{N}$. Suppose that $\omega(S) \leqslant \omega(f^{-1}(S))$ for any set $S \subset \mathbb{N}$. Show that $f(n) = n$ for ω-almost all $n \in \mathbb{N}$.

Classical definition. More commonly, a nonprincipal ultrafilter is defined as a collection, say $\mathfrak{F}$, of sets in $\mathbb{N}$ such that

(1) if $P \in \mathfrak{F}$ and $Q \supset P$, then $Q \in \mathfrak{F}$,

(2) if $P, Q \in \mathfrak{F}$, then $P \cap Q \in \mathfrak{F}$,

(3) for any subset $P \subset \mathbb{N}$, either P or its complement is an element of $\mathfrak{F}$,

(4) if $F \subset \mathbb{N}$ is finite, then $F \notin \mathfrak{F}$.

Setting

$$P \in \mathfrak{F} \iff \omega(P) = 1$$

makes these two definitions equivalent.

A nonempty collection of sets $\mathfrak{F}$ that does not include the empty set and satisfies only conditions (1) and (2) is called a *filter*; if in addition $\mathfrak{F}$ satisfies condition (3) it is called an *ultrafilter*. From Zorn's lemma, it follows that every filter is contained in an ultrafilter. Thus there is an ultrafilter $\mathfrak{F}$ contained in the filter of all complements of finite sets; clearly this $\mathfrak{F}$ is nonprincipal.

The existence of a selective ultrafilter follows from the continuum hypothesis; this was proved by Walter Rudin [**144**].

Stone–Čech compactification. Given a set $S \subset \mathbb{N}$, consider the subset Ω_S of all ultrafilters ω such that $\omega(S) = 1$. It is straightforward to check that the sets Ω_S for all $S \subset \mathbb{N}$ form a topology on the set of ultrafilters on $\mathbb{N}$. The resulting space is called the *Stone–Čech compactification* of $\mathbb{N}$; it is usually denoted by $\beta\mathbb{N}$.

There is a natural embedding $\mathbb{N} \hookrightarrow \beta\mathbb{N}$ defined by $n \mapsto \omega_n$, where ω_n is the principal ultrafilter such that $\omega_n(S) = 1$ if and only if $n \in S$. Using this embedding, we can (and will) consider $\mathbb{N}$ as a subset of $\beta\mathbb{N}$.

The space $\beta\mathbb{N}$ is the maximal compact Hausdorff space that contains $\mathbb{N}$ as an everywhere dense subset. More precisely, for any compact Hausdorff space $\mathcal{X}$ and a map $f : \mathbb{N} \to \mathcal{X}$, there is a unique continuous map $\bar{f} : \beta\mathbb{N} \to X$ such that the restriction $\bar{f}|_{\mathbb{N}}$ coincides with f.

B. Ultralimits of points

Fix an ultrafilter ω. Assume x_n is a sequence of points in a metric space $\mathcal{X}$. Define an *ω-limit* of x_n to be a point x_ω such that for any $\varepsilon > 0$, ω-almost all elements of x_n lie in $\mathrm{B}(x_\omega, \varepsilon)$; that is,

$$\omega\{\, n \in \mathbb{N} : |x_\omega - x_n| < \varepsilon \,\} = 1.$$

In this case, we write

$$x_\omega = \lim_{n\to\omega} x_n \quad \text{or} \quad x_n \to x_\omega \quad \text{as} \quad n \to \omega.$$

Also, if $\mathcal{X} = \mathbb{R}$ we write $\lim_{n\to\omega} x_n = \pm\infty$ if

$$\omega\{\, n \in \mathbb{N} : \pm x_n > L \,\} = 1$$

for any $L \in \mathbb{R}$.

It easily follows from the definition that ω-limits are unique if they exist. For example, if ω is the principal ultrafilter such that $\omega(\{n\}) = 1$ for some $n \in \mathbb{N}$, then $x_\omega = x_n$.

Note that ω-limits of a sequence and its subsequences may differ. For example, in general

$$\lim_{n\to\omega} x_n \neq \lim_{n\to\omega} x_{2\cdot n}.$$

The sequence x_n can be regarded as a map $\mathbb{N} \to \mathcal{X}$. If $\mathcal{X}$ is compact, then this map can be uniquely extended to a continuous map to the Stone–Čech compactification $\beta\mathbb{N}$ of $\mathbb{N}$. Then x_ω is the image of ω.

4.3. Proposition. *Let ω be a nonprincipal ultrafilter. Assume x_n is a sequence of points in a metric space $\mathcal{X}$ and $x_n \to x_\omega$ as $n \to \omega$. Then there is a subsequence of x_n that converges to x_ω in the usual sense.*

Moreover, if ω is selective, then the subsequence $(x_n)_{n\in S}$ can be chosen so that $\omega(S) = 1$.

Proof. Given $\varepsilon > 0$, let $S_\varepsilon = \{\, n \in \mathbb{N} : |x_n - x_\omega| < \varepsilon \,\}$.

Note that $\omega(S_\varepsilon) = 1$ for any $\varepsilon > 0$. Since ω is nonprincipal, the set S_ε is infinite. Therefore we can choose an increasing sequence n_k such that $n_k \in S_{\frac{1}{k}}$ for each $k \in \mathbb{N}$. Clearly $x_{n_k} \to x_\omega$ as $k \to \infty$.

Now assume that ω is selective. Consider the sets

$$C_k = \left\{\, n \in \mathbb{N} : \tfrac{1}{k} < |x_n - x_\omega| \leqslant \tfrac{1}{k-1} \,\right\},$$

where we assume $\frac{1}{0} = \infty$, and the set

$$C_\infty = \{\, n \in \mathbb{N} : x_n = x_\omega \,\}.$$

Note that $\omega(C_k) = 0$ for any $k \in \mathbb{N}$.

If $\omega(C_\infty) = 1$, we can take the subsequence consisting of the x_n, $n \in C_\infty$.

Otherwise, discarding all empty sets among C_k and C_∞ gives a partition of $\mathbb{N}$ into a countable collection of ω-small sets. Since ω is selective, we can choose a set $S \subset \mathbb{N}$ such that S meets each set of the partition at one point and $\omega(S) = 1$. Clearly the subsequence consisting of the x_n, $n \in S$, converges to x_ω in the usual sense. □

The following proposition is analogous to the statement that any sequence in a compact metric space has a convergent subsequence; it can be proved in the same way.

4.4. Proposition. *Let $\mathcal{X}$ be a compact metric space. Then any sequence of points x_n in $\mathcal{X}$ has a unique ω-limit x_ω.*

In particular, a bounded sequence of real numbers has a unique ω-limit.

The following lemma is an ultralimit analogue of the Cauchy convergence test.

4.5. Lemma. *Let x_n be a sequence of points in a complete metric space $\mathcal{X}$. If for each subsequence y_n of x_n, the ω-limit*

$$y_\omega = \lim_{n \to \omega} y_n \in \mathcal{X}$$

is defined and does not depend on the choice of a subsequence, then the sequence x_n converges in the usual sense.

Proof. Assume the contrary. Then for some $\varepsilon > 0$, there is a subsequence y_n of x_n such that $|x_n - y_n| \geqslant \varepsilon$ for all n.

It follows that $|x_\omega - y_\omega| \geqslant \varepsilon$, a contradiction. □

4.6. Exercise. Recall that ℓ^∞ denotes the space of bounded sequences of real numbers. Show that there is a linear functional $L\colon \ell^\infty \to \mathbb{R}$ such that for any sequence $\boldsymbol{s} = (s_1, s_2, \dots) \in S$ the image $L(\boldsymbol{s})$ is a partial limit of $s_1, s_2, \dots$.

4.7. Exercise. Suppose that $f\colon \mathbb{N} \to \mathbb{N}$ is a map such that

$$\lim_{n \to \omega} x_n = \lim_{n \to \omega} x_{f(n)}$$

for any bounded sequence x_n of real numbers. Show that $f(n) = n$ for ω-almost all $n \in \mathbb{N}$.

C. Ultralimits of spaces

Fix a selective ultrafilter ω on the set of natural numbers.

Let $\mathcal{X}_n$ be a sequence of metric spaces. Consider all sequences $x_n \in \mathcal{X}_n$. On the set of all such sequences, define a pseudometric by the formula

$$|(x_n) - (y_n)| = \lim_{n \to \omega} |x_n - y_n|. \tag{1}$$

Note that the ω-limit on the right-hand side is always defined and takes value in $[0, \infty]$.

Let $\mathcal{X}_\omega$ be the corresponding metric space; that is, the underlying set of $\mathcal{X}_\omega$ is formed by equivalence classes of sequences of points $x_n \in \mathcal{X}_n$ defined by the relation

$$(x_n) \sim (y_n) \quad \Longleftrightarrow \quad \lim_{n \to \omega} |x_n - y_n| = 0,$$

and the distance is defined as in (1).

The space $\mathcal{X}_\omega$ is called the *ω-limit* of $\mathcal{X}_n$. Typically $\mathcal{X}_\omega$ will denote the ω-limit of a sequence $\mathcal{X}_n$; we may also write

$$\mathcal{X}_n \to \mathcal{X}_\omega \quad \text{as} \quad n \to \omega \quad \text{or} \quad \mathcal{X}_\omega = \lim_{n \to \omega} \mathcal{X}_n.$$

Given a sequence $x_n \in \mathcal{X}_n$, we will denote by x_ω its equivalence class, which is a point in $\mathcal{X}_\omega$; in this case, we may write

$$x_n \to x_\omega \quad \text{as} \quad n \to \omega \quad \text{or} \quad x_\omega = \lim_{n \to \omega} x_n.$$

4.8. Observation. The ω-limit of any sequence of metric spaces is complete.

Proof. Let $\mathcal{X}_n$ be a sequence of metric spaces, and let $\mathcal{X}_n \to \mathcal{X}_\omega$ as $n \to \omega$. Choose a Cauchy sequence x_n in $\mathcal{X}_\omega$. Passing to a subsequence, we can assume that $|x_k - x_m|_{\mathcal{X}_\omega} < \frac{1}{k}$ for any $k < m$.

Let us choose points $x_{n,m} \in \mathcal{X}_n$ such that for any fixed m we have $x_{n,m} \to x_m$ as $n \to \omega$. Note that for any $k < m$ the inequality $|x_{n,k} - x_{n,m}| < \frac{1}{k}$ holds for ω-almost all n. It follows that we can choose a nested sequence of sets

$$\mathbb{N} = S_1 \supset S_2 \supset \cdots,$$

such that

- $\omega(S_m) = 1$ for each m,
- $\bigcap_m S_m = \varnothing$, and
- $|x_{n,k} - x_{n,l}| < \frac{1}{k}$ for $k < l \leqslant m$ and $n \in S_m$.

Consider the sequence $y_n = x_{n,m(n)}$, where $m(n)$ is the largest value such that $n \in S_{m(n)}$. Denote by $y_\omega \in \mathcal{X}_\omega$ the ω-limit of y_n.

Observe that $|y_m - x_{n,m}| < \frac{1}{m}$ for ω-almost all n. It follows that $|x_m - y_\omega| \leqslant \frac{1}{m}$ for any m. Therefore, $x_n \to y_\omega$ as $n \to \infty$. That is, any Cauchy sequence in $\mathcal{X}_\omega$ converges. $\square$

4.9. Observation. The ω-limit of any sequence of length spaces is geodesic.

Proof. If $\mathcal{X}_n$ is a sequence of length spaces, then for any sequence of pairs (x_n, y_n) in $\mathcal{X}_n$ there is a sequence of $\frac{1}{n}$-midpoints z_n.

Let $x_n \to x_\omega$, $y_n \to y_\omega$, and $z_n \to z_\omega$ as $n \to \omega$. Note that z_ω is a midpoint between x_ω and y_ω in $\mathcal{X}_\omega$.

By Observation 4.8, $\mathcal{X}_\omega$ is complete. Applying Lemma 2.13, we obtain the statement. □

A geodesic space $\mathcal{T}$ is called a *metric tree* if any pair of points in $\mathcal{T}$ are connected by a unique geodesic, and the union of any two geodesics $[xy]_\mathcal{T}$, and $[yz]_\mathcal{T}$ contains the geodesic $[xz]_\mathcal{T}$. The latter means that any triangle in $\mathcal{T}$ is a tripod; that is, for any three points x, y, and z there is a point p such that

$$[xy] \cup [yz] \cup [zx] = [px] \cup [py] \cup [pz].$$

4.10. Exercise. Let $\mathcal{T}$ be a metric component of the ultralimit of $\mathbb{M}^2(n)$ as $n \to$ $\to \omega$.

(a) Show that $\mathcal{T}$ is a complete metric tree.

(b) Show that $\mathcal{T}$ is homogeneous; that is, given two points $s, t \in \mathcal{T}$ there is an isometry of $\mathcal{T}$ that maps s to t.

(c) Show that $\mathcal{T}$ has *continuum degree* at any point; that is, for any point $t \in \mathcal{T}$ the set of connected components of the complement $\mathcal{T} \setminus \{t\}$ has cardinality continuum.

Ultrapower. If all the metric spaces in a sequence are identical, $\mathcal{X}_n = \mathcal{X}$, the ω-limit $\lim_{n\to\omega} \mathcal{X}_n$ is denoted by $\mathcal{X}^\omega$ and called the ω-power of $\mathcal{X}$.

By Theorem 5.16, there is a distance-preserving map $\iota\colon \mathcal{X} \hookrightarrow \mathcal{X}^\omega$, where $\iota(y)$ is the equivalence class of the constant sequence $y_n = y$.

The image $\iota(\mathcal{X})$ might be a proper subset of $\mathcal{X}^\omega$. For example, $\mathbb{R}^\omega$ has pairs of points at distance ∞ from each other, while each metric component of $\mathbb{R}^\omega$ is isometric to $\mathbb{R}$.

According to Theorem 5.16, if $\mathcal{X}$ is compact, then $\iota(\mathcal{X}) = \mathcal{X}^\omega$; in particular, $\mathcal{X}^\omega$ is isometric to $\mathcal{X}$. If $\mathcal{X}$ is proper, then $\iota(\mathcal{X})$ forms a metric component of $\mathcal{X}^\omega$.

The embedding ι allows us to treat $\mathcal{X}$ as a subset of its ultrapower $\mathcal{X}^\omega$.

4.11. Observation. Let $\mathcal{X}$ be a complete metric space. Then $\mathcal{X}^\omega$ is a geodesic space if and only if $\mathcal{X}$ is a length space.

Proof. Assume $\mathcal{X}^\omega$ is a geodesic space. Then any pair of points $x, y \in \mathcal{X}$ has a midpoint $z_\omega \in \mathcal{X}^\omega$. Fix a sequence of points $z_n \in \mathcal{X}$ such that $z_n \to z_\omega$ as $n \to \omega$.

Note that $|x - z_n|_{\mathcal{X}} \to \frac{1}{2}\cdot|x - y|_{\mathcal{X}}$ and $|y - z_n|_{\mathcal{X}} \to \frac{1}{2}\cdot|x - y|_{\mathcal{X}}$ as $n \to \omega$. In particular, for any $\varepsilon > 0$, the point z_n is an ε-midpoint between x and y for ω-almost all n. It remains to apply Lemma 2.13.

The "if" part follows from Observation 4.9. □

Note that the proof above together with Lemma 4.5 imply the following:

4.12. Corollary. *Assume $\mathcal{X}$ is a complete length space and $p, q \in \mathcal{X}$ cannot be joined by a geodesic in $\mathcal{X}$. Then there are at least continuum distinct geodesics between p and q in the ultrapower $\mathcal{X}^\omega$.*

4.13. Exercise. Let $\mathcal{X}$ be a countable set with discrete metric; that is $|x-y|_{\mathcal{X}} = 1$ if $x \neq y$. Show that

(a) $\mathcal{X}^\omega$ is not isometric to $\mathcal{X}$.

(b) $\mathcal{X}^\omega$ is isometric to $(\mathcal{X}^\omega)^\omega$.

4.14. Exercise. Given a nonprincipal ultrafilter ω, construct an ultrafilter ω_1 such that

$$\mathcal{X}^{\omega_1} \stackrel{\text{iso}}{=\!=} (\mathcal{X}^\omega)^\omega$$

for any metric space $\mathcal{X}$.

4.15. Exercise. Construct a proper metric space $\mathcal{X}$ such that $\mathcal{X}^\omega$ is not proper; that is, there is a point $p \in \mathcal{X}^\omega$ and $R < \infty$ such that the closed ball $\overline{\mathrm{B}}[p, R]_{\mathcal{X}^\omega}$ is not compact.

D. Ultralimits of sets

Let $\mathcal{X}_n$ be a sequence of metric spaces, and let $\mathcal{X}_n \to \mathcal{X}_\omega$ as $n \to \omega$.

For a sequence of sets $\Omega_n \subset \mathcal{X}_n$, consider the maximal set $\Omega_\omega \subset \mathcal{X}_\omega$ such that for any $x_\omega \in \Omega_\omega$ and any sequence $x_n \in \mathcal{X}_n$ such that $x_n \to x_\omega$ as $n \to \omega$, we have $x_n \in \Omega_n$ for ω-almost all n.

The set Ω_ω is called the *open ω-limit* of Ω_n; we could also write $\Omega_n \to \Omega_\omega$ as $n \to \omega$ or $\Omega_\omega = \lim_{n\to\omega} \Omega_n$.

Applying Observation 4.8 to the sequence of complements $\mathcal{X}_n \setminus \Omega_n$, we see that Ω_ω is open for any sequence Ω_n.

This definition can be applied to arbitrary sequences of sets, but we will apply it only for sequences of open sets.

E. Ultralimits of functions

Recall that a family of submaps (see Section 3.A) between metric spaces $\{f_\alpha\colon \mathcal{X} \multimap \mathcal{Y}\}_{\alpha\in\mathcal{A}}$ is called *equicontinuous* if for any $\varepsilon > 0$ there is $\delta > 0$ such that for any $p, q \in \mathcal{X}$ with $|p-q| < \delta$ and any $\alpha \in \mathcal{A}$, we have $|f_\alpha(p)-f_\alpha(q)| < \varepsilon$.

Let $f_n : \mathcal{X}_n \multimap \mathbb{R}$ be a sequence of subfunctions.

Set $\Omega_n = \operatorname{Dom} f_n$. Consider the open ω-limit set $\Omega_\omega \subset \mathcal{X}_\omega$ of Ω_n.

Assume there is a subfunction $f_\omega : \mathcal{X}_\omega \multimap \mathbb{R}$ that satisfies the following conditions:

(1) $\operatorname{Dom} f_\omega = \Omega_\omega$,

(2) if $x_n \to x_\omega \in \Omega_\omega$ for a sequence of points $x_n \in \mathcal{X}_n$, then $f_n(x_n) \to f_\omega(x_\omega)$ as $n \to \omega$.

In this case, the subfunction $f_\omega : \mathcal{X}_\omega \to \mathbb{R}$ is said to be the ω-limit of $f_n : \mathcal{X}_n \to \mathbb{R}$.

The following lemma gives a mild condition on a sequence of functions f_n guaranteeing the existence of its ω-limit.

4.16. Lemma. *Let $\mathcal{X}_n$ be a sequence of metric spaces, and let $f_n : \mathcal{X}_n \multimap \mathbb{R}$ be a sequence of subfunctions.*

Assume that for any positive integer k, there is an open set $\Omega_n(k) \subset \operatorname{Dom} f_n$ such that the restrictions $f_n|_{\Omega_n(k)}$ are uniformly bounded and equicontinuous and the open ω-limit of $\Omega_n(n)$ coincides with the open ω-limit of $\operatorname{Dom} f_n$. *Then the ω-limit f_ω of f_n is defined; moreover f_ω is a continuous subfunction.*

In particular, if the functions f_n are uniformly bounded and equicontinuous, then its ω-limit f_ω is defined, bounded and uniformly continuous.

The proof is straightforward.

4.17. Exercise. Construct a sequence of compact length spaces $\mathcal{X}_n$, with a converging sequence of ℓ-Lipschitz concave (see Definition 3.17) functions $f_n : \mathcal{X}_n \to \mathbb{R}$ such that the ω-limit $\mathcal{X}_\omega$ of $\mathcal{X}_n$ is compact and the ω-limit $f_\omega : \mathcal{X}_\omega \to \mathbb{R}$ of f_n is not concave.

If $f : \mathcal{X} \multimap \mathbb{R}$ is a subfunction, the ω-limit of the constant sequence $f_n = f$ is called the ω-power of f and is denoted by f^ω. So

$$f^\omega : \mathcal{X} \multimap \mathbb{R}, \quad f^\omega(x_\omega) = \lim_{n\to\omega} f(x_n).$$

Evidently, if f^ω is defined, then f is continuous.

Recall that we treat $\mathcal{X}$ as a subset of its ω-power $\mathcal{X}^\omega$. Note that $\operatorname{Dom} f = \mathcal{X} \cap \operatorname{Dom} f^\omega$. Moreover, if $\mathrm{B}(x,\varepsilon)_{\mathcal{X}} \subset \operatorname{Dom} f$ then $\mathrm{B}(x,\varepsilon)_{\mathcal{X}^\omega} \subset \operatorname{Dom} f^\omega$.

Chapter 5

Space of spaces

In this chapter we discuss the Gromov–Hausdorff convergence of metric spaces.

To the best of our knowledge, Hausdorff convergence of subsets of a fixed metric space was first introduced by Felix Hausdorff [**81**], and a couple of years later an equivalent definition was given by Wilhelm Blaschke [**29**]. A further refinement of this definition was introduced by Zdeněk Frolík [**66**] and then rediscovered by Robert Wijsman [**160**]. However this refinement was a step in the direction of the so-called *closed convergence* introduced by Hausdorff in the original book. For that reason we call it Hausdorff convergence instead of *Hausdorff–Blashcke–Frolík–Wijsman convergence.*

Gromov–Hausdorff convergence was first introduced by David Edwards [**62**] and rediscovered later by Michael Gromov [**72**]. It was an essential tool in Gromov's proof that any group of polynomial growth has a nilpotent subgroup of finite index. Other versions of convergence of metric spaces were considered earlier, but each time the definition was limited to very specific types of problems.

The definition of Gromov–Hausdorff convergence of metric spaces uses the notion of Hausdorff convergence. Gromov–Hausdorff convergence means that a sequence of metric spaces admits a sequence of distance-preserving embeddings into a common ambient metric space so that their images converge in the Hausdorff sense. Our definition of Gromov–Hausdorff convergence and Gromov–Hausdorff distance differ somewhat from the standard definition.

A. Convergence of subsets

Let $\mathcal{X}$ be a metric space and $A \subset \mathcal{X}$. Recall that the distance from A to a point x in $\mathcal{X}$ is given by

$$\operatorname{dist}_A x := \inf\{\, |a - x| \,:\, a \in A\,\}.$$

By this definition, we have $\operatorname{dist}_\varnothing x = \infty$ for any x.

5.1. Definition of Hausdorff convergence. *Given a sequence of closed sets A_n in a metric space $\mathcal{X}$, a closed set $A_\infty \subset \mathcal{X}$ is called the Hausdorff limit of A_n, briefly $A_n \underset{\mathrm{H}}{\longrightarrow} A_\infty$, if*

$$\operatorname{dist}_{A_n} x \to \operatorname{dist}_{A_\infty} x \quad \textit{as} \quad n \to \infty$$

for any fixed $x \in \mathcal{X}$.

In this case the sequence of closed sets A_n is said to be converging *or* converging in the sense of Hausdorff.

5.2. Selection theorem. *Let $\mathcal{X}$ be a proper metric space, and let A_n be a sequence of closed sets in $\mathcal{X}$. Then A_n has a converging subsequence in the sense of Hausdorff.*

Proof. Since $\mathcal{X}$ is proper, we can choose a countable dense set $\{x_1, x_2, \dots\}$ in $\mathcal{X}$.

If the sequence $a_n = \operatorname{dist}_{A_n} x_k$ is unbounded for some k, then we can pass to a subsequence of A_n such that $\operatorname{dist}_{A_n} x_k \to \infty$ as $n \to \infty$ for *any* k. The obtained sequence converges to the empty set.

Now suppose that $a_n = \operatorname{dist}_{A_n} x_k$ is bounded for each k. In this case, passing to a subsequence of A_n, we can assume that $\operatorname{dist}_{A_n} x_k$ converges as $n \to \infty$ for any fixed k.

Note that for each n, the function $\operatorname{dist}_{A_n} : \mathcal{X} \to \mathbb{R}$ is 1-Lipschitz and nonnegative. Therefore the sequence $\operatorname{dist}_{A_n}$ converges pointwise to a 1-Lipschitz nonnegative function $f : \mathcal{X} \to \mathbb{R}$.

Set $A_\infty = f^{-1}(0)$. Since f is 1-Lipschitz, $\operatorname{dist}_{A_\infty} y \geqslant f(y)$ for any $y \in \mathcal{X}$. It remains to show that $\operatorname{dist}_{A_\infty} y \leqslant f(y)$ for any y.

Assume the contrary, that is, $f(z) < R < \operatorname{dist}_{A_\infty} z$ for $z \in \mathcal{X}$ and $R > 0$. Then for any sufficiently large n there is a point $z_n \in A_n$ such that $|x - z_n| \leqslant R$. Since $\mathcal{X}$ is proper, we can pass to a partial limit z_∞ of z_n as $n \to \infty$.

Clearly $f(z_\infty) = 0$, that is, $z_\infty \in A_\infty$. On the other hand,

$$\operatorname{dist}_{A_\infty} y \leqslant |z_\infty - y| \leqslant R < \operatorname{dist}_{A_\infty} y,$$

a contradiction. □

B. Convergence of spaces

5.3. Definition. Let $\{\mathcal{X}_\alpha : \alpha \in \mathcal{A}\}$ be a set of metric spaces. A metric space $\boldsymbol{X}$ is called a *common space* of $\{\mathcal{X}_\alpha : \alpha \in \mathcal{A}\}$ if its underlying set is formed by the disjoint union

$$\bigsqcup_{\alpha\in\mathcal{A}} \mathcal{X}_\alpha$$

and each inclusion $\iota_\alpha : \mathcal{X}_\alpha \hookrightarrow \boldsymbol{X}$ is distance-preserving.

5.4. Definition. Let $\boldsymbol{X}$ be a common space for proper metric spaces $\mathcal{X}_1, \mathcal{X}_2, \dots$, and $\mathcal{X}_\infty$. Assume that $\mathcal{X}_n$ forms an open set in $\boldsymbol{X}$ for each $n < \infty$ and $\mathcal{X}_n \underset{\mathrm{H}}{\to} \mathcal{X}_\infty$ in $\boldsymbol{X}$ as $n \to \infty$.

Then the topology τ of $\boldsymbol{X}$ is called a *Gromov–Hausdorff convergence* and we write $\mathcal{X}_n \underset{\mathrm{GH}}{\longrightarrow} \mathcal{X}_\infty$ or $\mathcal{X}_n \xrightarrow{\tau} \mathcal{X}_\infty$; the latter notation is used if we need to consider the specific Gromov–Hausdorff convergence τ. The space $\mathcal{X}_\infty$ is called the *limit space* of the sequence $\mathcal{X}_n$ along τ.

When we write $\mathcal{X}_n \underset{\mathrm{GH}}{\longrightarrow} \mathcal{X}_\infty$, we mean that we made a choice of a Gromov–Hausdorff convergence.

Note that for a fixed sequence $\mathcal{X}_n$ of metric spaces, one may construct different Gromov–Hausdorff convergences, say $\mathcal{X}_n \xrightarrow{\tau} \mathcal{X}_\infty$ and $\mathcal{X}_n \xrightarrow{\tau'} \mathcal{X}'_\infty$, and their limit spaces $\mathcal{X}_\infty$ and $\mathcal{X}'_\infty$ need not be isometric to each other. For example, for the constant sequence $\mathcal{X}_n \overset{\mathrm{iso}}{=\!=} \mathbb{R}_{\geqslant 0}$, one may take $\mathcal{X}_\infty \overset{\mathrm{iso}}{=\!=} \mathbb{R}_{\geqslant 0}$. In this case, a point in the disjoint space $\boldsymbol{X}$ can be regarded as a pair $(x,n) \in \mathbb{R}_{\geqslant 0} \times (\mathbb{Z}_{>} \cup \{\infty\})$ and the metric on $\boldsymbol{X}$ can be defined by

$$|(x,n)-(y,m)|_{\boldsymbol{X}} := |\tfrac{1}{n}+\tfrac{1}{m}| + |x-y|,$$

where we assume that $0 = \frac{1}{\infty}$. On the other hand, one can take $\mathcal{X}'_\infty \overset{\mathrm{iso}}{=\!=} \mathbb{R}$, and consider the metric

$$\begin{aligned}
|(x,n)-(y,m)|_{\boldsymbol{X}'} &= |\tfrac{1}{n}-\tfrac{1}{m}| + |(x-n)-(y-m)|,\\
|(x,n)-(y,\infty)|_{\boldsymbol{X}'} &= \tfrac{1}{n} + |(x-n)-y|,\\
|(x,\infty)-(y,\infty)|_{\boldsymbol{X}'} &= |x-y|,
\end{aligned}$$

where $n, m < \infty$.

5.5. Induced convergences. *Suppose $\mathcal{X}_n \xrightarrow{\tau} \mathcal{X}_\infty$ as in Definition* 5.4, *and $\iota_n : \mathcal{X}_n \hookrightarrow \boldsymbol{X}$, $\iota_\infty : \mathcal{X}_\infty \hookrightarrow \boldsymbol{X}$ are the corresponding inclusions.*

(a) *A sequence of points $x_n \in \mathcal{X}_n$ converges to $x_\infty \in \mathcal{X}_\infty$ (briefly, $x_n \to x_\infty$ or $x_n \xrightarrow{\tau} x_\infty$) if $|x_n - x_\infty|_{\boldsymbol{X}} \to 0$.*

(b) *A sequence of closed sets* $\mathfrak{C}_n \subset \mathcal{X}_n$ *converges to a closed set* $\mathfrak{C}_\infty \subset \mathcal{X}_\infty$ *(briefly,* $\mathfrak{C}_n \to \mathfrak{C}_\infty$ *or* $\mathfrak{C}_n \xrightarrow{\tau} \mathfrak{C}_\infty$*) if* $\mathfrak{C}_n \underset{\mathrm{H}}{\longrightarrow} \mathfrak{C}_\infty$ *as subsets of* $\mathbf{X}$.

(c) *A sequence of open sets* $\Omega_n \subset \mathcal{X}_n$ *converges to an open set* $\Omega_\infty \subset \mathcal{X}_\infty$ *(briefly,* $\Omega_n \to \Omega_\infty$ *or* $\Omega_n \xrightarrow{\tau} \Omega_\infty$*) if the complements* $\mathcal{X}_n \setminus \Omega_n$ *converge to the complement* $\mathcal{X}_\infty \setminus \Omega_\infty$ *as closed sets.*

(d) *Let* $\mathcal{X}_n \xrightarrow{\tau} \mathcal{X}_\infty$ *and* $\mathcal{Y}_n \xrightarrow{\theta} \mathcal{Y}_\infty$*. A sequence of submaps (where a submap is a map defined on a subset; see Section* 3.A*)* $\Phi_n : \mathcal{X}_n \multimap \mathcal{Y}_n$ *converges to a submap* $\Phi_\infty : \mathcal{X}_\infty \multimap \mathcal{Y}_\infty$ *if the following conditions hold:*
- $\operatorname{Dom}\Phi_n \to \operatorname{Dom}\Phi_\infty$ *as a sequence of open sets.*
- *for any* $x_\infty \in \operatorname{Dom}\Phi_\infty$ *and any sequence* $x_n \in \mathcal{X}_n$ *such that* $x_n \to \to x_\infty$*, we have*

$$\mathcal{Y}_n \ni \Phi_n(x_n) \xrightarrow{\theta} \Phi_\infty(x_\infty) \in \mathcal{Y}_\infty$$

as $n \to \infty$.

(e) *Given a sequence of measures* μ_n *on* $\mathcal{X}_n$*, we say that* μ_n *weakly converges to a measure* μ_∞ *on* $\mathcal{X}_\infty$ *(briefly,* $\mu_n \to \mu_\infty$ *or* $\mu_n \xrightarrow{\tau} \mu_\infty$*) if the push-forward measures of* μ_n *weakly converge to the push-forward measure of* μ_∞.

In other words, if for any continuous function $\phi : \mathbf{X} \to \mathbb{R}$ *with a compact support, we have*

$$\int\limits_{\mathcal{X}_n} \phi \circ \iota_n \cdot \mu_n \to \int\limits_{\mathcal{X}_\infty} \phi \circ \iota_\infty \cdot \mu_\infty$$

as $n \to \infty$.

Liftings. Given a Gromov–Hausdorff convergence $\mathcal{X}_n \underset{\mathrm{GH}}{\longrightarrow} \mathcal{X}_\infty$ and a point $p_\infty \in \mathcal{X}_\infty$, any sequence of points $p_n \in \mathcal{X}_n$ such that $p_n \underset{\mathrm{GH}}{\longrightarrow} p_\infty$ will be called a *lifting* of p_∞. The point $p_n \in \mathcal{X}_n$ will be called a *lifting* of p_∞ to $\mathcal{X}_n$. We will also say that $\operatorname{dist}_{p_n} : \mathcal{X}_n \to \mathbb{R}$ is a *lifting* of the distance function $\operatorname{dist}_{p_\infty} : \mathcal{X}_\infty \to \mathbb{R}$. Clearly $\operatorname{dist}_{p_n} \underset{\mathrm{GH}}{\longrightarrow} \operatorname{dist}_{p_\infty}$.

Note that liftings are not uniquely defined.

Similarly, we may refer to liftings of a point array $\boldsymbol{p}_\infty = (p^1_\infty, p^2_\infty, \dots, p^k_\infty)$ and of the corresponding distance map $\operatorname{dist}_{\boldsymbol{p}_\infty} : \mathcal{X}_\infty \to \mathbb{R}^k$,

$$\operatorname{dist}_{\boldsymbol{p}_\infty} : x \mapsto \left(|p^1_\infty - x|, |p^2_\infty - x|, \dots, |p^k_\infty - x|\right).$$

C. Gromov's selection theorem

5.6. Gromov's selection theorem. *Let $\mathcal{X}_n$ be a sequence of proper metric spaces with marked points $x_n \in \mathcal{X}_n$. Assume that for any $R > 0$, $\varepsilon > 0$, there is $N = N(R, \varepsilon) \in \mathbb{Z}_{>0}$ such that for each n the ball $\overline{\mathrm{B}}[x_n, R] \subset \mathcal{X}_n$ admits a finite ε-net with at most N points. Then a subsequence of $\mathcal{X}_n$ admits a Gromov–Hausdorff convergence such that the sequence of marked points $x_n \in \mathcal{X}_n$ converges.*

Proof. By the main assumption, there is a sequence of integers $M_1 < M_2 < \cdots$ such that in each space $\mathcal{X}_n$ there is a sequence of points $z_{i,n} \in \mathcal{X}_n$ for which

$$|z_{i,n} - x_n|_{\mathcal{X}_n} \leqslant k + 1 \quad \text{if } i \leqslant M_k$$

and $\{z_{1,n}, \ldots, z_{M_k,n}\}$ is a $\frac{1}{k}$-net in $\overline{\mathrm{B}}[x_n, k]_{\mathcal{X}_n}$.

Passing to a subsequence, we may assume that the sequence

$$\ell_n = |z_{i,n} - z_{j,n}|_{\mathcal{X}_n}$$

converges for any i and j.

Let us consider a countable set of points $\mathcal{W} = \{w_1, w_2, \ldots\}$ equipped with the pseudometric defined by

$$|w_i - w_j|_{\mathcal{W}} = \lim_{n\to\infty} |z_{i,n} - z_{j,n}|_{\mathcal{X}_n}.$$

Let $\hat{\mathcal{W}}$ be the metric space corresponding to $\mathcal{W}$. Denote by $\mathcal{X}_\infty$ the completion of $\hat{\mathcal{W}}$.

It remains to construct a metric on the disjoint union of

$$\boldsymbol{X} = \mathcal{X}_\infty \sqcup \mathcal{X}_1 \sqcup \mathcal{X}_2 \sqcup \cdots$$

satisfying Definitions 5.3 and 5.4.

Such a metric can be defined as follows. Fix a sequence $\varepsilon_k \to 0+$ and let N_k be the minimal integer such that

$$|w_i - w_j|_{\mathcal{W}} \lesseqgtr |z_{i,n} - z_{j,n}|_{\mathcal{X}_n} \pm \varepsilon_k$$

if $i, j \leqslant N_k$ and $n \geqslant N_k$. Let us equip $\boldsymbol{X}$ with the maximal metric such that all the inclusions $\iota_n : \mathcal{X}_n \to \boldsymbol{X}$ and $\iota_\infty : \mathcal{X}_\infty \to \boldsymbol{X}$ are isometric and $|z_{i,n} - w_i| \leqslant \varepsilon_k$ for $i \leqslant N_k$ and $n \geqslant N_k$. It is easy to verify that such a metric on $\boldsymbol{X}$ satisfies Definitions 5.3 and 5.4. □

D. Convergence of compact spaces

5.7. Definition. Let $\mathcal{X}$ and $\mathcal{Y}$ be metric spaces. A map $f : \mathcal{X} \to \mathcal{Y}$ is called an *ε-isometry* if the following two conditions hold:

(a) $\operatorname{Im} f$ is an ε-net in $\mathcal{Y}$.

(b) $||f(x) - f(x')|_{\mathcal{Y}} - |x - x'|_{\mathcal{X}}| \leqslant \varepsilon$ for any $x, x' \in \mathcal{X}$.

5.8. Lemma. *Let $\mathcal{X}_1, \mathcal{X}_2, \dots,$ and $\mathcal{X}_\infty$ be metric spaces and $\varepsilon_n \to 0+$ as $n \to \infty$. Suppose that either*

(a) *for each n there is an ε_n-isometry $f_n : \mathcal{X}_n \to \mathcal{X}_\infty$, or*

(b) *for each n there is an ε_n-isometry $h_n : \mathcal{X}_\infty \to \mathcal{X}_n$.*

Then there is a Gromov–Hausdorff convergence $\mathcal{X}_n \xrightarrow[\text{GH}]{} \mathcal{X}_\infty$.

Proof. To prove part (a), let us construct a common space $\boldsymbol{X}$ for the spaces $\mathcal{X}_1, \mathcal{X}_2, \dots,$ and $\mathcal{X}_\infty$ by taking the metric ρ on the disjoint union $\mathcal{X}_\infty \sqcup \mathcal{X}_1 \sqcup \mathcal{X}_2 \sqcup \cdots$ that is defined the following way:

$$\rho(x_n, y_n) = |x_n - y_n|_{\mathcal{X}_n}, \quad \rho(x_\infty, y_\infty) = |x_\infty - y_\infty|_{\mathcal{X}_\infty},$$
$$\rho(x_n, x_\infty) = \inf\{\, |x_n - y_n|_{\mathcal{X}_n} + \varepsilon_n + |x_\infty - f(y_n)|_{\mathcal{X}_\infty} : y_n \in \mathcal{X}_n \,\},$$
$$\rho(x_n, x_m) = \inf\{\, \rho(x_n, y_\infty) + \rho(x_m, y_\infty) : y_\infty \in \mathcal{X}_\infty \,\},$$

where we assume that $x_m \in \mathcal{X}_m$, $x_n \in \mathcal{X}_n$, and $x_\infty \in \mathcal{X}_\infty$.

It remains to observe that ρ is indeed a metric and $\mathcal{X}_n \xrightarrow[\text{H}]{} \mathcal{X}_\infty$ in $\boldsymbol{X}$.

The proof of the second part is analogous; one only needs to change one line in the definition of ρ to the following:

$$\rho(x_n, x_\infty) = \inf\{\, |x_n - h(y_\infty)|_{\mathcal{X}_n} + \varepsilon_n + |x_\infty - y_\infty|_{\mathcal{X}_\infty} : y_\infty \in \mathcal{X}_\infty \,\}. \qquad \square$$

5.9. Definition. Given two compact spaces $\mathcal{X}$ and $\mathcal{Y}$, we will write

- $\mathcal{X} \leqslant \mathcal{Y}$ if there is a noncontracting map $\Phi : \mathcal{X} \to \mathcal{Y}$.
- $\mathcal{X} \leqslant \mathcal{Y} + \varepsilon$ if there is a map $\Phi : \mathcal{X} \to \mathcal{Y}$ such that for any $x, x' \in \mathcal{X}$ we have

$$|x - x'| \leqslant |\Phi(x) - \Phi(x')| + \varepsilon.$$

5.10. Lemma. *Let $\mathcal{X}$ and $\mathcal{Y}$ be two metric spaces, and let $\mathcal{X}$ be compact. Then*

$$\mathcal{X} \geqslant \mathcal{Y} \geqslant \mathcal{X} \quad \Longleftrightarrow \quad \mathcal{X} \overset{\text{iso}}{=\!=} \mathcal{Y}.$$

The following proof was suggested by Travis Morrison.

Proof. Let $f : \mathcal{X} \to \mathcal{Y}$ and $g : \mathcal{Y} \to \mathcal{X}$ be noncontracting mappings. It is sufficient to prove that $h = g \circ f : \mathcal{X} \to \mathcal{X}$ is an isometry.

Given any pair of points $x, y \in \mathcal{X}$, let $x_n = h^{\circ n}(x)$ and $y_n = h^{\circ n}(y)$. Since $\mathcal{X}$ is compact, one can choose an increasing sequence of integers n_k such that both sequences x_{n_i} and y_{n_i} converge. In particular, both of these sequences are Cauchy, that is,

$$|x_{n_i} - x_{n_j}|, |y_{n_i} - y_{n_j}| \to 0$$

as $\min\{i, j\} \to \infty$. Since h is noncontracting, we have

$$|x - x_{|n_i - n_j|}| \leqslant |x_{n_i} - x_{n_j}|.$$

It follows that there is a sequence $m_i \to \infty$ such that

$$x_{m_i} \to x \quad \text{and} \quad y_{m_i} \to y \quad \text{as} \quad i \to \infty. \tag{1}$$

Let $\ell_n = |x_n - y_n|$. Since h is noncontracting, the sequence ℓ_n is nondecreasing. On the other hand, from (1) it follows that $\ell_{m_i} \to |x - y| = \ell_0$ as $m_i \to \infty$; that is, ℓ_n is a constant sequence. In particular, $\ell_0 = \ell_1$ for any x and y in $\mathcal{X}$, so h is a distance-preserving map.

Thus $h(\mathcal{X})$ is isometric to $\mathcal{X}$. From (1), $h(\mathcal{X})$ is everywhere dense. Since $\mathcal{X}$ is compact, $h(\mathcal{X}) = \mathcal{X}$. □

The *Gromov–Hausdorff distance* between isometry classes of compact metric spaces $\mathcal{X}$ and $\mathcal{Y}$, is defined by

$$\mathrm{D_{GH}}(\mathcal{X}, \mathcal{Y}) := \inf\{\,\varepsilon > 0 : \mathcal{X} \leqslant \mathcal{Y} + \varepsilon \text{ and } \mathcal{Y} \leqslant \mathcal{X} + \varepsilon\,\}.$$

The Gromov–Hausdorff distance turns the set of all isometry classes of compact metric spaces into a metric space. The following theorem shows that convergence in this space coincides with the Gromov–Hausdorff convergence defined above.

5.11. Theorem. *Let $\mathcal{X}_1, \mathcal{X}_2, \ldots,$ and $\mathcal{X}_\infty$ be compact metric spaces. Then there is a convergence $\mathcal{X}_n \xrightarrow[\mathrm{GH}]{} \mathcal{X}_\infty$ if and only if $\mathrm{D_{GH}}(\mathcal{X}_n, \mathcal{X}_\infty) \to 0$ as $n \to \infty$.*

Proof.

"If" part. Suppose $a_n : \mathcal{X}_\infty \to \mathcal{X}_n$ and $b_n : \mathcal{X}_n \to \mathcal{X}_\infty$ are sequences of maps such that

$$|a_n(x) - a_n(y)|_{\mathcal{X}_\infty} \geqslant |x - y|_{\mathcal{X}_n} - \delta_n,$$
$$|b_n(v) - b_n(w)|_{\mathcal{X}_n} \geqslant |v - w|_{\mathcal{X}_\infty} - \delta_n$$

for any $x, y \in \mathcal{X}_n$, $v, w \in \mathcal{X}_\infty$, and a sequence $\delta_n \to 0+$.

Fix $\varepsilon > 0$ and choose a maximal ε-packing $\{x^1, x^2, \ldots, x^k\}$ in $\mathcal{X}_\infty$ such that $\sum_{i<j} |x^i - x^j|$ is maximal. Note that

$$|a_n \circ b_n(x^i) - a_n \circ b_n(x^j)| \geqslant |x^i - x^j| - 2\cdot\delta_n.$$

Since $\sum_{i<j} |x^i - x^j|$ is maximal,

$$|a_n \circ b_n(x^i) - a_n \circ b_n(x^j)| \to |x^i - x^j|$$

for all i and j as $n \to \infty$. For all large n, we have $2\cdot\delta_n < |x^i - x^j| - \varepsilon$, and so

$$|b_n(x^i) - b_n(x^j)|_{\mathcal{X}_n} > \varepsilon \quad \text{and} \quad |a_n \circ b_n(x^i) - a_n \circ b_n(x^j)|_{\mathcal{X}_\infty} > \varepsilon$$

for all $i \neq j$. Therefore for each large n, the set $\{a_n \circ b_n(x^i)\}$ is a maximal ε-packing and hence an ε-net in $\mathcal{X}_\infty$.

Since $\{a_n \circ b_n(x^i)\}$ is an ε-net in $\mathcal{X}_\infty$, we have that for any $y_n \in \mathcal{X}_n$ there is x^i such that $|a_n \circ b_n(x^i) - a_n(y_n)| < \varepsilon$. Thus $|b_n(x^i) - y_n| < \varepsilon + \delta_n$, that is, $\{b_n(x^i)\}$ is a $(\varepsilon + \delta_n)$-net in $\mathcal{X}_n$.

Given $y \in \mathcal{X}_n$, choose x^i so that $|b_n(x^i) - y_n| < \varepsilon + \delta_n$ and define $h_n(y) = a_n \circ b_n(x^i)$. Observe that h_n is a $3\cdot\varepsilon$-isometry for all large n. Since $\varepsilon > 0$ is arbitrary, there is a sequence of ε_n-isometries $\mathcal{X}_n \to \mathcal{X}_\infty$ such that $\varepsilon_n \to 0+$ as $n \to \infty$. It remains to apply Lemma 5.8.

"Only-if" part. Assume $\mathcal{X}_n \xrightarrow{\tau} \mathcal{X}_\infty$. Fix $\varepsilon > 0$, and choose a maximal ε-packing $\{x^1, x^2, \dots, x^k\}$ in $\mathcal{X}_\infty$. For each x^i, choose a sequence $x_n^i \in \mathcal{X}_n$ such that $x_n^i \to x^i$. Define a map $a_n : \mathcal{X}_n \to \mathcal{X}_\infty$ such that $a_n(x_n^i) = x_n$. Note that for all large n, we have $|x_n^i - x_n^j| > \varepsilon$. For each point $z \in \mathcal{X}_\infty$, choose x^i so that $|z - x^i| < \varepsilon$. Define a map $b_n : \mathcal{X}_\infty \to \mathcal{X}_n$ by setting $b_n(z) = x_n^i$. Observe that

$$|b_n(y) - b_n(z)|_{\mathcal{X}_n} + 3\cdot\varepsilon > |y - z|_{\mathcal{X}_\infty}$$

for all large n.

In the same way we can construct a map $a_n : \mathcal{X}_n \to \mathcal{X}_\infty$ such that

$$|a_n(y) - a_n(z)|_{\mathcal{X}_\infty} + 3\cdot\varepsilon > |y - z|_{\mathcal{X}_n}.$$

Hence $\mathrm{D_{GH}}(\mathcal{X}_n, \mathcal{X}_\infty) \to 0$ as $n \to \infty$. □

The following theorem states that the isometry class of a Gromov–Hausdorff limit is uniquely defined if it is compact.

5.12. Theorem. *Let $\mathcal{X}_1$, $\mathcal{X}_2$, ..., and $\mathcal{X}_\infty$ and $\bar{\mathcal{X}}_\infty$ be metric spaces such that $\mathcal{X}_n \xrightarrow{\tau} \mathcal{X}_\infty$, $\mathcal{X}_n \xrightarrow{\bar{\tau}} \bar{\mathcal{X}}_\infty$.*

Assume that $\bar{\mathcal{X}}_\infty$ is compact. Then $\mathcal{X}_\infty \overset{\text{iso}}{=\!=} \bar{\mathcal{X}}_\infty$.

Proof. For each point $x_\infty \in \mathcal{X}_\infty$, choose liftings $x_n \in \mathcal{X}_n$.

Choose a nonprincipal ultrafilter ω on $\mathbb{N}$. Define $\bar{x}_\infty \in \bar{\mathcal{X}}_\infty$ as the ω-limit of x_n with respect to $\bar{\tau}$. We claim that the map $x_\infty \to \bar{x}_\infty$ is an isometry.

Indeed, by the definition of Gromov–Hausdorff convergence,

$$|\bar{x}_\infty - \bar{y}_\infty|_{\bar{\mathcal{X}}_\infty} = \lim_{n\to\omega} |x_n - y_n|_{\mathcal{X}_n} = |x_\infty - y_\infty|_{\mathcal{X}_\infty}.$$

Thus the map $x_\infty \to \bar{x}_\infty$ gives a distance-preserving map $\Phi : \mathcal{X}_\infty \hookrightarrow \bar{\mathcal{X}}_\infty$. In particular, $\mathcal{X}_\infty$ is compact. Switching $\mathcal{X}_\infty$ and $\bar{\mathcal{X}}_\infty$ and applying the same argument, we get an isometric embedding $\bar{\mathcal{X}}_\infty \hookrightarrow \mathcal{X}_\infty$. Now the result follows from Lemma 5.10. □

5.13. Exercise.

(a) Show that a sequence of compact simply connected length spaces cannot converge to a circle.

(b) Construct a sequence of compact simply connected length spaces that converges to a compact nonsimply connected space.

5.14. Exercise.

(a) Show that a sequence of length metrics on the 2-sphere cannot converge to the unit disk.

(b) Construct a sequence of length metrics on the 3-sphere that converges to a unit 3-ball.

5.15. Exercise. Let $\mathcal{X}_n$ be a sequence of metric spaces that admits two Gromov–Hausdorff convergences τ and τ'. Assume $\mathcal{X}_n \xrightarrow{\tau} \mathcal{X}_\infty$ and $\mathcal{X}_n \xrightarrow{\tau'} \mathcal{X}'_\infty$. Show that if $\mathcal{X}_\infty$ is proper and there is a sequence of points $x_n \in \mathcal{X}_n$ that converges in both τ and τ', then $\mathcal{X}_\infty \overset{\text{iso}}{=\!=} \mathcal{X}'_\infty$.

E. Ultralimits revisited

Recall that ω denotes an ultrafilter of the set of natural numbers.

5.16. Theorem. *Assume $\mathcal{X}_n$ is a sequence of complete metric spaces. Let $\mathcal{X}_n \to \mathcal{X}_\omega$ as $n \to \omega$, and let $\mathcal{Y}_n$ be a sequence of subspaces of $\mathcal{X}_n$ such that $\mathcal{Y}_n \underset{\text{GH}}{\longrightarrow} \mathcal{Y}_\infty$. Then there is a distance-preserving map $\iota : \mathcal{Y}_\infty \to \mathcal{X}_\omega$. Moreover:*

(a) *If $\mathcal{X}_n \underset{\text{GH}}{\longrightarrow} \mathcal{X}_\infty$ and $\mathcal{X}_\infty$ is compact, then $\mathcal{X}_\infty$ is isometric to $\mathcal{X}_\omega$.*

(b) *If $\mathcal{X}_n \underset{\text{GH}}{\longrightarrow} \mathcal{X}_\infty$ and $\mathcal{X}_\infty$ is proper, then $\mathcal{X}_\infty$ is isometric to a metric component of $\mathcal{X}_\omega$.*

Proof. For each point $y_\infty \in \mathcal{Y}_\infty$ choose a lifting $y_n \in \mathcal{Y}_n$. Pass to the ω-limit $y_\omega \in \mathcal{X}_\omega$ of y_n. Clearly for any $y_\infty, z_\infty \in \mathcal{Y}_\infty$, we have

$$|y_\infty - z_\infty|_{\mathcal{Y}_\infty} = |y_\omega - z_\omega|_{\mathcal{X}_\omega};$$

that is, the map $y_\infty \mapsto y_\omega$ gives a distance-preserving map $\iota : \mathcal{Y}_\infty \to \mathcal{X}_\omega$.

(a) + (b). Fix $x_\omega \in \mathcal{X}_\omega$. Choose a sequence x_n of points in $\mathcal{X}_n$, such that $x_n \to \to x_\omega$ as $n \to \omega$.

Denote by $\boldsymbol{X} = \mathcal{X}_\infty \sqcup \mathcal{X}_1 \sqcup \mathcal{X}_2 \sqcup \cdots$ the common space for the convergence $\mathcal{X}_n \underset{\text{GH}}{\longrightarrow} \mathcal{X}_\infty$, as in the definition of Gromov–Hausdorff convergence. Note that x_n is a sequence of points in $\boldsymbol{X}$.

If the ω-limit x_∞ of x_n in $\boldsymbol{X}$ exists, it must lie in $\mathcal{X}_\infty$.

The point x_∞, if defined, does not depend on the choice of x_n. Indeed, if $y_n \in \mathcal{X}_n$ is another sequence such that $y_n \to x_\omega$ as $n \to \omega$, then

$$|y_\infty - x_\infty| = \lim_{n \to \omega} |y_n - x_n| = 0;$$

therefore, $x_\infty = y_\infty$.

This way we obtain a map $\nu\colon x_\omega \to x_\infty$, defined on $\operatorname{Dom}\nu \subset \mathcal{X}_\omega$. By construction of ι, we have $\iota\circ\nu(x_\omega) = x_\omega$ for any $x_\omega \in \operatorname{Dom}\nu$.

Finally note that if $\mathcal{X}_\infty$ is compact, then ν is defined on all of $\mathcal{X}_\omega$; this proves (a).

If $\mathcal{X}_\infty$ is proper, choose any point $z_\infty \in \mathcal{X}_\infty$ and set $z_\omega = \iota(z_\infty)$. For any point $x_\omega \in \mathcal{X}_\omega$ at finite distance from z_ω, for the sequence x_n as above we have that $|z_n - x_n|$ is bounded for ω-almost all n. Since $\mathcal{X}_\infty$ is proper, $\nu(x_\omega)$ is defined; in other words, ν is defined on the metric component of z_ω. Hence (b) follows. □

Chapter 6

The ghost of Euclid

A. Geodesics, triangles and hinges

Geodesics and their relatives. Let $\mathcal{X}$ be a metric space, and let $\mathbb{I} \subset \mathbb{R}$ be an interval. A globally distance-preserving map $\gamma \colon \mathbb{I} \to \mathcal{X}$ is called a *unit-speed geodesic*. (Various authors call it differently: *shortest path*, *minimizing geodesic*.) In other words, $\gamma \colon \mathbb{I} \to \mathcal{X}$ is a unit-speed geodesic if the equality

$$|\gamma(s) - \gamma(t)|_{\mathcal{X}} = |s - t|$$

holds for any pair $s, t \in \mathbb{I}$.

A unit-speed geodesic between p and q in $\mathcal{X}$ will be denoted by $\mathrm{geod}_{[pq]}$. We will always assume $\mathrm{geod}_{[pq]}$ is parametrized starting at p; that is, $\mathrm{geod}_{[pq]}(0) = p$ and $\mathrm{geod}_{[pq]}(|p - q|) = q$. The image of $\mathrm{geod}_{[pq]}$ will be denoted by $[pq]$ and called a *geodesic*. The term *geodesic* will also be used for a linear reparametrization of a unit-speed geodesic. With a slight abuse of notation, we will use the notation $[pq]$ also for the class of all linear reparametrizations of $\mathrm{geod}_{[pq]}$.

A unit-speed geodesic $\gamma \colon \mathbb{R}_{\geqslant 0} \to \mathcal{X}$ is called a *half-line*.

A unit-speed geodesic $\gamma \colon \mathbb{R} \to \mathcal{X}$ is called a *line*.

A piecewise geodesic curve is called a *polygonal line*; we may say a *polygonal line* $p_1, \dots, p_n$ meaning a polygonal line with edges $[p_1 p_2], \dots, [p_{n-1} p_n]$. A closed polygonal line will be also called a polygon.

We may write $[pq]_{\mathcal{X}}$ to emphasize that the geodesic $[pq]$ is in the space $\mathcal{X}$. Also, we use the following short-cut notation:

$$]pq[= [pq] \setminus \{p, q\}, \qquad]pq] = [pq] \setminus \{p\}, \qquad [pq[= [pq] \setminus \{q\}.$$

In general, a geodesic between p and q need not exist, and if it does exist, it need not be unique. However, once we write $\mathrm{geod}_{[pq]}$ or $[pq]$, we mean that we have fixed a choice of a geodesic.

A constant-speed geodesic $\gamma\colon [0,1] \to \mathcal{X}$ is called a *geodesic path*. Given a geodesic $[pq]$, we denote by $\mathrm{path}_{[pq]}$ the corresponding geodesic path; that is,

$$\mathrm{path}_{[pq]}(t) \equiv \mathrm{geod}_{[pq]}(t\cdot|p-q|).$$

A curve $\gamma\colon \mathbb{I} \to \mathcal{X}$ is called a *local geodesic* if for any $t \in \mathbb{I}$ there is a neighborhood $U \ni t$ in $\mathbb{I}$ such that the restriction $\gamma|_U$ is a constant-speed geodesic. If $\mathbb{I} = [0,1]$, then γ is called a *local geodesic path*.

6.1. Proposition. *Suppose $\mathcal{X}$ is a metric space and $\gamma\colon [0,\infty) \to \mathcal{X}$ is a half-line. Then the* Busemann function $\mathrm{bus}_\gamma\colon \mathcal{X} \to \mathbb{R}$

$$\mathrm{bus}_\gamma(x) = \lim_{t\to\infty} |\gamma(t) - x| - t \tag{1}$$

is defined and is 1*-Lipschitz.*

Proof. By the triangle inequality, the function $t \mapsto |\gamma(t)-x|-t$ is nonincreasing. Clearly $|\gamma(t)-x|-t \geqslant -|\gamma(0)-x|$. Thus the limit in (1) is defined, and it is 1-Lipschitz as a limit of 1-Lipschitz functions. □

6.2. Example. If $\mathcal{X}$ is a Euclidean space and $\gamma(t) = p + t\cdot v$ where v is a unit vector, then

$$\mathrm{bus}_\gamma(x) = \langle x-p, v\rangle.$$

Triangles. For a triple of points $p, q, r \in \mathcal{X}$, a choice of a triple of geodesics $([qr],[rp],[pq])$ will be called a *triangle*, and we will use the short notation $[pqr] = ([qr],[rp],[pq])$. Again, given a triple $p, q, r \in \mathcal{X}$, there may be no triangle $[pqr]$, simply because one of the pairs of these points cannot be joined by a geodesic. Or there may be many different triangles, any of which can be denoted by $[pqr]$. Once we write $[pqr]$, it means we have chosen such a triangle; that is, made a choice of each $[qr]$, $[rp]$, and $[pq]$.

The value $|p-q|+|q-r|+|r-p|$ will be called the *perimeter of triangle* $[pqr]$; it obviously coincides with perimeter of the triple p, q, r as defined below.

Hinges. Let $p, x, y \in \mathcal{X}$ be a triple of points such that p is distinct from x and y. A pair of geodesics $([px],[py])$ will be called a *hinge*, and will be denoted by $[p\,{}^x_y] = ([px],[py])$.

B. Model angles and triangles

Let $\mathcal{X}$ be a metric space, $p, q, r \in \mathcal{X}$, and $\kappa \in \mathbb{R}$. Let us define the *model triangle* $[\tilde p\tilde q\tilde r]$ (briefly, $[\tilde p\tilde q\tilde r] = \tilde\triangle^\kappa(pqr)$) to be a triangle in the model plane $\mathbb{M}^2(\kappa)$ such that

$$|\tilde p-\tilde q| = |p-q|,\quad |\tilde q-\tilde r| = |q-r|,\quad |\tilde r-\tilde p| = |r-p|.$$

In the notation of Section 1.A, $\tilde\triangle^\kappa(pqr) = \tilde\triangle^\kappa\{|q-r|,|r-p|,|p-q|\}$.

If $\kappa \leqslant 0$, the model triangle is always defined, that is, it exists and is unique up to an isometry of $\mathbb{M}^2(\kappa)$. If $\kappa > 0$, the model triangle is said to be defined if in addition

$$|p-q| + |q-r| + |r-p| < 2\cdot\varpi^\kappa;$$

here ϖ^κ denotes the diameter of the model space $\mathbb{M}^2(\kappa)$. In this case, the model triangle also exists and is unique up to an isometry of $\mathbb{M}^2(\kappa)$. The value $|p-q| + |q-r| + |r-p|$ will be called the *perimeter of the triple* p, q, r.

If for $p, q, r \in \mathcal{X}$, $[\tilde p\tilde q\tilde r] = \tilde\triangle^\kappa(pqr)$ is defined and $|p-q|, |p-r| > 0$, the angle measure of $[\tilde p\tilde q\tilde r]$ at $\tilde p$ will be called the *model angle* of the triple p, q, r, and will be denoted by $\tilde\measuredangle^\kappa\left(p\,{}^q_r\right)$.

In the notation of Section 1.A,

$$\tilde\measuredangle^\kappa\left(p\,{}^q_r\right) = \tilde\measuredangle^\kappa\{|q-r|;|p-q|,|p-r|\}.$$

6.3. Alexandrov's lemma. *Let p, q, r, z be distinct points in a metric space such that $z \in \,]pr[$ and*

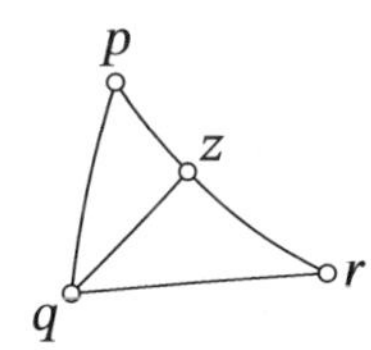

$$|p-q| + |q-r| + |r-p| < 2\cdot\varpi^\kappa.$$

Then the following expressions have the same sign:

(a) $\tilde\measuredangle^\kappa\left(p\,{}^q_r\right) - \tilde\measuredangle^\kappa\left(p\,{}^q_z\right)$,

(b) $\tilde\measuredangle^\kappa\left(z\,{}^q_p\right) + \tilde\measuredangle^\kappa\left(z\,{}^q_r\right) - \pi$.

Moreover,

$$\tilde\measuredangle^\kappa\left(q\,{}^p_r\right) \geqslant \tilde\measuredangle^\kappa\left(q\,{}^p_z\right) + \tilde\measuredangle^\kappa\left(q\,{}^z_r\right),$$

with equality if and only if the expressions in (a) *and* (b) *vanish.*

Proof. By the triangle inequality,

$$|p-q| + |q-z| + |z-p| \leqslant |p-q| + |q-r| + |r-p| < 2\cdot\varpi^\kappa.$$

Therefore the model triangle $[\tilde p\tilde q\tilde z] = \tilde\triangle^\kappa pqz$ is defined. Take a point $\tilde r$ on the extension of $[\tilde p\tilde z]$ beyond $\tilde z$ so that $|\tilde p-\tilde r| = |p-r|$ (and therefore $|\tilde p-\tilde z| = |p-z|$).

From monotonicity of the function $a \mapsto \tilde\measuredangle^\kappa\{a;b,c\}$ (see (1.1(c))), the following expressions have the same sign:

(i) $\measuredangle\left[\tilde p\,{}^{\tilde q}_{\tilde r}\right] - \tilde\measuredangle^\kappa\left(p\,{}^q_r\right)$;

(ii) $|\tilde p-\tilde r| - |p-r|$;

(iii) $\measuredangle\left[\tilde z\,{}^{\tilde q}_{\tilde r}\right] - \tilde\measuredangle^\kappa\left(z\,{}^q_r\right)$.

Since

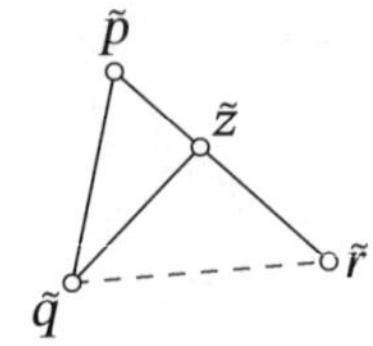

$$\measuredangle\left[\tilde p\,{}^{\tilde q}_{\tilde r}\right] = \measuredangle\left[\tilde p\,{}^{\tilde q}_{\tilde z}\right] = \tilde\measuredangle^\kappa\left(p\,{}^q_z\right)$$

and

$$\measuredangle\left[\tilde z\,{}^{\tilde q}_{\tilde r}\right] = \pi - \measuredangle\left[\tilde z\,{}^{\tilde p}_{\tilde q}\right] = \pi - \tilde\measuredangle^\kappa\left(z\,{}^p_q\right),$$

the first statement follows.

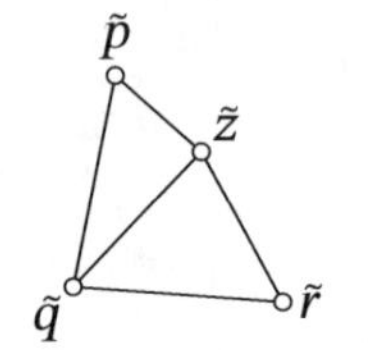

For the second statement, let us redifine $\tilde r$; construct $[\tilde q\tilde z\tilde r] = \tilde\triangle^\kappa qzr$ on the opposite side of $[\tilde q\tilde z]$ from $[\tilde p\tilde q\tilde z]$. Since

$$\begin{aligned}|\tilde p - \tilde r| &\leqslant |\tilde p - \tilde z| + |\tilde z - \tilde r| = \\ &= |p - z| + |z - r| = \\ &= |p - r|,\end{aligned}$$

we have

$$\begin{aligned}\tilde\measuredangle^\kappa\left(q\,{}^p_z\right) + \tilde\measuredangle^\kappa\left(q\,{}^z_r\right) &= \measuredangle\left[\tilde q\,{}^{\tilde p}_{\tilde z}\right] + \measuredangle\left[\tilde q\,{}^{\tilde z}_{\tilde r}\right] = \\ &= \measuredangle\left[\tilde q\,{}^{\tilde p}_{\tilde r}\right] \leqslant \\ &\leqslant \tilde\measuredangle^\kappa\left(q\,{}^p_r\right).\end{aligned}$$

Equality holds if and only if $|\tilde p - \tilde r| = |p - r|$, as required. □

C. Angles and the first variation

Given a hinge $\left[p\,{}^x_y\right]$, we define its *angle* to be

$$\measuredangle\left[p\,{}^x_y\right] := \lim_{\bar x,\bar y\to p} \tilde\measuredangle^\kappa\left(p\,{}^{\bar x}_{\bar y}\right), \tag{1}$$

for $\bar x \in\,]px]$ and $\bar y \in\,]py]$, if this limit exists.

Similarly to $\tilde\measuredangle^\kappa\left(p\,{}^x_y\right)$, we will use the short notation

$$\tilde\curlyvee^\kappa\left[p\,{}^x_y\right] = \tilde\curlyvee^\kappa\left\{\measuredangle\left[p\,{}^x_y\right]; |p-x|, |p-y|\right\},$$

where the right-hand side is defined in Section 1.A. The value $\tilde\curlyvee^\kappa\left[p\,{}^x_y\right]$ will be called the *model side* of the hinge $\left[p\,{}^x_y\right]$.

6.4. Lemma. *Let p, x, y be a triple of points in a metric space with perimeter ℓ. Then for any $\kappa, \mathrm{K} \in \mathbb{R}$,*

$$\left|\tilde\measuredangle^{\mathrm{K}}\left(p\,{}^x_y\right) - \tilde\measuredangle^\kappa\left(p\,{}^x_y\right)\right| \leqslant 100(|\mathrm{K}| + |\kappa|)\cdot\ell^2, \tag{2}$$

whenever the left-hand side is defined.

Lemma 6.4 implies that the definition of angle is independent of κ. In particular, one can take $\kappa = 0$ in (1); thus the angle can be calculated from the cosine law:

$$\cos\tilde\measuredangle^0\left(p\,{}^x_y\right) = \frac{|p-x|^2 + |p-y|^2 - |x-y|^2}{2\cdot|p-x|\cdot|p-y|}.$$

Proof. The function $\kappa \mapsto \tilde{\measuredangle}^{\kappa}(p\,{}^{x}_{y})$ is nondecreasing (in Section 1.1(d)). Thus, for $\mathrm{K} > \kappa$, we have

$$\begin{aligned}0 \leqslant \tilde{\measuredangle}^{\mathrm{K}}(p\,{}^{x}_{y}) - \tilde{\measuredangle}^{\kappa}(p\,{}^{x}_{y}) &\leqslant \tilde{\measuredangle}^{\mathrm{K}}(p\,{}^{x}_{y}) + \tilde{\measuredangle}^{\mathrm{K}}(x\,{}^{p}_{y}) + \tilde{\measuredangle}^{\mathrm{K}}(y\,{}^{p}_{x}) - \\ &\quad - \tilde{\measuredangle}^{\kappa}(p\,{}^{x}_{y}) - \tilde{\measuredangle}^{\kappa}(x\,{}^{p}_{y}) - \tilde{\measuredangle}^{\kappa}(y\,{}^{p}_{x}) = \\ &= \mathrm{K}\cdot \operatorname{area}\tilde{\triangle}^{\mathrm{K}}(pxy) - \kappa\cdot \operatorname{area}\tilde{\triangle}^{\kappa}(pxy).\end{aligned}$$

Note that for $\kappa \geqslant 0$, a triangle of perimeter ℓ in $\mathbb{M}^2(\kappa)$ lies in a ball of radius $2\cdot\ell$, which easily implies that $\operatorname{area}\tilde{\triangle}^{\kappa}(pxy) \leqslant 100\cdot\ell^2$. For $\kappa < 0$ one gets the same estimate by a direct computation in the hyperbolic plane.

Therefore

$$\operatorname{area}\tilde{\triangle}^{\kappa}(pxy) \leqslant 100\cdot\ell^2, \qquad \operatorname{area}\tilde{\triangle}^{\mathrm{K}}(pxy) \leqslant 100\cdot\ell^2.$$

Thus (2) follows. □

6.5. Triangle inequality for angles. *Let $[px^1]$, $[px^2]$, and $[px^3]$ be three geodesics in a metric space. If all of the angles $\alpha^{ij} = \measuredangle\left[p\,{}^{x^i}_{x^j}\right]$ are defined, then they satisfy the triangle inequality*

$$\alpha^{13} \leqslant \alpha^{12} + \alpha^{23}.$$

Proof. Since $\alpha^{13} \leqslant \pi$, we can assume that $\alpha^{12} + \alpha^{23} < \pi$. Set $\gamma^i = \operatorname{geod}_{[px^i]}$. Given any $\varepsilon > 0$, for all sufficiently small $t, \tau, s \in \mathbb{R}_{\geqslant 0}$ we have

$$\begin{aligned}|\gamma^1(t) - \gamma^3(\tau)| &\leqslant |\gamma^1(t) - \gamma^2(s)| + |\gamma^2(s) - \gamma^3(\tau)| < \\ &< \sqrt{t^2 + s^2 - 2\cdot t\cdot s\cdot \cos(\alpha^{12} + \varepsilon)} + \\ &\quad + \sqrt{s^2 + \tau^2 - 2\cdot s\cdot \tau\cdot \cos(\alpha^{23} + \varepsilon)} \leqslant\end{aligned}$$

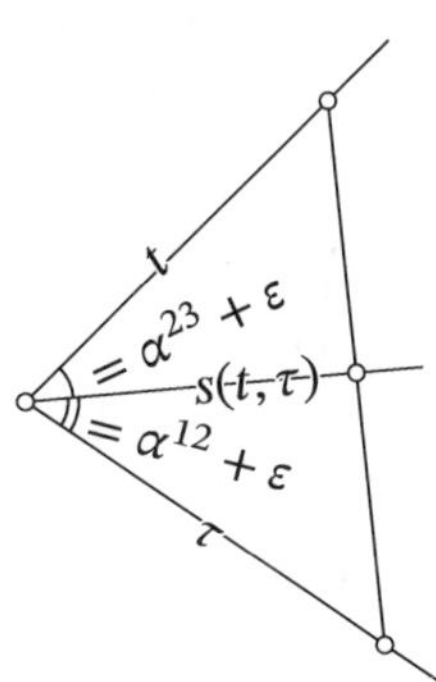

Below we define $s(t,\tau)$ so that for $s = s(t,\tau)$, this chain of inequalities can be continued as follows:

$$\leqslant \sqrt{t^2 + \tau^2 - 2\cdot t\cdot \tau\cdot \cos(\alpha^{12} + \alpha^{23} + 2\cdot\varepsilon)}.$$

Thus for any $\varepsilon > 0$,

$$\alpha^{13} \leqslant \alpha^{12} + \alpha^{23} + 2\cdot\varepsilon.$$

Hence the result follows.

To define $s(t,\tau)$, consider three half-lines $\tilde{\gamma}^1$, $\tilde{\gamma}^2$, $\tilde{\gamma}^3$ on a Euclidean plane starting at one point, such that $\measuredangle(\tilde{\gamma}^1,\tilde{\gamma}^2) = \alpha^{12} + \varepsilon$, $\measuredangle(\tilde{\gamma}^2,\tilde{\gamma}^3) = \alpha^{23} + \varepsilon$, and $\measuredangle(\tilde{\gamma}^1,\tilde{\gamma}^3) = \alpha^{12} + \alpha^{23} + 2\cdot\varepsilon$. We parametrize each half-line by the distance from the starting point. Given two positive numbers $t, \tau \in \mathbb{R}_{\geqslant 0}$, let $s = s(t,\tau)$ be the number such that $\tilde{\gamma}^2(s) \in [\tilde{\gamma}^1(t)\,\tilde{\gamma}^3(\tau)]$. Clearly $s \leqslant \max\{t,\tau\}$, so t, τ, s may be taken sufficiently small. □

6.6. Exercise. Prove that the sum of adjacent angles is at least π.

More precisely, suppose that the hinges $[p\,{}^x_z]$ and $[p\,{}^y_z]$ are *adjacent*; that is, the side $[pz]$ is shared and the union of two sides $[px]$ and $[py]$ is a geodesic $[xy]$. Then

$$\measuredangle[p\,{}^x_z] + \measuredangle[p\,{}^y_z] \geqslant \pi$$

whenever each angle on the left-hand side is defined.

The above inequality can be strict. For example in a metric tree angles between any two different edges coming out of the same vertex are all equal to π.

6.7. First variation inequality. *Assume that for a hinge $[q\,{}^p_x]$, the angle $\alpha = \measuredangle[q\,{}^p_x]$ is defined. Then*

$$|p - \mathrm{geod}_{[qx]}(t)| \leqslant |q - p| - t\cdot\cos\alpha + o(t).$$

Proof. Take a sufficiently small $\varepsilon > 0$. For all sufficiently small $t > 0$, we have

$$\begin{aligned}|\mathrm{geod}_{[qp]}(t/\varepsilon) - \mathrm{geod}_{[qx]}(t)| &\leqslant \tfrac{t}{\varepsilon}\cdot\sqrt{1+\varepsilon^2 - 2\cdot\varepsilon\cdot\cos\alpha} + o(t) \leqslant \\ &\leqslant \tfrac{t}{\varepsilon} - t\cdot\cos\alpha + t\cdot\varepsilon.\end{aligned}$$

Applying the triangle inequality, we get

$$\begin{aligned}|p - \mathrm{geod}_{[qx]}(t)| &\leqslant |p - \mathrm{geod}_{[qp]}(t/\varepsilon)| + |\mathrm{geod}_{[qp]}(t/\varepsilon) - \mathrm{geod}_{[qx]}(t)| \leqslant \\ &\leqslant |p - q| - t\cdot\cos\alpha + t\cdot\varepsilon\end{aligned}$$

for any $\varepsilon > 0$ and all sufficiently small t. Hence the result. □

D. Space of directions

Let $\mathcal{X}$ be a metric space. If the angle $\measuredangle[p\,{}^x_y]$ is defined for any hinge $[p\,{}^x_y]$ in $\mathcal{X}$, then we will say that the space $\mathcal{X}$ has *defined angles*.

Let us note that this is a strong condition. For example a Banach space which has defined angles must be Hilbert.

Let $\mathcal{X}$ be a space with defined angles. For $p \in \mathcal{X}$, consider the set $\mathfrak{S}_p$ of all nontrivial unit-speed geodesics starting at p. By Theorem 6.5, the triangle inequality holds for $\measuredangle$ on $\mathfrak{S}_p$, that is, $(\mathfrak{S}_p, \measuredangle)$ forms a pseudometric space.

The metric space corresponding to $(\mathfrak{S}_p, \measuredangle)$ is called the *space of geodesic directions* at p, denoted by Σ'_p or $\Sigma'_p\mathcal{X}$. The elements of Σ'_p are called *geodesic directions* at p. Each geodesic direction is formed by an equivalence class of geodesics starting from p for the equivalence relation

$$[px] \sim [py] \quad\Longleftrightarrow\quad \measuredangle[p\,{}^x_y] = 0;$$

the direction of $[px]$ is denoted by $\uparrow_{[px]}$.

The completion of Σ'_p is called the *space of directions* at p and is denoted by Σ_p or $\Sigma_p\mathcal{X}$. The elements of Σ_p are called *directions* at p.

E. Tangent space

The *Euclidean cone* $\mathcal{Y} = \operatorname{Cone}\mathcal{X}$ over a metric space $\mathcal{X}$ is defined as the metric space whose underlying set consists of equivalence classes in $[0,\infty)\times\mathcal{X}$ with the equivalence relation "$\sim$" given by $(0,p)\sim(0,q)$ for any points $p,q\in\mathcal{X}$, and whose metric is given by the cosine rule

$$|(s,p)-(t,q)|_{\mathcal{Y}} = \sqrt{s^2+t^2-2\cdot s\cdot t\cdot\cos\theta},$$

where $\theta = \min\{\pi, |p-q|_{\mathcal{X}}\}$. The point in $\mathcal{Y}$ that corresponds $(t,x)\in[0,\infty)\times\mathcal{X}$ will be denoted by $t\cdot x$.

The point in $\operatorname{Cone}\mathcal{X}$ formed by the equivalence class of $\{0\}\times\mathcal{X}$ is called the *tip of the cone* and is denoted by 0 or $0_{\mathcal{Y}}$. For $v\in\mathcal{Y}$ the distance $|0-v|_{\mathcal{Y}}$ is called the norm of v and is denoted by $|v|$ or $|v|_{\mathcal{Y}}$.

The *scalar product* $\langle v,w\rangle$ of two vectors $v=s\cdot p$ and $w=t\cdot q$ is defined by

$$\langle v,w\rangle := |v|\cdot|w|\cdot\cos\theta;$$

we set $\langle v,w\rangle := 0$ if $v=0$ or $w=0$.

6.8. Example. $\operatorname{Cone}\mathbb{S}^n$ is isometric to $\mathbb{R}^{n+1}$. If $G<\mathrm{O}(n+1)$ is a closed subgroup, then $\operatorname{Cone}(\mathbb{S}^n/G)$ is isometric to $\mathbb{R}^{n+1}/G$.

The Euclidean cone $\operatorname{Cone}\Sigma_p$ over the space of directions Σ_p is called the *tangent space* at p and is denoted by T_p or $\mathrm{T}_p\mathcal{X}$. The elements of $\mathrm{T}_p\mathcal{X}$ will be called *tangent vectors* at p (despite the fact that T_p is only a cone—not a vector space).

The tangent space T_p could be also defined directly, without introducing the space of directions. To do so, consider the set $\mathfrak{T}_p$ of all geodesics starting at p, with arbitrary speed. Given $\alpha,\beta\in\mathfrak{T}_p$, set

$$(1)\qquad |\alpha-\beta|_{\mathfrak{T}_p} = \lim_{\varepsilon\to 0}\frac{|\alpha(\varepsilon)-\beta(\varepsilon)|_{\mathcal{X}}}{\varepsilon}.$$

If the angles in $\mathcal{X}$ are defined, then so is the limit in (1), and we obtain a pseudometric on $\mathfrak{T}_p$.

The corresponding metric space admits a natural isometric identification with the cone $\mathrm{T}'_p = \operatorname{Cone}\Sigma'_p$. The vectors of T'_p are the equivalence classes for the relation

$$\alpha\sim\beta \quad\Longleftrightarrow\quad |\alpha(t)-\beta(t)|_{\mathcal{X}} = o(t).$$

The completion of T'_p is therefore naturally isometric to T_p. A vector in T'_p that corresponds to the geodesic path $\operatorname{geod}_{[pq]}$ is called the *logarithm of* $[pq]$ and denoted by $\log[pq]$.

F. Velocity of curves

6.9. Definition. Let $\mathcal{X}$ be a metric space, let $a > 0$, and let $\alpha\colon [0,a) \to \mathcal{X}$ be a function, not necessarily continuous, such that $\alpha(0) = p$. We say that $v \in \mathrm{T}_p$ is the *right derivative* of α at 0, briefly $\alpha^+(0) = v$, if for some (and therefore any) sequence of vectors $v_n \in \mathrm{T}'_p$ such that $v_n \to v$ as $n \to \infty$, and corresponding geodesics γ_n, we have

$$\overline{\lim_{\varepsilon\to 0+}} \frac{|\alpha(\varepsilon) - \gamma_n(\varepsilon)|_{\mathcal{X}}}{\varepsilon} \to 0 \quad \text{as} \quad n \to \infty.$$

We define right and left derivatives $\alpha^+(t_0)$ and $\alpha^-(t_0)$ of α at $t_0 \in \mathbb{I}$ by

$$\alpha^{\pm}(t_0) = \check{\alpha}^+(0),$$

where $\check{\alpha}(t) = \alpha(t_0 \pm t)$.

The sign convention is not quite standard; if α is a smooth curve in a Riemannian manifold, then we have $\alpha^+(t) = -\alpha^-(t)$.

Note that if γ is a geodesic starting at p and the tangent vector $v \in \mathrm{T}'_p$ corresponds to γ, then $\gamma^+(0) = v$.

6.10. Exercise. Assume $\mathcal{X}$ is a metric space with defined angles, and let $\alpha, \beta\colon [0,a) \to \mathcal{X}$ be two maps such that the right derivatives $\alpha^+(0)$, $\beta^+(0)$ are defined and $\alpha^+(0) = \beta^+(0)$. Show that

$$|\alpha(t) - \beta(t)|_{\mathcal{X}} = o(t).$$

6.11. Proposition. *Let $\mathcal{X}$ be a metric space with defined angles, and let $p \in \mathcal{X}$. Then for any tangent vector $v \in \mathrm{T}_p\mathcal{X}$ there is a map $\alpha\colon [0,\varepsilon) \to \mathcal{X}$ such that $\alpha^+(0) = v$.*

Proof. If $v \in \mathrm{T}'_p$, then for the corresponding geodesic α we have $\alpha^+(0) = v$.

Given $v \in \mathrm{T}_p$, construct a sequence $v_n \in \mathrm{T}'_p$ such that $v_n \to v$, and let γ_n be a sequence of corresponding geodesics.

The needed map α can be found among the maps such that $\alpha(0) = p$ and

$$\alpha(t) = \gamma_n(t) \quad \text{if } \varepsilon_{n+1} \leqslant t < \varepsilon_n,$$

where ε_n is a decreasing sequence converging to 0 as $n \to \infty$. In order to satisfy the conclusion of Proposition 6.11, one has to choose the sequence ε_n converging to 0 very fast. Note that in this construction α is not continuous. □

6.12. Definition. Let $\mathcal{X}$ be a metric space, and let $\alpha\colon \mathbb{I} \to \mathcal{X}$ be a curve.

For $t_0 \in \mathbb{I}$, if $\alpha^+(t_0)$ or $\alpha^-(t_0)$ or both are defined, we say respectively that α is *right* or *left* or *both-sided differentiable* at t_0. In the exceptional cases where t_0 is the left (respectively, right) end of $\mathbb{I}$, α is by definition left (respectively, right) differentiable at t_0.

If α is both-sided differentiable at t and

$$|\alpha^+(t)| = |\alpha^-(t)| = \tfrac{1}{2}\cdot|\alpha^+(t) - \alpha^-(t)|_{\mathrm{T}_{\alpha(t)}},$$

then we say that α is *differentiable* at t.

6.13. Exercise. Assume $\mathcal{X}$ is a metric space with defined angles. Show that any geodesic $\gamma\colon \mathbb{I} \to \mathcal{X}$ is differentiable everywhere.

Recall that the speed of a curve is defined in Theorem 3.10.

6.14. Exercise. Let α be a curve in a metric space with defined angles. Suppose that $\operatorname{speed}_t \alpha$, $\alpha^+(t)$, and $\alpha^-(t)$ are defined.

Show that α is differentiable at t.

G. Differential

6.15. Definition. Let $\mathcal{X}$ be a metric space with defined angles, and let $f\colon \mathcal{X} \circ\!\!\to \mathbb{R}$ be a subfunction. For $p \in \operatorname{Dom} f$, a function $\phi\colon \mathrm{T}_p \to \mathbb{R}$ is called the *differential* of f at p (briefly $\phi = \mathbf{d}_p f$) if for any map $\alpha\colon \mathbb{I} \to \mathcal{X}$ such that $\mathbb{I}$ is a real interval, $\alpha(0) = p$, and $\alpha^+(0)$ is defined, we have

$$(f \circ \alpha)^+(0) = \phi(\alpha^+(0)).$$

6.16. Proposition. *Let $f\colon \mathcal{X} \circ\!\!\to \mathbb{R}$ be a locally Lipschitz semiconcave subfunction on a metric space $\mathcal{X}$ with defined angles. Then the differential $\mathbf{d}_p f$ is uniquely defined for any $p \in \operatorname{Dom} f$. Moreover,*

(a) *The differential $\mathbf{d}_p f\colon \mathrm{T}_p \to \mathbb{R}$ is Lipschitz and*

$$\operatorname{lip} \mathbf{d}_p f \leqslant \operatorname{lip}_p f;$$

that is, the Lipschitz constant of $\mathbf{d}_p f$ does not exceed the Lipschitz constant of f in any neighborhood of p.

(b) *$\mathbf{d}_p f\colon \mathrm{T}_p \to \mathbb{R}$ is a positive homogeneous function; that is, for any $r \geqslant 0$ and $v \in \mathrm{T}_p$ we have*

$$r\cdot \mathbf{d}_p f(v) = \mathbf{d}_p f(r\cdot v).$$

(c) *The differential $\mathbf{d}_p f\colon \mathrm{T}_p \to \mathbb{R}$ is the restriction of ultradifferential defined in Section 6.I; that is,*

$$\mathbf{d}_p f = \mathbf{d}_p^\omega f|_{\mathrm{T}_p}.$$

Proof. Passing to a subdomain of f if necessary, we can assume that f is ℓ-Lipschitz and λ-concave for some $\ell, \lambda \in \mathbb{R}$.

Take a geodesic γ in $\operatorname{Dom} f$ starting at p. Since $f \circ \gamma$ is λ-concave, the right derivative $(f \circ \gamma)^+(0)$ is defined. Since f is ℓ-Lipschitz, we have

$$(1)\qquad |(f \circ \gamma)^+(0) - (f \circ \gamma_1)^+(0)| \leqslant \ell\cdot|\gamma^+(0) - \gamma_1^+(0)|$$

for any other geodesic γ_1 starting at p.

Define $\phi\colon \mathrm{T}'_p \to \mathbb{R}\colon \gamma^+(0) \mapsto (f\circ\gamma)^+(0)$. From (1), ϕ is an ℓ-Lipschtz function defined on T'_p. Thus we can extend ϕ to all of T_p as an ℓ-Lipschitz function.

It remains to check that ϕ is the differential of f at p. Assume $\alpha\colon [0,a) \to \to \mathcal{X}$ is a map such that $\alpha(0) = p$ and $\alpha^+(0) = v \in \mathrm{T}_p$. Let $\gamma_n \in \Gamma_p$ be a sequence of geodesics as in the definition 6.9; that is, if

$$v_n = \gamma_n^+(0) \quad \text{and} \quad a_n = \varlimsup_{t\to 0+} |\alpha(t) - \gamma_n(t)|/t,$$

then $a_n \to 0$ and $v_n \to v$ as $n \to \infty$. Then

$$\phi(v) = \lim_{n\to\infty} \phi(v_n),$$

$$f\circ\gamma_n(t) = f(p) + \phi(v_n)\cdot t + o(t),$$

$$|f\circ\alpha(t) - f\circ\gamma_n(t)| \leqslant \ell\cdot|\alpha(t) - \gamma_n(t)|.$$

Hence

$$f\circ\alpha(t) = f(p) + \phi(v)\cdot t + o(t).$$

The last part follows from the definitions of differential and ultradifferential; see Section 4.E. ☐

H. Ultratangent space

Fix a selective ultrafilter ω on the set of natural numbers.

For a metric space $\mathcal{X}$ and $r > 0$, we will denote by $r\cdot\mathcal{X}$ its *r-blowup*, which is a metric space with the same underlying set as $\mathcal{X}$ and the metric multiplied by r. The tautological bijection $\mathcal{X} \to r\cdot\mathcal{X}$ will be denoted by $x \mapsto x^r$, so

$$|x^r - y^r| = r\cdot|x - y|$$

for any $x, y \in \mathcal{X}$.

The *ω-blowup* $\omega\cdot\mathcal{X}$ of $\mathcal{X}$ is defined to be the ω-limit of the n-blowups $n\cdot\mathcal{X}$; that is,

$$\omega\cdot\mathcal{X} := \lim_{n\to\omega} n\cdot\mathcal{X}.$$

Given a point $x \in \mathcal{X}$, we can consider the sequence x^n, where $x^n \in n\cdot\mathcal{X}$ is the image of x under n-blowup. Note that if $x \neq y$, then

$$|x^\omega - y^\omega|_{\omega\cdot\mathcal{X}} = \infty;$$

that is, x^ω and y^ω belong to different metric components of $\omega\cdot\mathcal{X}$.

The metric component of x^ω in $\omega\cdot\mathcal{X}$ is called the *ultratangent space* of $\mathcal{X}$ at x and is denoted by $\mathrm{T}^\omega_x\mathcal{X}$ or T^ω_x.

Equivalently, the ultratangent space $\mathrm{T}_x^\omega \mathcal{X}$ can be defined as follows. Consider all the sequences of points $x_n \in \mathcal{X}$ such that the sequence $(n\cdot|x - x_n|_{\mathcal{X}})$ is bounded. Define the pseudodistance between two such sequences as

$$|(x_n) - (y_n)| = \lim_{n\to\omega} n\cdot|x_n - y_n|_{\mathcal{X}}.$$

Then $\mathrm{T}_x^\omega \mathcal{X}$ is the corresponding metric space.

Tangent spaces (see Section 6.E) as well as ultratangent spaces generalize the notion of tangent spaces on Riemannian manifolds. In the simplest cases these two notions define the same space. However in general they are different and are both useful—often a lack of a property in one is compensated by the other. It is clear from the definition that a tangent space has a cone structure. On the other hand, in general an ultratangent space does not have a cone structure. Hilbert's cube $\prod_{n=1}^{\infty}[0, 2^{-n}]$ is an example. We remark that Hilbert's cube is a CBB(0) as well as a CAT(0) Alexandrov space.

The next theorem shows that the tangent space T_p can be (and often will be) considered as a subset of T_p^ω.

6.17. Theorem. *Let $\mathcal{X}$ be a metric space with defined angles. Then for any $p \in$ $\in \mathcal{L}$, there is a distance-preserving map*

$$\iota : \mathrm{T}_p \hookrightarrow \mathrm{T}_p^\omega$$

such that for any geodesic γ starting at p we have

$$\gamma^+(0) \overset{\iota}{\mapsto} \lim_{n\to\omega} [\gamma(\tfrac{1}{n})]^n.$$

Proof. Given $v \in \mathrm{T}_p'$, choose a geodesic γ that starts at p and such that $\gamma^+(0) =$ $= v$. Set $v^n = [\gamma(\frac{1}{n})]^n \in n\cdot\mathcal{X}$ and

$$v^\omega = \lim_{n\to\omega} v^n.$$

Note that the value $v^\omega \in \mathrm{T}_p^\omega$ does not depend on the choice of γ; that is, if γ_1 is another geodesic starting at p such that $\gamma_1^+(0) = v$, then

$$\lim_{n\to\omega} v^n = \lim_{n\to\omega} v_1^n,$$

where $v_1^n = [\gamma_1(\frac{1}{n})]^n \in n\cdot\mathcal{X}$. The latter follows since

$$|\gamma(t) - \gamma_1(t)|_{\mathcal{X}} = o(t),$$

and therefore $|v^n - v_1^n|_{n\cdot\mathcal{X}} \to 0$ as $n \to \infty$.

Set $\iota(v) = v^\omega$. Since angles between geodesics in $\mathcal{X}$ are defined, for any v, $w \in \mathrm{T}_p'$, we have $n\cdot|v_n - w_n| \to |v - w|$. Thus $|v_\omega - w_\omega| = |v - w|$; that is, $\iota\colon \mathrm{T}_p' \to \mathrm{T}_p$ is a distance-preserving map.

Since T'_p is dense in T_p, we can extend ι to a distance-preserving map $\mathrm{T}_p \to \to \mathrm{T}^\omega_p$. □

I. Ultradifferential

Given a function $f\colon \mathcal{L} \to \mathbb{R}$, consider the sequence of functions $f_n\colon n\cdot\mathcal{L} \to \mathbb{R}$ defined by

$$f_n(x^n) = n\cdot(f(x) - f(p)),$$

where $x \mapsto x^n$ denotes the natural map $\mathcal{L} \to n\cdot\mathcal{L}$. While $n\cdot(\mathcal{L}, p) \to (\mathrm{T}^\omega, 0)$ as $n \to \omega$, the functions f_n converge to the *ω-differential* of f at p. It will be denoted by $\boldsymbol{d}^\omega_p f$:

$$\boldsymbol{d}^\omega_p f\colon \mathrm{T}^\omega_p \to \mathbb{R}, \quad \boldsymbol{d}^\omega_p f = \lim_{n\to\omega} f_n.$$

Clearly, the ω-differential $\boldsymbol{d}^\omega_p f$ of a locally Lipschitz subfunction f is defined and Lipschitz at each point $p \in \operatorname{Dom} f$.

J. Remarks

Spaces with defined angles include CAT and CBB spaces; see 8.14(c) and 9.15(b).

For general metric spaces, angles may not exist, and given a hinge $[p\,{}^x_y]$ it is more natural to consider the *upper angle* defined by

$$\measuredangle^{\mathrm{up}}[p\,{}^x_y] := \varlimsup_{\bar{x},\bar{y}\to p} \tilde{\measuredangle}^\kappa(p\,{}^{\bar{x}}_{\bar{y}}),$$

where $\bar{x} \in\,]px]$ and $\bar{y} \in\,]py]$. The triangle inequality (see 6.5) holds for upper angles as well.

Chapter 7

Dimension theory

A. Definitions

In this section, we give definitions of different types of dimension-like invariants of metric spaces and state general relations between them. The proofs of most of the statements in this section can be found in the book [**85**] by Witold Hurewicz and Henry Wallman; the rest follow directly from the definitions.

7.1. Hausdorff dimension. *Let $\mathcal{X}$ be a metric space. Its Hausdorff dimension is defined as*

$$\operatorname{HausDim}\mathcal{X} = \sup\{\,\alpha \in \mathbb{R} : \operatorname{HausMes}_{\alpha}(\mathcal{X}) > 0\,\},$$

where $\operatorname{HausMes}_{\alpha}$ *denotes the α-dimensional Hausdorff measure.*

Let $\mathcal{X}$ be a metric space, and let $\{V_{\beta}\}_{\beta\in\mathcal{B}}$ be an open cover of $\mathcal{X}$. Let us recall two notions in general topology:

- The *order* of $\{V_{\beta}\}$ is the supremum of all integers n such that there is a collection of $n+1$ elements of $\{V_{\beta}\}$ with nonempy intersection.
- An open cover $\{W_{\alpha}\}_{\alpha\in\mathcal{A}}$ of $\mathcal{X}$ is called a *refinement* of $\{V_{\beta}\}_{\beta\in\mathcal{B}}$ if for any $\alpha \in \mathcal{A}$ there is $\beta \in \mathcal{B}$ such that $W_{\alpha} \subset V_{\beta}$.

7.2. Topological dimension. *Let $\mathcal{X}$ be a metric space. The topological dimension of $\mathcal{X}$ is defined to be the minimum of nonnegative integers n such that for any open cover of $\mathcal{X}$ there is a finite open refinement with order n.*

If no such n exists, the topological dimension of $\mathcal{X}$ is infinite.

The topological dimension of $\mathcal{X}$ will be denoted by $\operatorname{TopDim}\mathcal{X}$.

The invariants satisfying the following two statements, 7.3 and 7.4, are commonly called "dimension"; for that reason we call these statements axioms.

7.3. Normalization axiom. *For any $m \in \mathbb{Z}_{\geqslant 0}$,*

$$\operatorname{TopDim} \mathbb{E}^m = \operatorname{HausDim} \mathbb{E}^m = m.$$

7.4. Cover axiom. *If $\{A_n\}_{n=1}^{\infty}$ is a countable closed cover of $\mathcal{X}$, then*

$$\operatorname{TopDim} \mathcal{X} = \sup_n \{\operatorname{TopDim} A_n\},$$
$$\operatorname{HausDim} \mathcal{X} = \sup_n \{\operatorname{HausDim} A_n\}.$$

On product spaces. Recall that the direct product $\mathcal{X} \times \mathcal{Y}$ of metric spaces $\mathcal{X}$ and $\mathcal{Y}$ is defined in Section 2.A. The direct product satisfies the following two inequalities:

$$\operatorname{TopDim}(\mathcal{X} \times \mathcal{Y}) \leqslant \operatorname{TopDim} \mathcal{X} + \operatorname{TopDim} \mathcal{Y}$$

and

$$\operatorname{HausDim}(\mathcal{X} \times \mathcal{Y}) \geqslant \operatorname{HausDim} \mathcal{X} + \operatorname{HausDim} \mathcal{Y}.$$

These inequalities might be strict. For topological dimension, strict inequality holds for a pair of Pontryagin surfaces [**140**]. For Hausdorff dimension, an example was constructed by Abram Besicovitch and Pat Moran [**26**].

The following theorem follows from [**85**, theorems V 8 and VII 2].

7.5. Szpilrajn's theorem. *Let $\mathcal{X}$ be a separable metric space, and assume* $\operatorname{TopDim} \mathcal{X} \geqslant m$. *Then* $\operatorname{HausMes}_m \mathcal{X} > 0$.

In particular, $\operatorname{TopDim} \mathcal{X} \leqslant \operatorname{HausDim} \mathcal{X}$.

In fact it is true that for any separable metric space $\mathcal{X}$, we have

$$\operatorname{TopDim} \mathcal{X} = \inf\{\operatorname{HausDim} \mathcal{Y}\},$$

where the infimum is taken over all metric spaces $\mathcal{Y}$ homeomorphic to $\mathcal{X}$.

7.6. Definition. Let $\mathcal{X}$ be a metric space, and let $F : \mathcal{X} \to \mathbb{R}^m$ be a continuous map. A point $z \in \operatorname{Im} F$ is called a *stable value* of F if there is $\varepsilon > 0$ such that $z \in \operatorname{Im} F'$ for any *ε-close* to F continuous map $F' : \mathcal{X} \to \mathbb{R}^m$, that is, $|F'(x) - F(x)| < \varepsilon$ for all $x \in \mathcal{X}$.

The next theorem follows from [**85**, theorems VI 1&2]. (This theorem also holds for nonseparable metric spaces [**120**], [**63**, 3.2.10].)

7.7. Stable value theorem. *Let $\mathcal{X}$ be a separable metric space. Then* $\operatorname{TopDim} \mathcal{X} \geqslant m$ *if and only if there is a map $F : \mathcal{X} \to \mathbb{R}^m$ with a stable value.*

7.8. Proposition. *Suppose $\mathcal{X}$ and $\mathcal{Y}$ are metric spaces and $\Phi : \mathcal{X} \to \mathcal{Y}$ satisfies*

$$|\Phi(x) - \Phi(x')| \geqslant \varepsilon \cdot |x - x'|$$

for fixed $\varepsilon > 0$ *and any pair* $x, x' \in \mathcal{X}$. *Then*

$$\operatorname{HausDim} \mathcal{X} \leqslant \operatorname{HausDim} \mathcal{Y}.$$

In particular, if there is a Lipschitz onto map $\mathcal{Y} \to \mathcal{X}$, *then*

$$\operatorname{HausDim} \mathcal{X} \leqslant \operatorname{HausDim} \mathcal{Y}.$$

B. Linear dimension

In addition to HausDim and TopDim, we will use the so-called linear dimension. It will be applied only to Alexandrov spaces and to their open subsets (in cases both of curvature bounded below and curvature bounded above). As we shall see, in all these cases LinDim behaves nicely and is easy to work with.

Recall that a *cone map* is a map between cones respecting the cone multiplication.

7.9. Definition of linear dimension. *Let* $\mathcal{X}$ *be a metric space with defined angles. The* linear dimension *of* $\mathcal{X}$ *(denoted by* $\operatorname{LinDim} \mathcal{X}$*) is defined as the exact upper bound on* $m \in \mathbb{Z}_{\geqslant 0}$ *such that there is a distance-preserving cone embedding* $\mathbb{E}^m \hookrightarrow \mathrm{T}_p \mathcal{X}$ *for some* $p \in \mathcal{X}$; *here* $\mathbb{E}^m$ *denotes the* m*-dimensional Euclidean space and* $\mathrm{T}_p \mathcal{X}$ *denotes the tangent space of* $\mathcal{X}$ *at* p *(defined in Section* 6.D*).*

Note that LinDim takes values in $\mathbb{Z}_{\geqslant 0} \cup \{\infty\}$.

The linear dimension LinDim has no immediate relations to HausDim and TopDim. Also, LinDim does not satisfy Cover Axiom 7.4. Note that

$$\operatorname{LinDim}(\mathcal{X} \times \mathcal{Y}) = \operatorname{LinDim} \mathcal{X} + \operatorname{LinDim} \mathcal{Y} \tag{1}$$

for any two metric spaces $\mathcal{X}$ and $\mathcal{Y}$ with defined angles.

The following exercise is based on a construction of Thomas Foertsch and Viktor Schroeder [**147**]; it shows that the condition on existence of angles in (1) cannot be removed.

7.10. Exercise. Construct metrics ρ_1 and ρ_2 on $\mathbb{R}^{10}$ defined by norms, such that $(\mathbb{R}^{10}, \rho_i)$ do *not* contain an isometric copy of $\mathbb{E}^2$ but $(\mathbb{R}^{10}, \rho_1) \times (\mathbb{R}^{10}, \rho_2)$ has an isometric copy of $\mathbb{E}^{10}$.

Remarks. Linear dimension was first introduced by Conrad Plaut [**139**] under the name *local dimension*. *Geometric dimension*, introduced by Bruce Kleiner [**96**] is closely related; it coincides with the linear dimension for CBB and CAT spaces.

One can extend the definition to arbitrary metric spaces. To do this, one should modify the definition of tangent space and take an arbitrary n-dimensional Banach space instead of the Euclidean n-space. For Alexandrov spaces (either CBB or CAT) this modification is equivalent to our definition.

Part 2

Fundamentals

Chapter 8

Fundamentals of curvature bounded below

A. Four-point comparison

Recall (Section 6.B) that the model angle $\tilde\measuredangle^\kappa\left(p\,{}^x_y\right)$ is defined if

$$|p-x|+|p-y|+|x-y|<\varpi^\kappa;$$

here ϖ^κ denotes the diameter of model space $\mathbb{M}^2(\kappa)$.

8.1. Four-point comparison. *A quadruple of points p, x^1, x^2, x^3 in a metric space satisfies* CBB(κ) comparison *if*

(1) $$\tilde\measuredangle^\kappa\left(p\,{}^{x^1}_{x^2}\right)+\tilde\measuredangle^\kappa\left(p\,{}^{x^2}_{x^3}\right)+\tilde\measuredangle^\kappa\left(p\,{}^{x^3}_{x^1}\right)\leqslant 2\cdot\pi,$$

or at least one of the model angles $\tilde\measuredangle^\kappa\left(p\,{}^{x^i}_{x^j}\right)$ is not defined.

8.2. Definition. Let $\mathcal{L}$ be a metric space.

(a) $\mathcal{L}$ is CBB(κ) if any quadruple in $\mathcal{L}$ satisfies CBB(κ) comparison.

(b) $\mathcal{L}$ is *locally* CBB(κ) if any point $q\in\mathcal{L}$ admits a neighborhood $\Omega\ni q$ such that any quadruple in Ω satisfies CBB(κ) comparison.

(c) $\mathcal{L}$ is a CBB space if $\mathcal{L}$ is CBB(κ) for some $\kappa\in\mathbb{R}$.

Remarks.

- CBB(κ) length spaces are often called *spaces with curvature* $\geqslant\kappa$ *in the sense of Alexandrov.* These spaces will usually be denoted by $\mathcal{L}$, for $\mathcal{L}$ower curvature bound.

- In the definition of CBB(κ), when $\kappa > 0$ most authors assume in addition that the diameter is at most the model diameter ϖ^κ. For a complete length space, the latter means that it is not isometric to one of the exceptional spaces; see 8.44. We do not make this assumption. In particular, we consider the real line to have curvature $\geqslant 1$.
- If $\kappa < \mathrm{K}$, then any complete length CBB(K) space is CBB(κ). Moreover directly from the definition it follows that if $\mathrm{K} \leqslant 0$, then any CBB(K) space is CBB(κ). However, in the case $\mathrm{K} > 0$ the latter statement does not hold and the former statement is not trivial; it will be proved in Corollary 8.33.

8.3. Exercise. Let $\mathcal{L}$ be a metric space, and let $\kappa \leqslant 0$. Show that $\mathcal{L}$ is CBB(κ) if for any quadruple of points $p, x^1, x^2, x^3 \in \mathcal{L}$ there is a quadruple of points q, $y^1, y^2, y^3 \in \mathbb{M}^2(\kappa)$ such that

$$|p - x^i| = |q - y^i| \quad \text{and} \quad |x^i - x^j| \leqslant |y^i - y^j|$$

for all i and j.

The exercise above is a special case of $(1 + n)$-point comparison (see 10.8).

Recall that ω denotes a selective ultrafilter on $\mathbb{N}$, which is fixed once and for all. The following proposition follows directly from the definition of CBB(κ) comparison and the definitions of ω-limit and ω-power given in Section 4.B.

8.4. Proposition. *Let $\mathcal{L}_n$ be a* CBB(κ_n) *space for each n. Assume $\mathcal{L}_n \to \mathcal{L}_\omega$ and $\kappa_n \to \kappa_\omega$ as $n \to \omega$. Then $\mathcal{L}_\omega$ is* CBB(κ_ω).

Moreover, a metric space $\mathcal{L}$ is CBB(κ) *if and only if so is its ultrapower $\mathcal{L}^\omega$.*

8.5. Theorem. *Let $\mathcal{L}$ be a* CBB(κ) *space, let $\mathcal{M}$ be a metric space, and let $\sigma : \mathcal{L} \to \to \mathcal{M}$ be a submetry. Assume p, x^1, x^2, x^3 is a quadruple of points in $\mathcal{M}$ such that $|p - x^i| < \frac{\varpi^\kappa}{2}$ for any i. Then the quadruple satisfies* CBB(κ) *comparison.*

In particular,

(a) *The space $\mathcal{M}$ is locally* CBB(κ)*. Moreover, any open ball of radius $\frac{\varpi^\kappa}{4}$ in $\mathcal{M}$ is* CBB(κ).

(b) *If $\kappa \leqslant 0$, then $\mathcal{M}$ is* CBB(κ).

Corollary 8.34 gives a stronger statement; it states that if $\mathcal{L}$ is a complete length space, then $\mathcal{M}$ is always CBB(κ). The last theorem together with Proposition 3.8 imply the following:

8.6. Corollary. *Assume that the group G acts isometrically on a* CBB(κ) *space $\mathcal{L}$ and has closed orbits. Then the quotient space $\mathcal{L}/G$ is locally* CBB(κ).

8.7. Example. If $G < \mathrm{O}(n+1)$ is a closed subgroup, then $\mathbb{S}^n/G$ is CBB(1) and $\mathbb{R}^{n+1}/G$ is CBB(0).

Proof of Theorem 8.5. Fix a quadruple of points $p, x^1, x^2, x^3 \in \mathcal{M}$ such that $|p - x^i| < \frac{\varpi^\kappa}{2}$ for any i. Choose an arbitrary $\hat{p} \in \mathcal{L}$ such that $\sigma(\hat{p}) = p$.

Since σ is submetry, we can choose the points $\hat{x}^1, \hat{x}^2, \hat{x}^3 \in \mathcal{L}$ such that $\sigma(\hat{x}_i) = x_i$ and

$$|p - x^i|_{\mathcal{M}} \lesseqgtr |\hat{p} - \hat{x}^i|_{\mathcal{L}} \pm \delta$$

for all i and any fixed $\delta > 0$.

Note that

$$|x^i - x^j|_{\mathcal{M}} \leqslant |\hat{x}^i - \hat{x}^j|_{\mathcal{L}} \leqslant |p - x^i|_{\mathcal{M}} + |p - x^j|_{\mathcal{M}} + 2\cdot\delta$$

for all i and j.

Since $|p - x^i| < \frac{\varpi^\kappa}{2}$, we can choose $\delta > 0$ above so that the angles $\tilde{\measuredangle}^\kappa\left(\hat{p}\,{}^{\hat{x}^i}_{\hat{x}^j}\right)$ are defined. Moreover, given $\varepsilon > 0$, the value δ can be chosen in such a way that the inequality

$$\tilde{\measuredangle}^\kappa\left(p\,{}^{x^i}_{x^j}\right) < \tilde{\measuredangle}^\kappa\left(\hat{p}\,{}^{\hat{x}^i}_{\hat{x}^j}\right) + \varepsilon \tag{2}$$

holds for all i and j.

By CBB(κ) comparison in $\mathcal{L}$, we have

$$\tilde{\measuredangle}^\kappa\left(\hat{p}\,{}^{\hat{x}^1}_{\hat{x}^2}\right) + \tilde{\measuredangle}^\kappa\left(\hat{p}\,{}^{\hat{x}^2}_{\hat{x}^3}\right) + \tilde{\measuredangle}^\kappa\left(\hat{p}\,{}^{\hat{x}^3}_{\hat{x}^1}\right) \leqslant 2\cdot\pi.$$

Applying (2), we get

$$\tilde{\measuredangle}^\kappa\left(p\,{}^{x^1}_{x^2}\right) + \tilde{\measuredangle}^\kappa\left(p\,{}^{x^2}_{x^3}\right) + \tilde{\measuredangle}^\kappa\left(p\,{}^{x^3}_{x^1}\right) < 2\cdot\pi + 3\cdot\varepsilon.$$

Since $\varepsilon > 0$ is arbitrary, we have

$$\tilde{\measuredangle}^\kappa\left(p\,{}^{x^1}_{x^2}\right) + \tilde{\measuredangle}^\kappa\left(p\,{}^{x^2}_{x^3}\right) + \tilde{\measuredangle}^\kappa\left(p\,{}^{x^3}_{x^1}\right) \leqslant 2\cdot\pi;$$

that is, the CBB(κ) comparison holds for this quadruple in $\mathcal{M}$. □

B. Geodesics

Recall that general complete length spaces might have no geodesics; see Exercise 2.10.

8.8. Exercise. Construct a complete length CBB(0) space that is not geodesic.

We are going to show that all complete length CBB spaces have plenty of geodesics in the following sense. Recall that a subset of a topological space is called G-delta if it is a countable intersection of open sets.

8.9. Definition. A metric space $\mathcal{X}$ is called *G-delta geodesic* if for any point $p \in \mathcal{X}$ there is a dense G-delta set $W_p \subset \mathcal{X}$ such that for any $q \in W_p$ there is a geodesic $[pq]$.

A metric space $\mathcal{X}$ is called *locally G-delta geodesic* if for any point $p \in \mathcal{X}$ there is a G-delta set $W_p \subset \mathcal{X}$ such that W_p is dense in a neighborhood of p and for any $q \in W_p$ there is a geodesic $[pq]$.

8.10. Definition. Let $\mathcal{X}$ be a metric space, and let $p \in \mathcal{X}$. A point $q \in \mathcal{X}$ is called *p-straight* (briefly, $q \in \operatorname{Str}(p)$) if

$$\varlimsup_{r \to q} \frac{|p-r| - |p-q|}{|q-r|} = 1.$$

For an array of points $x^1, x^2, \dots, x^k$, we use the notation

$$\operatorname{Str}(x^1, x^2, \dots, x^k) = \bigcap_i \operatorname{Str}(x^i).$$

8.11. Theorem. *Let $\mathcal{L}$ be a complete length* CBB *space, and let $p \in \mathcal{L}$. Then the set* $\operatorname{Str}(p)$ *is a dense G-delta set. Moreover, for any $q \in \operatorname{Str}(p)$ there is a unique geodesic $[pq]$.*

In particular, $\mathcal{L}$ is G-delta geodesic.

This theorem was proved by by Conrad Plaut [**139**, Th. 27].

Proof. Given a positive integer n, consider the set Ω_n of all points $q \in \mathcal{L}$ such that

$$(1 - \tfrac{1}{n})\cdot|q-r| < |p-r| - |p-q| < \tfrac{1}{n}$$

for some $r \in \mathcal{L}$. Clearly Ω_n is open; let us show that Ω_n is dense in $\mathcal{L}$.

Assuming the contrary, there is a point $x \in \mathcal{L}$ such that

$$\mathrm{B}(x, \varepsilon) \cap \Omega_n = \varnothing$$

for $\varepsilon > 0$. Since $\mathcal{L}$ is a length space, for any $\delta > 0$, there exists a point $y \in \mathcal{L}$ such that

$$|x-y| < \tfrac{\varepsilon}{2} + \delta \quad \text{and} \quad |p-y| < |p-x| - \tfrac{\varepsilon}{2} + \delta.$$

If ε and δ are sufficiently small, then

$$(1 - \tfrac{1}{n})\cdot|y-x| < |p-x| - |p-y| < \tfrac{1}{n};$$

that is, $y \in \Omega_n$, a contradiction.

Note that $\operatorname{Str}(p) = \bigcap_n \Omega_n$; therefore, $\operatorname{Str}(p)$ is a dense G-delta set.

Assuming $q \in \operatorname{Str}(p)$, let us show that there is a unique geodesic connecting p and q. Note that it is sufficient to show that for all sufficiently small $t > 0$ there is a unique point z such that

(1) $$t = |q-z| = |p-q| - |p-z|.$$

First let us show uniqueness. Assume z and z' both satisfy (1). Take a sequence $r_n \to q$ such that

$$\frac{|p-r_n| - |p-q|}{|q-r_n|} \to 1.$$

By the triangle inequality,

$$|z-r| - |z-q|, \quad |z'-r| - |z'-q| \geqslant |p-r| - |p-q|;$$

thus, as $n \to \infty$,

$$\frac{|z - r_n| - |z - q|}{|q - r_n|}, \quad \frac{|z' - r_n| - |z' - q|}{|q - r_n|} \to 1.$$

Therefore $\tilde{\measuredangle}^\kappa(q\,{}^{z}_{r_n}) \to \pi$ and $\tilde{\measuredangle}^\kappa(q\,{}^{z'}_{r_n}) \to \pi$. (Here we use that t is small, otherwise if $\kappa > 0$ the angles might be undefined.)

From CBB(κ) comparison (Definition 8.2), $\tilde{\measuredangle}^\kappa(q\,{}^{z}_{z'}) = 0$ and thus $z = z'$.

The proof of existence is similar. Choose a sequence r_n as above. Since $\mathcal{L}$ is a complete length space, there is a sequence $z_k \in \mathcal{L}$ such that $|q - z_k| \to t$ and $|p - q| - |p - z_k| \to t$ as $k \to \infty$. Then

$$\lim_{n\to\infty} \lim_{k\to\infty} \tilde{\measuredangle}^\kappa(q\,{}^{z_k}_{r_n}) = \pi.$$

Thus, for any $\varepsilon > 0$ and sufficiently large n, k, we have $\tilde{\measuredangle}^\kappa(q\,{}^{z_k}_{r_n}) > \pi - \varepsilon$. From CBB($\kappa$) comparison (Definition 8.2), for all large k and j, we have $\tilde{\measuredangle}^\kappa(q\,{}^{z_k}_{z_j}) <$ $< 2\cdot\varepsilon$ and thus

$$|z_k - z_j| < \varepsilon\cdot c(\kappa, t);$$

that is, z_n is a Cauchy sequence, and its limit z satisfies (1). □

8.12. Exercise. Let $\mathcal{L}$ be a complete length CBB space, and let $A \subset \mathcal{L}$ be a closed subset. Show that there is a dense G-delta set $W \subset \mathcal{L}$ such that for any $q \in W$, there is a unique geodesic $[pq]$ with $p \in A$ that realizes the distance from q to A; that is, $|p - q| = \operatorname{dist}_A q$.

8.13. Exercise. Construct a complete length CBB space $\mathcal{L}$ with an everywhere dense G-delta set A such that $A \cap\,]xy[\, = \varnothing$ for any geodesic $[xy]$ in $\mathcal{L}$.

C. More comparisons

The following theorem makes it easier to use Euclidean intuition in the Alexandrov setting.

8.14. Theorem. *If $\mathcal{L}$ is a* CBB(κ) *space, then the following conditions hold for all $p, x, y \in \mathcal{L}$, provided the model triangle $\tilde{\triangle}^\kappa(pxy)$ is defined.*

(a) Adjacent angle comparison. *For any geodesic $[xy]$ and $z \in\,]xy[$, $z \neq p$ we have*

$$\tilde{\measuredangle}^\kappa(z\,{}^{p}_{x}) + \tilde{\measuredangle}^\kappa(z\,{}^{p}_{y}) \leqslant \pi.$$

(b) Point-on-side comparison. *For any geodesic $[xy]$ and $z \in\,]xy[$, we have*

$$\tilde{\measuredangle}^\kappa(x\,{}^{p}_{y}) \leqslant \tilde{\measuredangle}^\kappa(x\,{}^{p}_{z}),$$

or, equivalently,

$$|\tilde{p} - \tilde{z}| \leqslant |p - z|,$$

where $[\tilde{p}\tilde{x}\tilde{y}] = \tilde{\triangle}^\kappa(pxy)$, $\tilde{z} \in\,]\tilde{x}\tilde{y}[$, $|\tilde{x} - \tilde{z}| = |x - z|$.

(c) Hinge comparison. *For any hinge* $\left[x\,{}^{p}_{y}\right]$, *the angle* $\measuredangle\left[x\,{}^{p}_{y}\right]$ *is defined and*

$$\measuredangle\left[x\,{}^{p}_{y}\right] \geqslant \tilde\measuredangle^{\kappa}\left(x\,{}^{p}_{y}\right),$$

or equivalently

$$\tilde{\curlyvee}^{\kappa}\left[x\,{}^{p}_{y}\right] \geqslant |p-y|.$$

Moreover,

$$\measuredangle\left[z\,{}^{p}_{y}\right] + \measuredangle\left[z\,{}^{p}_{x}\right] \leqslant \pi$$

for any two adjacent hinges $\left[z\,{}^{p}_{y}\right]$ *and* $\left[z\,{}^{p}_{x}\right]$.

Moreover, in each case, the converse holds if $\mathcal{L}$ *is G-delta geodesic. That is, if one of the conditions* ((a)), (b), *or* (c) *holds in a G-delta geodesic space* $\mathcal{L}$, *then* $\mathcal{L}$ *is* CBB(κ).

A slightly stronger form of (c) is given in 8.29. See also Open Question 8.51.

Proof. Since $z \in \,]xy[$, we have $\tilde\measuredangle^{\kappa}\left(z\,{}^{x}_{y}\right) = \pi$. Thus, CBB($\kappa$) comparison

$$\tilde\measuredangle^{\kappa}\left(z\,{}^{x}_{y}\right) + \tilde\measuredangle^{\kappa}\left(z\,{}^{p}_{x}\right) + \tilde\measuredangle^{\kappa}\left(z\,{}^{p}_{y}\right) \leqslant 2\cdot\pi$$

implies

$$\tilde\measuredangle^{\kappa}\left(z\,{}^{p}_{x}\right) + \tilde\measuredangle^{\kappa}\left(z\,{}^{p}_{y}\right) \leqslant \pi.$$

(a)⇔(b). Follows from Alexandrov's lemma 6.3.

(a) + (b) ⇒ (c). From (b) we get that for $\bar p \in \,]xp]$ and $\bar y \in \,]xy]$, the function $(|x-\bar p|, |x-\bar y|) \mapsto \tilde\measuredangle^{\kappa}\left(x\,{}^{\bar p}_{\bar y}\right)$ is nonincreasing in each argument. In particular, $\measuredangle\left[x\,{}^{p}_{y}\right] = \sup\{\tilde\measuredangle^{\kappa}\left(x\,{}^{\bar p}_{\bar y}\right)\}$ is defined and is at least $\tilde\measuredangle^{\kappa}\left(x\,{}^{p}_{y}\right)$.

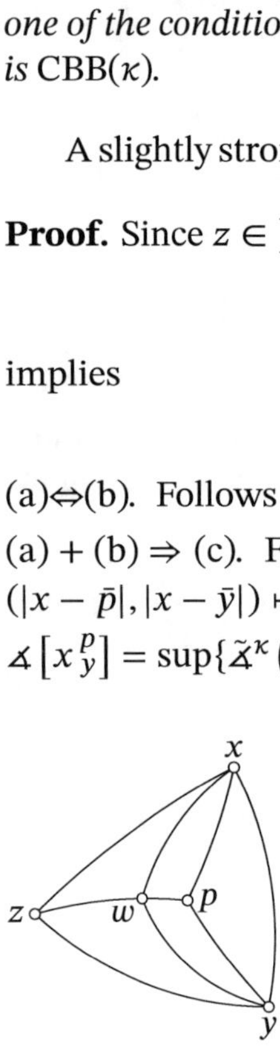

From above and (a), it follows that

$$\measuredangle\left[z\,{}^{p}_{y}\right] + \measuredangle\left[z\,{}^{p}_{x}\right] \leqslant \pi.$$

Converse. Assume first that $\mathcal{L}$ is geodesic. Consider a point $w \in \,]pz[$ close to p. From (c), it follows that

$$\measuredangle\left[w\,{}^{x}_{z}\right] + \measuredangle\left[w\,{}^{x}_{p}\right] \leqslant \pi \quad \text{and} \quad \measuredangle\left[w\,{}^{y}_{z}\right] + \measuredangle\left[w\,{}^{y}_{p}\right] \leqslant \pi.$$

Since $\measuredangle\left[w\,{}^{x}_{y}\right] \leqslant \measuredangle\left[w\,{}^{x}_{p}\right] + \measuredangle\left[w\,{}^{y}_{p}\right]$ (see Theorem 6.5), we get

$$\measuredangle\left[w\,{}^{x}_{z}\right] + \measuredangle\left[w\,{}^{y}_{z}\right] + \measuredangle\left[w\,{}^{x}_{y}\right] \leqslant 2\cdot\pi.$$

Applying the first inequality in (c),

$$\tilde\measuredangle^{\kappa}\left(w\,{}^{x}_{z}\right) + \tilde\measuredangle^{\kappa}\left(w\,{}^{y}_{z}\right) + \tilde\measuredangle^{\kappa}\left(w\,{}^{x}_{y}\right) \leqslant 2\cdot\pi.$$

Passing to the limits $w \to p$, we have

$$\tilde\measuredangle^{\kappa}\left(p\,{}^{x}_{z}\right) + \tilde\measuredangle^{\kappa}\left(p\,{}^{y}_{z}\right) + \tilde\measuredangle^{\kappa}\left(p\,{}^{x}_{y}\right) \leqslant 2\cdot\pi.$$

If $\mathcal{L}$ is only G-delta geodesic, we can apply the above arguments to sequences of points $p_n, w_n \to p$, $x_n \to x$, $y_n \to y$ such that $[p_n z]$ exists, $w_n \in$ $\in\,]zp_n[$ and $[x_n w_n]$, $[y_n w_n]$ exist, and then pass to the limit as $n \to \infty$. □

8.15. Exercise. Let $\mathcal{L}$ be $\mathbb{R}^m$ with a metric defined by a norm. Show that $\mathcal{L}$ is a complete length CBB space if and only if $\mathcal{L} \overset{\text{iso}}{=\!=} \mathbb{E}^m$.

8.16. Exercise. Assume $\mathcal{L}$ is a complete length CBB space, and $[px]$, $[py]$ be two geodesics in the same geodesic direction $\xi \in \Sigma'_p$. Show that

$$[px] \subset [py] \quad \text{or} \quad [px] \supset [py].$$

8.17. Angle-sidelength monotonicity. *Let p, x, y be points in a complete length* CBB(κ) *space $\mathcal{L}$. Suppose that the model triangle $\tilde{\triangle}^\kappa(pxy)$ is defined and there is a geodesic $[xy]$. Then for $\bar{y} \in\,]xy]$ the function*

$$|x - \bar{y}| \mapsto \tilde{\measuredangle}^\kappa \left(x\,{}^{p}_{\bar{y}}\right)$$

is nonincreasing.

In particular, if a geodesic $[xp]$ exists and $\bar{p} \in\,]xp]$, then

(a) *the function*

$$(|x - \bar{y}|, |x - \bar{p}|) \mapsto \tilde{\measuredangle}^\kappa \left(x\,{}^{\bar{p}}_{\bar{y}}\right)$$

is nonincreasing in each argument;

(b) *the angle $\measuredangle\left[x\,{}^{p}_{y}\right]$ is defined and*

$$\measuredangle\left[x\,{}^{p}_{y}\right] = \sup\left\{\, \tilde{\measuredangle}^\kappa \left(x\,{}^{\bar{p}}_{\bar{y}}\right) \,:\, \bar{p} \in\,]xp],\ \bar{y} \in\,]xy] \,\right\}.$$

The proof is contained in the first part of (a) + (b) ⇒ (c) of the proof above.

8.18. Exercise. Let $\mathcal{L}$ be a CBB(κ) space, p, x, $y \in \mathcal{L}$ and v, $w \in\,]xy[$. Prove that

$$\tilde{\measuredangle}^\kappa \left(x\,{}^{y}_{p}\right) = \tilde{\measuredangle}^\kappa \left(x\,{}^{v}_{p}\right) \quad \Longleftrightarrow \quad \tilde{\measuredangle}^\kappa \left(x\,{}^{y}_{p}\right) = \tilde{\measuredangle}^\kappa \left(x\,{}^{w}_{p}\right).$$

8.19. Advanced exercise. Construct a geodesic space $\mathcal{X}$ that is not CBB(0), but meets the following condition: for any three points p, x, $y \in \mathcal{X}$, there is a geodesic $[xy]$ such that for any $z \in\,]xy[$

$$\tilde{\measuredangle}^0 \left(z\,{}^{p}_{x}\right) + \tilde{\measuredangle}^0 \left(z\,{}^{p}_{y}\right) \leqslant \pi.$$

8.20. Advanced exercise. Let $\mathcal{L}$ be a complete length space such that for any quadruple p, x, y, $z \in \mathcal{L}$ the following inequality holds;

(1) $$|p - x|^2 + |p - y|^2 + |p - z|^2 \geqslant \tfrac{1}{3} \cdot \left[|x - y|^2 + |y - z|^2 + |z - x|^2\right].$$

Prove that $\mathcal{L}$ is CBB(0).

Construct a four-point metric space $\mathcal{X}$ that satisfies inequality (1) for any relabeling of its points by p, x, y, z, such that $\mathcal{X}$ is not CBB(0).

Assume that for a given triangle $[x^1x^2x^3]$ in a metric space, its κ-model triangle $[\tilde{x}^1\tilde{x}^2\tilde{x}^3] = \tilde{\triangle}^\kappa(x^1x^2x^3)$ is defined. We say the triangle $[x^1x^2x^3]$ is *κ-thick* if the natural map (see 9.19) $[\tilde{x}^1\tilde{x}^2\tilde{x}^3] \to [x^1x^2x^3]$ is distance noncontracting.

8.21. Exercise. Prove that any triangle with perimeter $< \varpi^\kappa$ in a CBB(κ) space is κ-*thick*.

8.22. Exercise.

(a) Show that any CBB(0) space $\mathcal{L}$ satisfies the following condition: for any three points $p, q, r \in \mathcal{L}$, if $\bar{q}$ and $\bar{r}$ are midpoints of geodesics $[pq]$ and $[pr]$, respectively, then $2\cdot|\bar{q} - \bar{r}| \geqslant |q - r|$.

(b) Show that there is a metric on $\mathbb{R}^2$ defined by a norm that satisfies the above condition, but is not CBB(0).

Remarks. Monotonicity of the model angle with respect to adjacent side-lengths (see 8.17) was named the *convexity property* by Alexandrov.

D. Function comparison

In this section we will translate the angle comparison definitions (see 8.14) to a concavity-like property of the distance functions as defined in Section 3.F. This is a conceptual step—we reformulate a global geometric condition into an infinitesimal condition on distance functions.

8.23. Theorem. *Let $\mathcal{L}$ be a complete length space. Then the following statements are equivalent:*

(a) *$\mathcal{L}$ is* CBB(κ).

(b) Function comparison. *$\mathcal{L}$ is G-delta geodesic and for any $p \in \mathcal{L}$, the function $f = \mathrm{md}^\kappa \circ \mathrm{dist}_p$ satisfies the differential inequality*

$$f'' \leqslant 1 - \kappa\cdot f$$

in $\mathrm{B}(p, \varpi^\kappa)$.

8.24. Corollary. *A complete G-delta geodesic space $\mathcal{L}$ is* CBB(0) *if and only if for any $p \in \mathcal{L}$, the function $\mathrm{dist}_p^2 : \mathcal{L} \to \mathbb{R}$ is 2-concave.*

Proof of Theorem 8.23. Let $[xy]$ be a geodesic in $\mathrm{B}(p, \varpi^\kappa)$, and let $\ell = |x-y|$. Consider the model triangle $[\tilde{p}\tilde{x}\tilde{y}] = \tilde{\triangle}^\kappa(pxy)$. Set

$$\tilde{r}(t) = |\tilde{p} - \mathrm{geod}_{[\tilde{x}\tilde{y}]}(t)|, \qquad r(t) = |p - \mathrm{geod}_{[xy]}(t)|.$$

Clearly $\tilde{r}(0) = r(0)$ and $\tilde{r}(\ell) = r(\ell)$. Set $\tilde{f} = \mathrm{md}^\kappa \circ \tilde{r}$ and $f = \mathrm{md}^\kappa \circ r$. From 1.1(a) we get that $\tilde{f}'' = 1 - \kappa\cdot\tilde{f}$.

Note that the point-on-side comparison (see 8.14(b)) for point p and geodesic $[xy]$ is equivalent to $\tilde{r} \leqslant r$. Since md^κ is increasing on $[0, \varpi^\kappa)$, $\tilde{r} \leqslant r$

is equivalent to $\tilde{f} \leqslant f$. The latter is Jensen's inequality (see 3.14(c)) for the function $t \mapsto \operatorname{md}^\kappa |p - \operatorname{geod}_{[xy]}(t)|$ on the interval $[0, \ell]$. Hence the result. □

Recall that Busemann functions are defined in Proposition 6.1.

8.25. Exercise. Let $\mathcal{L}$ be a complete length CBB(κ) space, and let $\operatorname{bus}_\gamma : \mathcal{L} \to \to \mathbb{R}$ be the Busemann function for a half-line $\gamma : [0, \infty) \to \mathcal{L}$.

(a) If $\kappa = 0$, then the Busemann function $\operatorname{bus}_\gamma$ is concave.

(b) If $\kappa = -1$, then the function $f = \exp \circ \operatorname{bus}_\gamma$ satisfies

$$f'' - f \leqslant 0.$$

Exercise 9.28 is an analogous statement for upper curvature bound.

E. Development

In this section we reformulate the function comparison using a more geometric language based on the definition of development given below.

This definition appears in [**17**] and an earlier form of it can be found in [**106**]. The definition is somewhat lengthy, but it defines a useful comparison object for a curve. Often it is easier to write proofs in terms of function comparison but to think in terms of developments.

8.26. Lemma-definition. *Let $\kappa \in \mathbb{R}$, $\mathcal{X}$ be a metric space, let $\gamma : \mathbb{I} \to \mathcal{X}$ be a 1-Lipschitz curve, let $p \in \mathcal{X}$, and let $\tilde{p} \in \mathbb{M}^2(\kappa)$. Assume $0 < |p - \gamma(t)| < \varpi^\kappa$ for all $t \in \mathbb{I}$. Then there exists a unique up to rotation curve $\tilde{\gamma} : \mathbb{I} \to \mathbb{M}^2(\kappa)$, parametrized by arc-length, such that $|\tilde{p} - \tilde{\gamma}(t)| = |p - \gamma(t)|$ for all t and the direction of $[\tilde{p}\tilde{\gamma}(t)]$ monotonically turns around $\tilde{p}$ counterclockwise as t increases.*

*If p, $\tilde{p}$, γ, and $\tilde{\gamma}$ are as above, then $\tilde{\gamma}$ is called the κ-*development *of γ with respect to p; the point $\tilde{p}$ is called the* basepoint *of the development. When we say that the κ-development of γ with respect to p is defined, we always assume that $0 < |p - \gamma(t)| < \varpi^\kappa$ for all $t \in \mathbb{I}$.*

Proof. Consider the functions $\rho, \theta : \mathbb{I} \to \mathbb{R}$ defined as

$$\rho(t) = |p - \gamma(t)|, \qquad \theta(t) = \int_{t_0}^{t} \frac{\sqrt{1 - (\rho')^2}}{\operatorname{sn}^\kappa \rho},$$

where $t_0 \in \mathbb{I}$ is a fixed number and $\int$ denotes Lebesgue integral. Since γ is 1-Lipshitz, so is $\rho(t)$, and thus the function θ is defined and nondecreasing.

It is straightforward to check that (ρ, θ) uniquely describe $\tilde{\gamma}$ in polar coordinates on $\mathbb{M}^2(\kappa)$ with center at $\tilde{p}$. □

We need the following analogues of sub- and super-graphs and convex/concave functions, adapted to polar coordinates in $\mathbb{M}^2(\kappa)$.

8.27. Definition. Let $\tilde{\gamma} : \mathbb{I} \to \mathbb{M}^2(\kappa)$ be a curve, and let $\tilde{p} \in \mathbb{M}^2(\kappa)$ be such that there is a unique geodesic $[\tilde{p}\,\tilde{\gamma}(t)]$ for any $t \in \mathbb{I}$ and the direction of $[\tilde{p}\,\tilde{\gamma}(t)]$ turns monotonically as t grows.

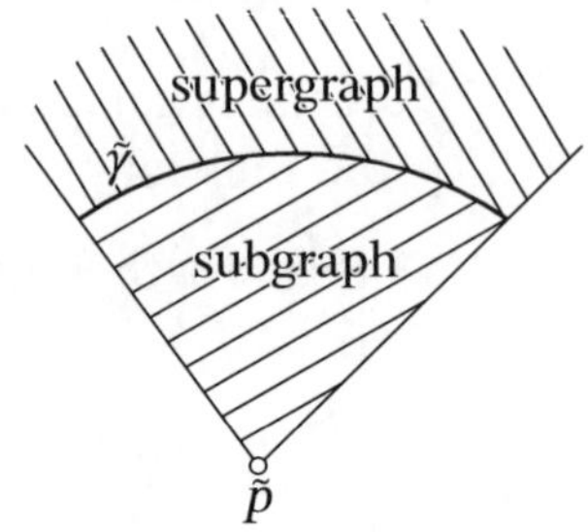

The set formed by all geodesics from $\tilde{p}$ to the points on $\tilde{\gamma}$ is called the *subgraph* of $\tilde{\gamma}$ with respect to $\tilde{p}$.

The set of all points $\tilde{x} \in \mathbb{M}^2(\kappa)$ such that a geodesic $[\tilde{p}\tilde{x}]$ intersects $\tilde{\gamma}$ is called the *supergraph* of $\tilde{\gamma}$ with respect to $\tilde{p}$.

The curve $\tilde{\gamma}$ is called *convex* (*resp., concave*) *with respect to* $\tilde{p}$ if the subgraph (resp., supergraph) of $\tilde{\gamma}$ with respect to $\tilde{p}$ is convex.

The curve $\tilde{\gamma}$ is called *locally convex* (*resp., concave*) *with respect to* $\tilde{p}$ if for any interior value t_0 in $\mathbb{I}$ there is a subsegment $(a,b) \subset \mathbb{I}$, $(a,b) \ni t_0$, such that the restriction $\tilde{\gamma}|_{(a,b)}$ is convex (resp., concave) with respect to $\tilde{p}$.

Note that if $\kappa > 0$, then the supergraph of a curve is the subgraph with respect to the opposite point.

For developments, all the notions above will be considered with respect to their basepoints. In particular, if $\tilde{\gamma}$ is a development, we will say it is (*locally*) *convex* if it is (locally) convex with respect to its basepoint.

8.28. Development comparison. *A complete G-delta geodesic space $\mathcal{L}$ is* CBB(κ) *if and only if for any point $p \in \mathcal{L}$ and any geodesic γ in* $\mathrm{B}(p, \varpi^\kappa) \setminus \{p\}$, *its κ-development with respect to p is convex.*

A simpler proof of the only-if part can be built on the adjacent angle comparison theorem (see 8.14(a)). We use a longer proof since it also implies the short hinge lemma (see 8.29).

Proof.

Only-if part. Let $\gamma : [0,T] \to \mathrm{B}(p, \varpi^\kappa) \setminus \{p\}$ be a unit-speed geodesic in $\mathcal{L}$.

Consider a fine partition

$$0 = t_0 < t_1 < \cdots < t_n = T.$$

Set $x_i = \gamma(t_i)$ and choose a point

$$p' \in \mathrm{Str}(x_0, x_1, \ldots, x_n)$$

sufficiently close to p; recall that geodesics $[p'x_i]$ exist for all i (see Theorem 8.11).

Let us construct a chain of model triangles $[\tilde p'\tilde x_{i-1}\tilde x_i] = \tilde\triangle^\kappa(p'x_{i-1}x_i)$ in such a way that direction $[\tilde p'\tilde x_i]$ turns counterclockwise as i grows. By the hinge comparison theorem (see 8.14(c)), we have

$$\begin{aligned}\measuredangle\left[\tilde x_i\,{}^{\tilde x_{i-1}}_{\tilde p'}\right]+\measuredangle\left[\tilde x_i\,{}^{\tilde x_{i+1}}_{\tilde p'}\right]&=\tilde\measuredangle^\kappa\left(x_i\,{}^{x_{i-1}}_{p'}\right)+\tilde\measuredangle^\kappa\left(x_i\,{}^{x_{i+1}}_{p'}\right)\leqslant\\ &\leqslant\measuredangle\left[x_i\,{}^{x_{i-1}}_{p'}\right]+\measuredangle\left[x_i\,{}^{x_{i+1}}_{p'}\right]\leqslant\\ &\leqslant\pi.\end{aligned}\tag{1}$$

Further, since γ is a unit-speed geodesic, we have

$$\sum_{i=1}^{n}|x_{i-1}-x_i|\leqslant|p'-x_0|+|p'-x_n|.\tag{2}$$

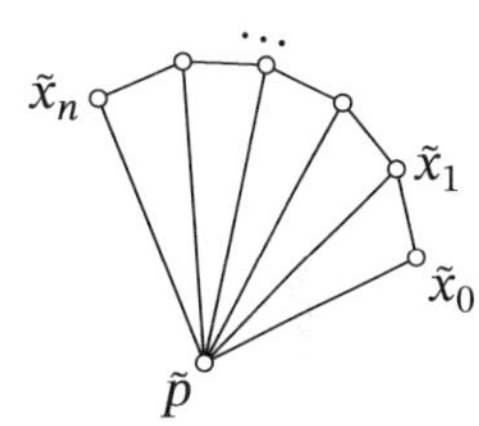

Since $p'\notin\gamma$, the development comparison implies that $\tilde p'$ does not lie on the polygonal line $\tilde x_0\cdots\tilde x_n$.

If $\kappa\leqslant 0$, then (2) implies that

$$\theta:=\sum_{i=1}^{n}\measuredangle\left[\tilde p'\,{}^{\tilde x_i}_{\tilde x_{i-1}}\right]\leqslant\pi.\tag{3}$$

In the case $\kappa>0$, the proof of (3) requires more work. Applying rescaling, we can assume that $\kappa=1$. Since γ lies in $B_\pi(p')$, the point-on-side comparison implies that the antipodal point of $\tilde p'$ does not lie on the polygonal line $\tilde x_0\cdots\tilde x_n$.

Consider the space L glued from the solid model triangles $[\tilde p'\tilde x_0\tilde x_1]$, ..., $[\tilde p'\tilde x_{n-1}\tilde x_n]$ along the corresponding sides. Note that θ is the total angle of L at $\tilde p'$. We can assume that L has nonempty interior. Otherwise all the triangles are degenerate, and therefore $\theta=0$; the latter holds since $\tilde p'$ is not on the polygonal line $\tilde x_0\cdots\tilde x_n$.

Consider a minimizing geodesic $[\tilde x_0\tilde x_n]_L$. By (2) we may assume that $\tilde p'\notin$ $\notin[\tilde x_0\tilde x_n]_L$. Further if the geodesic $[\tilde x_0\tilde x_n]_L$ contains one of the points $\tilde x_1$, ..., $\tilde x_{n-1}$, then it coincides with the polygonal line $\tilde x_0\cdots\tilde x_n$. (In particular, we have equality in (1) for each i.) In this case, the sum in the left-hand side of (2) must be at most π; otherwise $[\tilde x_0\tilde x_n]_L$ is not minimizing. Therefore (3) follows. In the remaining case $[\tilde x_0\tilde x_n]_L$ meets the boundary of L only at its ends. In this case, $|\tilde x_0-\tilde x_n|_L\leqslant\pi$; otherwise $[\tilde x_0\tilde x_n]_L$ is not minimizing. Whence (3) follows.

Inequalities (1) and (3) imply that the polygon $[\tilde p'\tilde x_0\tilde x_1\ldots\tilde x_n]$ is convex.

Let us take finer and finer partitions and pass to the limit of the polygon $\tilde p'\tilde x_0\tilde x_1\cdots\tilde x_n$ as $p'\to p$. We obtain a convex curvilinear triangle formed by a curve $\tilde\gamma\colon[0,T]\to\mathbb{M}^2(\kappa)$—the limit of polygonal line $\tilde x_0\tilde x_1\cdots\tilde x_n$ and two geodesics $[\tilde p'\,\tilde\gamma(0)]$, $[\tilde p'\,\tilde\gamma(T)]$. Since $[\tilde p'\tilde x_0\tilde x_1\ldots\tilde x_n]$ is convex, the natural parametrization of $\tilde x_0\tilde x_1\cdots\tilde x_n$ converges to the natural parametrization of $\tilde\gamma$. Thus $\tilde\gamma$ is the κ-development of γ with respect to p. This proves the only-if part of 8.28.

If part. Assuming convexity of the development, we will prove the point-on-side comparison theorem (see 8.14(b)). We can assume that $p \notin [xy]$; otherwise the statement is trivial.

Set $T = |x - y|$ and $\gamma(t) = \mathrm{geod}_{[xy]}(t)$; note that γ is a geodesic in $\mathrm{B}(p, \varpi^\kappa) \setminus \{p\}$. Let $\tilde\gamma\colon [0, T] \to \mathbb{M}^2(\kappa)$ be the κ-development with base $\tilde p$ of γ with respect to p. Take a partition $0 = t_0 < t_1 < \dots < t_n = T$, and set

$$\tilde y_i = \tilde\gamma(t_i) \quad \text{and} \quad \tau_i = |\tilde y_0 - \tilde y_1| + |\tilde y_1 - \tilde y_2| + \dots + |\tilde y_{i-1} - \tilde y_i|.$$

Since $\tilde\gamma$ is convex, for a fine partition we have that the polygonal line $\tilde y_0 \tilde y_1 \cdots \tilde y_n$ is also convex. Applying Alexandrov's Lemma 6.3 inductively to pairs of model triangles

$$\tilde\triangle^\kappa\{\tau_{i-1}, |\tilde p - \tilde y_0|, |\tilde p - \tilde y_{i-1}|\}$$

and

$$\tilde\triangle^\kappa\{|\tilde y_{i-1} - \tilde y_i|, |\tilde p - \tilde y_{i-1}|, |\tilde p - \tilde y_i|\}$$

we obtain that the sequence $\tilde\measuredangle^\kappa\{|\tilde p - \tilde y_i|; |\tilde p - \tilde y_0|, \tau_i\}$ is nonincreasing.

For finer and finer partitions we have

$$\max_i\{|\tau_i - t_i|\} \to 0.$$

Thus, the point-on-side comparison (see 8.14(b)) follows. □

Note that in the proof of the if part we could use a slightly weaker version of the hinge comparison (see 8.14(c)). Namely, we proved the following lemma, which will be needed later in the proof of the Globalization Theorem 8.31.

8.29. Short hinge lemma. *Let $\mathcal{L}$ be a complete G-delta geodesic space such that for any hinge $[x\,{}^p_y]$ in $\mathcal{L}$ the angle $\measuredangle[x\,{}^p_y]$ is defined, and*

$$\measuredangle[x\,{}^p_y] + \measuredangle[x\,{}^p_z] \leqslant \pi$$

for any two adjacent hinges.

Assume that for any hinge $[x\,{}^p_y]$ in $\mathcal{L}$ we have

$$|p - x| + |x - y| < \varpi^\kappa \quad \Rightarrow \quad \measuredangle[x\,{}^p_y] \geqslant \tilde\measuredangle^\kappa(x\,{}^p_y).$$

Then $\mathcal{L}$ is CBB(κ).

F. Local definitions and globalization

In this section we discuss locally CBB(κ) spaces. In particular, we prove the *globalization theorem*: equivalence of local and global definitions for complete length spaces.

The following theorem summarizes equivalent definitions of locally CBB(κ) spaces.

8.30. Theorem. *Let $\mathcal{X}$ be a complete length space, and let $p \in \mathcal{X}$. Then the following conditions are equivalent:*

(1) Local CBB(κ) comparison. *There is $R_1 > 0$ such that the comparison*

$$\tilde{\measuredangle}^\kappa\left(q\,{}^{x^1}_{x^2}\right) + \tilde{\measuredangle}^\kappa\left(q\,{}^{x^2}_{x^3}\right) + \tilde{\measuredangle}^\kappa\left(q\,{}^{x^3}_{x^1}\right) \leqslant 2\cdot\pi$$

holds for any q, x^1, x^2, $x^3 \in \mathrm{B}(p, R_1)$.

(2) Local Kirszbraun property. *There is $R_2 > 0$ such that for any three-point subset F_3 and any four-point subset $F_4 \supset F_3$ in $\mathrm{B}(p, R_2)$, any short map $f : F_3 \to \mathbb{M}^2(\kappa)$ can be extended to a short map $\bar{f} : F_4 \to \mathbb{M}^2(\kappa)$ (so $f = \bar{f}|_{F_3}$).*

(3) Local function comparison. *There is $R_3 > 0$ such that $\mathrm{B}(p, R_3)$ is G-delta geodesic and for any $q \in \mathrm{B}(p, R_3)$, the function $f = \mathrm{md}^\kappa \circ \mathrm{dist}_q$ satisfies $f'' \leqslant 1 - \kappa\cdot f$ in $\mathrm{B}(p, R_3)$.*

(4) Local adjacent angle comparison. *There is $R_4 > 0$ such that $\mathrm{B}(p, R_4)$ is G-delta geodesic, and if q and a geodesic $[xy]$ lie in $\mathrm{B}(p, R_4)$ and $z \in$ $\in\,]xy[$, then*

$$\tilde{\measuredangle}^\kappa\left(z\,{}^{q}_{x}\right) + \tilde{\measuredangle}^\kappa\left(z\,{}^{q}_{y}\right) \leqslant \pi.$$

(5) Local point-on-side comparison. *There is $R_5 > 0$ such that $\mathrm{B}(p, R_5)$ is G-delta geodesic and if q and a geodesic $[xy]$ lie in $\mathrm{B}(p, R_5)$ and $z \in$ $\in\,]xy[$, then*

$$\tilde{\measuredangle}^\kappa\left(x\,{}^{q}_{y}\right) \leqslant \tilde{\measuredangle}^\kappa\left(x\,{}^{q}_{z}\right),$$

or, equivalently,

$$|\tilde{p} - \tilde{z}| \leqslant |p - z|,$$

where $[\tilde{p}\tilde{x}\tilde{y}] = \tilde{\triangle}^\kappa(pxy)$, $\tilde{z} \in\,]\tilde{x}\tilde{y}[$, $|\tilde{x} - \tilde{z}| = |x - z|$.

(6) Local hinge comparison. *There is $R_6 > 0$ such that $\mathrm{B}(p, R_6)$ is G-delta geodesic, and if $x \in \mathrm{B}(p, R_6)$, then for any hinge $\left[x\,{}^{q}_{y}\right]$, the angle $\measuredangle\left[x\,{}^{q}_{y}\right]$ is defined, and*

$$\measuredangle\left[x\,{}^{q}_{y}\right] + \measuredangle\left[x\,{}^{q}_{z}\right] \leqslant \pi$$

for any two adjacent hinges. Moreover, if a hinge $\left[x\,{}^{q}_{y}\right]$ lies in $\mathrm{B}(p, R_6)$, then

$$\measuredangle\left[x\,{}^{q}_{y}\right] \geqslant \tilde{\measuredangle}^\kappa\left(x\,{}^{q}_{y}\right),$$

or, equivalently,

$$\tilde{\curlyvee}^\kappa\left[x\,{}^{q}_{y}\right] \geqslant |q - y|.$$

(7) Local development comparison. *There is $R_7 > 0$ such that $\mathrm{B}(p, R_7)$ is G-delta geodesic, and if a geodesic γ lies in $\mathrm{B}(p, R_7)$ and $q \in \mathrm{B}(p, R_7)\backslash\gamma$, then the κ-development $\tilde{\gamma}$ with respect to q is convex.*

Moreover, for each pair i, $j \in \{1, 2, \ldots, 7\}$ we can assume that

$$R_i > \tfrac{1}{9}\cdot R_j.$$

The proofs of each of these equivalences repeat the proofs of the corresponding global equivalences in localized form; see the proofs of Theorems 8.14, 8.23, and 10.1, and Lemma-definition 8.26.

8.31. Globalization theorem. *Any complete length locally* CBB(κ) *space is* CBB(κ).

In the two-dimensional case this theorem was proved by Paolo Pizzetti [**136**]; later it was reproved independently by Alexandr Alexandrov [**17**]. Victor Toponogov [**156**] proved it for Riemannian manifolds of all dimensions. In the above generality, the theorem first appears in the paper of Michael Gromov, Yuriy Burago, and Grigory Perelman [**44**]; simplifications and modifications were given by Conrad Plaut [**138**], Katsuhiro Shiohama [**150**], and in the book of Dmitry Burago, Yuriy Burago, and Sergei Ivanov [**37**]. A generalization for noncomplete but geodesic spaces was obtained by the third author [**129**]; namely it solves the following exercise.

8.32. Advanced exercise. Show that any locally CBB(κ) geodesic space is CBB(κ).

Corollary 8.33 of the globalization theorem says that the expression "space with curvature $\geqslant \kappa$" makes sense.

8.33. Corollary. *Let $\mathcal{L}$ be a complete length space. Then $\mathcal{L}$ is* CBB(K) *if and only if $\mathcal{L}$ is* CBB(κ) *for any $\kappa <$ K.*

Proof. Note that if K $\leqslant 0$, this statement follows directly from the definition of an Alexandrov space (Definition 8.2) and monotonicity of the function $\kappa \mapsto$ $\mapsto \tilde{\measuredangle}^{\kappa}\left(x\,{}^{y}_{z}\right)$ (see 1.1(d)).

The if part also follows directly from the definition.

For K > 0, the angle $\tilde{\measuredangle}^{\mathrm{K}}\left(x\,{}^{y}_{z}\right)$ might be undefined while $\tilde{\measuredangle}^{\kappa}\left(x\,{}^{y}_{z}\right)$ is defined. However, $\tilde{\measuredangle}^{\mathrm{K}}\left(x\,{}^{y}_{z}\right)$ is defined if x, y, and z are sufficiently close to each other. Thus, if K $> \kappa$, then any CBB(K) space is locally CBB(κ). It remains to apply the globalization theorem. □

8.34. Corollary. *Let $\mathcal{L}$ be a complete length* CBB(κ) *space. Assume that a space $\mathcal{M}$ is the target space of a submetry from $\mathcal{L}$. Then $\mathcal{M}$ is a complete length space* CBB(κ) *space.*

In particular, if $G \curvearrowright \mathcal{L}$ is an isometric group action with closed orbits, then the quotient space $\mathcal{L}/G$ is a complete length CBB(κ) *space.*

Proof. This follows from the globalization theorem and Theorem 8.5. □

Our proof of Globalization Theorem 8.31 is based on presentations in [**138**] and [**37**]; this proof was rediscovered independently by Urs Lang and Viktor Schroeder [**100**]. We will need Short Hinge Lemma 8.29 together with the following two lemmas. The following lemma says that if comparison holds for all small hinges, then it holds for slightly bigger hinges near the given point.

8.35. Key lemma. *Let $\kappa\in\mathbb{R}$, $0<\ell\leqslant\varpi^\kappa$, $\mathcal{X}$ be a complete geodesic space, and let $p\in\mathcal{X}$ be a point such that* $\mathrm{B}(p,2\cdot\ell)$ *is locally* $\mathrm{CBB}(\kappa)$.

Assume that for any point $q\in\mathrm{B}(p,\ell)$ the comparison

$$\measuredangle\big[x\,{}^{y}_{q}\big]\geqslant\tilde\measuredangle^\kappa\big(x\,{}^{y}_{q}\big)$$

holds for any hinge $\big[x\,{}^{y}_{q}\big]$ with $|x-y|+|x-q|<\tfrac{2}{3}\cdot\ell$. Then the comparison

$$\measuredangle\big[x\,{}^{p}_{q}\big]\geqslant\tilde\measuredangle^\kappa\big(x\,{}^{p}_{q}\big)$$

holds for any hinge $\big[x\,{}^{p}_{q}\big]$ with $|x-p|+|x-q|<\ell$.

Proof. It is sufficient to prove the inequality

$$\tilde\curlyvee^\kappa\big[x\,{}^{p}_{q}\big]\geqslant|p-q| \tag{1}$$

for any hinge $\big[x\,{}^{p}_{q}\big]$ with $|x-p|+|x-q|<\ell$.

Fix q. Consider a hinge $\big[x\,{}^{p}_{q}\big]$ such that

$$\tfrac{2}{3}\cdot\ell\leqslant|p-x|+|x-q|<\ell.$$

First we construct a new smaller hinge $\big[x'\,{}^{p}_{q}\big]$ with

$$|p-x|+|x-q|\geqslant|p-x'|+|x'-q|, \tag{2}$$

such that

$$\tilde\curlyvee^\kappa\big[x\,{}^{p}_{q}\big]\geqslant\tilde\curlyvee^\kappa\big[x'\,{}^{p}_{q}\big]. \tag{3}$$

Assume $|x-q|\geqslant|x-p|$; otherwise switch the roles of p and q in the following construction. Take $x'\in[xq]$ such that

$$|p-x|+3\cdot|x-x'|=\tfrac{2}{3}\cdot\ell. \tag{4}$$

Choose a geodesic $[x'p]$ and consider the hinge $\big[x'\,{}^{p}_{q}\big]$ formed by $[x'p]$ and $[x'q]\subset[xq]$. (In fact by Corollary 8.38 the condition $[x'q]\subset[xq]$ always holds.) Then (2) follows from the triangle inequality.

Further, note that we have $x,x'\in\mathrm{B}(p,\ell)\cap\mathrm{B}(q,\ell)$ and moreover

$$|p-x|+|x-x'|<\tfrac{2}{3}\cdot\ell,\qquad\qquad |p-x'|+|x'-x|<\tfrac{2}{3}\cdot\ell.$$

In particular,

$$\measuredangle\big[x\,{}^{p}_{x'}\big]\geqslant\tilde\measuredangle^\kappa\big(x\,{}^{p}_{x'}\big)\quad\text{and}\quad\measuredangle\big[x'\,{}^{p}_{x}\big]\geqslant\tilde\measuredangle^\kappa\big(x'\,{}^{p}_{x}\big). \tag{5}$$

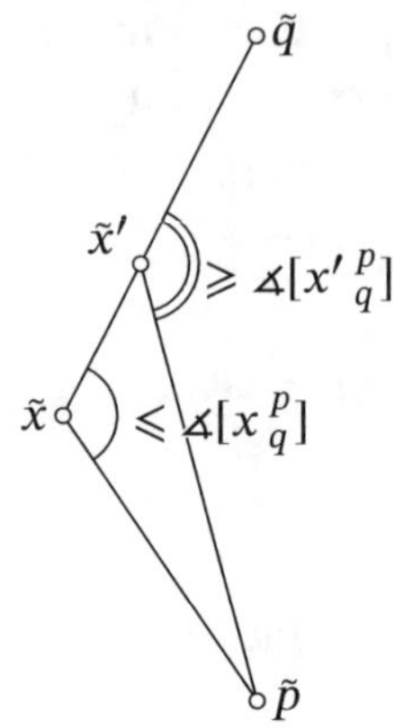

Now, let $[\tilde x\tilde x'\tilde p] = \tilde\triangle^\kappa(xx'p)$. Take $\tilde q$ on the extension of $[\tilde x\tilde x']$ beyond x' such that $|\tilde x-\tilde q| = |x-q|$ (and therefore $|\tilde x' - \tilde q| = |x' - q|$). From (5),

$$\measuredangle\left[x\,{}^{p}_{q}\right] = \measuredangle\left[x\,{}^{p}_{x'}\right] \geqslant \tilde\measuredangle^\kappa\left(x\,{}^{p}_{x'}\right) \quad\Rightarrow\quad \tilde\gamma^\kappa\left[x\,{}^{q}_{p}\right] \geqslant |\tilde p - \tilde q|.$$

Hence

$$\begin{aligned}\measuredangle\left[\tilde x'\,{}^{\tilde p}_{\tilde q}\right] &= \pi - \tilde\measuredangle^\kappa\left(x'\,{}^{p}_{x}\right) \geqslant\\ &\geqslant \pi - \measuredangle\left[x'\,{}^{p}_{x}\right] =\\ &= \measuredangle\left[x'\,{}^{p}_{q}\right],\end{aligned}$$

and (3) follows.

Let us continue the proof. Set $x_0 = x$. Let us apply inductively the above construction to get a sequence of hinges $\left[x_n\,{}^{p}_{q}\right]$ with $x_{n+1} = x'_n$. From (3), we have that the sequence $s_n = \tilde\gamma^\kappa\left[x_n\,{}^{p}_{q}\right]$ is nonincreasing.

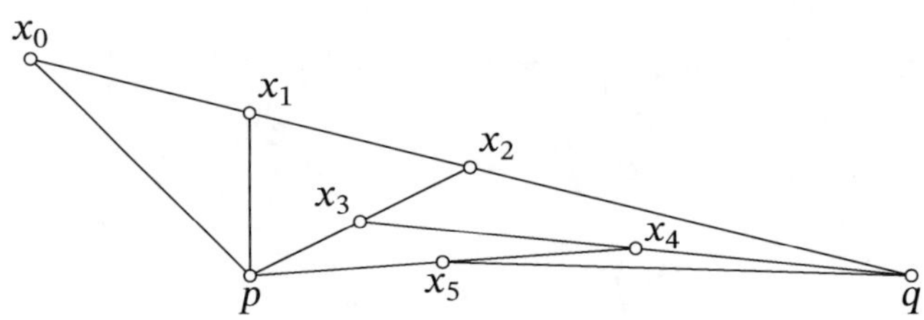

The sequence might terminate at x_n only if $|p-x_n|+|x_n-q| < \frac{2}{3}\cdot\ell$. In this case, by the assumptions of the lemma, $\tilde\gamma^\kappa\left[x_n\,{}^{p}_{q}\right] \geqslant |p-q|$. Since the sequence s_n is nonincreasing, inequality (1) follows.

Otherwise, the sequence $r_n = |p - x_n| + |x_n - q|$ is nonincreasing and $r_n \geqslant \frac{2}{3}\cdot\ell$ for all n. Note that by construction, the distances $|x_n - x_{n+1}|$, $|x_n - p|$, and $|x_n - q|$ are bounded away from zero for all large n. Indeed, since on each step we move x_n toward to the point p or q that is further away, the distances $|x_n - p|$ and $|x_n - q|$ become about the same. Namely, by (4), we have that $|p - x_n| - |x_n - q| \leqslant \frac{2}{9}\cdot\ell$ for all large n. Since $|p - x_n| + |x_n - q| \geqslant \frac{2}{3}\cdot\ell$, we have $|x_n - p| \geqslant \frac{\ell}{100}$ and $|x_n - q| \geqslant \frac{\ell}{100}$. Further, since $r_n \geqslant \frac{2}{3}\cdot\ell$, (4) implies that $|x_n - x_{n+1}| > \frac{\ell}{100}$.

Since the sequence r_n is nonincreasing, it converges. In particular, $r_n - r_{n+1} \to 0$ as $n \to \infty$. It follows that $\tilde\measuredangle^\kappa\left(x_n\,{}^{p_n}_{x_{n+1}}\right) \to \pi$, where $p_n = p$ if $x_{n+1} \in [x_n q]$, and otherwise $p_n = q$. Since $\measuredangle\left[x_n\,{}^{p_n}_{x_{n+1}}\right] \geqslant \tilde\measuredangle^\kappa\left(x_n\,{}^{p_n}_{x_{n+1}}\right)$, we have $\measuredangle\left[x_n\,{}^{p_n}_{x_{n+1}}\right] \to \pi$ as $n \to \infty$.

It follows that

$$r_n - s_n = |p - x_n| + |x_n - q| - \tilde\gamma^\kappa\left[x_n\,{}^{p}_{q}\right] \to 0.$$

(Here we used that $\ell \leqslant \varpi^\kappa$.) Together with the triangle inequality

$$|p - x_n| + |x_n - q| \geqslant |p - q|$$

this yields

$$\lim_{n\to\infty} \tilde{\gamma}^\kappa \left[x_n\,{}^p_q\right] \geqslant |p - q|.$$

Applying monotonicity of the sequence $s_n = \tilde{\gamma}^\kappa \left[x_n\,{}^p_q\right]$, we obtain (1). □

The final part of the proof above resembles the *cat's cradle construction* introduced by the first author and Richard Bishop [**5**].

The next lemma works in all complete spaces; it will be used as a substitute for the existence of a minimum point of a continuous function on a compact space.

8.36. Lemma on almost minimum. *Let $\mathcal{X}$ be a complete metric space. Suppose $r\colon \mathcal{X} \to \mathbb{R}$ is a function, $p \in \mathcal{X}$, and $\varepsilon > 0$. Assume that the function r is strictly positive in $\overline{\mathrm{B}}[p, \tfrac{1}{\varepsilon^2}\cdot r(p)]$ and $\underline{\lim}_n\, r(x_n) > 0$ for any convergent sequence $x_n \to x \in \overline{\mathrm{B}}[p, \tfrac{1}{\varepsilon^2}\cdot r(p)]$.*

Then there is a point $p^* \in \overline{\mathrm{B}}[p, \tfrac{1}{\varepsilon^2}\cdot r(p)]$ such that

(a) *$r(p^*) \leqslant r(p)$ and*

(b) *$r(x) > (1-\varepsilon)\cdot r(p^*)$ for any $x \in \overline{\mathrm{B}}[p^*, \tfrac{1}{\varepsilon}\cdot r(p^*)]$.*

Proof. Assume the statement is wrong. Then for any $x \in \mathrm{B}(p, \tfrac{1}{\varepsilon^2}\cdot r(p))$ with $r(x) \leqslant r(p)$, there is a point $x' \in \mathcal{X}$ such that

$$|x - x'| < \tfrac{1}{\varepsilon}\cdot r(x) \quad \text{and} \quad r(x') \leqslant (1-\varepsilon)\cdot r(x).$$

Take $x_0 = p$ and consider a sequence of points x_n such that $x_{n+1} = x_n'$. Clearly

$$|x_{n+1} - x_n| \leqslant \tfrac{r(p)}{\varepsilon}\cdot(1-\varepsilon)^n \quad \text{and} \quad r(x_n) \leqslant r(p)\cdot(1-\varepsilon)^n.$$

In particular, $|p - x_n| < \tfrac{1}{\varepsilon^2}\cdot r(p)$. Therefore the sequence x_n is Cauchy, $x_n \to$ $\to x \in \overline{\mathrm{B}}[p, \tfrac{1}{\varepsilon^2}\cdot r(p)]$ and $\lim_n r(x_n) = 0$, a contradiction. □

Proof of Globalization Theorem 8.31. Exactly the same argument as in the proof of Theorem 8.11 shows that $\mathcal{L}$ is G-delta geodesic. By Theorem 8.30(6), for any hinge $\left[x\,{}^p_y\right]$ in $\mathcal{L}$ the angle $\measuredangle\left[x\,{}^p_y\right]$ is defined and

$$\measuredangle\left[x\,{}^p_y\right] + \measuredangle\left[x\,{}^p_z\right] \leqslant \pi$$

for any two adjacent hinges.

Let us denote by $\operatorname{ComRad}(p, \mathcal{L})$ (which stands for *comparison radius* of $\mathcal{L}$ at p) the maximal value (possibly ∞) such that the comparison

$$\measuredangle\left[x\,{}^p_y\right] \geqslant \tilde{\measuredangle}^\kappa\left(x\,{}^p_y\right)$$

holds for any hinge $\left[x\,{}^p_y\right]$ with $|p - x| + |x - y| < \operatorname{ComRad}(p, \mathcal{L})$.

As follows from Theorem8.30(3), $\operatorname{ComRad}(p,\mathcal{L}) > 0$ for any $p \in \mathcal{L}$ and

$$\varliminf_{n\to\infty} \operatorname{ComRad}(p_n,\mathcal{L}) > 0$$

for any converging sequence of points $p_n \to p$. That makes it possible to apply the lemma on almost minimum (Lemma 8.36) to the function $p \mapsto \operatorname{ComRad}(p,\mathcal{L})$.

According to Short Hinge Lemma 8.29, it is sufficient to show that

$$s_0 = \inf_{p\in\mathcal{L}} \operatorname{ComRad}(p,\mathcal{L}) \geqslant \varpi^\kappa \quad \text{for any } p \in \mathcal{L}. \tag{6}$$

We argue by contradiction, assuming that (6) does not hold.

The rest of the proof is easier for geodesic spaces and easier still for compact spaces. Thus we give three different arguments for each of these cases.

Compact case. Assume $\mathcal{L}$ is compact. By Theorem 8.30(3), $s_0 > 0$. Take a point $p^* \in \mathcal{L}$ such that $r^* = \operatorname{ComRad}(p^*,\mathcal{L})$ is sufficiently close to s_0 (p^* such that $s_0 \leqslant r^* < \min\{\varpi^\kappa, \frac{3}{2}\cdot s_0\}$ will do). Then Key Lemma 8.35 applied to p^* and ℓ slightly bigger than r^* (say, such that $r^* < \ell < \min\{\varpi^\kappa, \frac{3}{2}\cdot s_0\}$) implies that

$$\measuredangle\left[x\,{}^{p^*}_{q}\right] \geqslant \tilde{\measuredangle}^\kappa\left(x\,{}^{p^*}_{q}\right)$$

for any hinge $\left[x\,{}^{p^*}_{q}\right]$ such that $|p^*-x|+|x-q| < \ell$. Thus $r^* \geqslant \ell$, a contradiction.

Geodesic case. Assume $\mathcal{L}$ is geodesic. Fix a small $\varepsilon > 0$ ($\varepsilon = 0.0001$ will do). Apply Lemma 8.36 on almost minimum to find a point $p^* \in \mathcal{L}$ such that

$$r^* = \operatorname{ComRad}(p^*,\mathcal{L}) < \varpi^\kappa$$

and

$$\operatorname{ComRad}(q,\mathcal{L}) > (1-\varepsilon)\cdot r^* \tag{7}$$

for any $q \in \overline{\mathrm{B}}[p^*, \frac{1}{\varepsilon}\cdot r^*]$.

Applying Key Lemma 8.35 for p^* and ℓ slightly bigger than r^* leads to a contradiction.

General case. Let us construct $p^* \in \mathcal{L}$ as in the previous case. Since $\mathcal{L}$ is not geodesic, we cannot apply the key lemma directly. Instead, let us pass to the ultrapower $\mathcal{L}^\omega$, which is a geodesic space (see Observation 4.9).

In Theorem 8.30, inequality (7) implies that condition 8.30(1) holds for some fixed $R_1 = \frac{r^*}{100} > 0$ at any point $q \in \overline{\mathrm{B}}[p^*, \frac{1}{2\cdot\varepsilon}\cdot r^*] \subset \mathcal{L}$. Therefore a similar statement is true in the ultrapower $\mathcal{L}^\omega$; that is, for any point $q_\omega \in$ $\in \overline{\mathrm{B}}[p^*, \frac{1}{2\cdot\varepsilon}\cdot r^*] \subset \mathcal{L}^\omega$, condition 8.30(1) holds for, say, $R_1 = \frac{r^*}{101}$.

Note that $r^* \geqslant \operatorname{ComRad}(p^*,\mathcal{L}^\omega)$. Therefore we can apply the lemma on almost minimum at the point p^* to the function $x \mapsto \operatorname{ComRad}(x,\mathcal{L}^\omega)$ and $\varepsilon' = \sqrt{\varepsilon} = 0.01$.

For the resulting point $p^{**} \in \mathcal{L}^{\omega}$, we have

$$r^{**} = \operatorname{ComRad}(p^{**}, \mathcal{L}) < \varpi^{\kappa}, \quad \text{and} \quad \operatorname{ComRad}(q_{\omega}, \mathcal{L}^{\omega}) > (1-\varepsilon')\cdot r^{**}$$

for any $q_{\omega} \in \overline{\mathrm{B}}[p^{**}, \frac{1}{\varepsilon'}\cdot r^{**}]$. Thus applying Key Lemma 8.35 for p^{**} and for ℓ slightly bigger than r^{**} leads to a contradiction. □

G. Properties of geodesics and angles

Remark. All proofs in this section can be easily modified to use only the local definition of CBB spaces without use of Globalization Theorem 8.31.

8.37. Geodesics do not split. *In a* CBB *space, geodesics do not bifurcate.*

More precisely, let $\mathcal{L}$ be a CBB *space, and let $[px]$, $[py]$ be two geodesics. Then:*

(a) *If there is $\varepsilon > 0$ such that $\operatorname{geod}_{[px]}(t) = \operatorname{geod}_{[py]}(t)$ for all $t \in [0, \varepsilon)$, then $[px] \subset [py]$ or $[py] \subset [px]$.*

(b) *If $\measuredangle\left[p\,{}^{x}_{y}\right] = 0$, then $[px] \subset [py]$ or $[py] \subset [px]$.*

8.38. Corollary. *Let $\mathcal{L}$ be a* CBB *space. Then the restriction of any geodesic in $\mathcal{L}$ to a proper segment is the unique minimal geodesic joining its endpoints.*

In case $\kappa \leqslant 0$, the proof is easier, since the model triangles are always defined. To deal with $\kappa > 0$, we have to argue locally.

Proof of 8.37.

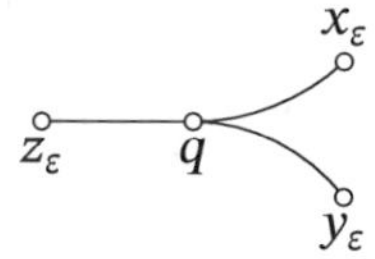

(a). Let $t_{\max}$ be the maximal value such that $\operatorname{geod}_{[px]}(t) = \operatorname{geod}_{[py]}(t)$ for all $t \in [0, t_{\max})$. Since geodesics are continuous, $\operatorname{geod}_{[px]}(t_{\max}) = \operatorname{geod}_{[py]}(t_{\max})$. Let

$$q = \operatorname{geod}_{[px]}(t_{\max}) = \operatorname{geod}_{[py]}(t_{\max}).$$

We must show that $t_{\max} = \min\{|p-x|, |p-y|\}$.

If that is not true, choose a sufficiently small $\varepsilon > 0$ such that the points

$$x_{\varepsilon} = \operatorname{geod}_{[px]}(t_{\max}+\varepsilon) \quad \text{and} \quad y_{\varepsilon} = \operatorname{geod}_{[py]}(t_{\max}+\varepsilon)$$

are distinct. Let

$$z_{\varepsilon} = \operatorname{geod}_{[px]}(t_{\max}-\varepsilon) = \operatorname{geod}_{[py]}(t_{\max}-\varepsilon).$$

Clearly, $\tilde{\measuredangle}^{\kappa}\left(q\,{}^{z_{\varepsilon}}_{x_{\varepsilon}}\right) = \tilde{\measuredangle}^{\kappa}\left(q\,{}^{z_{\varepsilon}}_{y_{\varepsilon}}\right) = \pi$. Thus from the CBB($\kappa$) comparison (Definition 8.2), $\tilde{\measuredangle}^{\kappa}\left(q\,{}^{x_{\varepsilon}}_{y_{\varepsilon}}\right) = 0$ and thus $x_{\varepsilon} = y_{\varepsilon}$, a contradiction.

(b). From the hinge comparison theorem (see 8.14(c)),

$$\measuredangle\left[p\,{}^{x}_{y}\right]=0 \quad\Rightarrow\quad \tilde\measuredangle^{\kappa}\left(p\,{}^{\operatorname{geod}_{[px]}(t)}_{\operatorname{geod}_{[py]}(t)}\right)=0$$

and thus $\operatorname{geod}_{[px]}(t)=\operatorname{geod}_{[py]}(t)$ for all small t. Therefore we can apply (a). □

8.39. Adjacent angle lemma. *Let $\mathcal{L}$ be a* CBB *space. Assume that two hinges $\left[z\,{}^{x}_{p}\right]$ and $\left[z\,{}^{y}_{p}\right]$ in $\mathcal{L}$ are adjacent. Then*

$$\measuredangle\left[z\,{}^{p}_{y}\right]+\measuredangle\left[z\,{}^{p}_{x}\right]=\pi.$$

Proof. From the hinge comparison theorem (see 8.14(c)) we have that both angles $\measuredangle\left[z\,{}^{p}_{y}\right]$ and $\measuredangle\left[z\,{}^{p}_{x}\right]$ are defined and

$$\measuredangle\left[z\,{}^{p}_{y}\right]+\measuredangle\left[z\,{}^{p}_{x}\right]\leqslant\pi.$$

Clearly $\measuredangle\left[z\,{}^{x}_{y}\right]=\pi$. Thus the result follows from the triangle inequality for angles (see 6.5). □

8.40. Angle semicontinuity. *Suppose $\mathcal{L}_n$ is a sequence of* CBB(κ) *spaces and $\mathcal{L}_n\to\mathcal{L}_\omega$ as $n\to\omega$. Assume that a sequence of hinges $\left[p_n\,{}^{x_n}_{y_n}\right]$ in $\mathcal{L}_n$ converges to a hinge $\left[p_\omega\,{}^{x_\omega}_{y_\omega}\right]$ in $\mathcal{L}_\omega$. Then*

$$\measuredangle\left[p_\omega\,{}^{x_\omega}_{y_\omega}\right]\leqslant\lim_{n\to\omega}\measuredangle\left[p_n\,{}^{x_n}_{y_n}\right].$$

Proof. From 8.17,

$$\measuredangle\left[p_\omega\,{}^{x_\omega}_{y_\omega}\right]=\sup\left\{\tilde\measuredangle^{\kappa}\left(p_\omega\,{}^{\bar x_\omega}_{\bar y_\omega}\right)\,:\,\bar x_\omega\in\,]p_\omega x_\omega],\ \bar y_\omega\in\,]p_\omega x_\omega]\right\}.$$

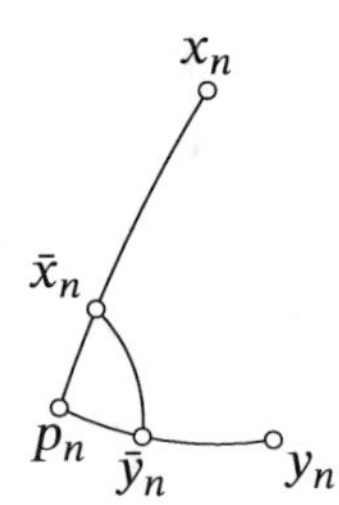

For fixed $\bar x_\omega\in\,]p_\omega x_\omega]$ and $\bar y_\omega\in\,]p_\omega x_\omega]$, choose $\bar x_n\in$ $\in\,]px_n]$ and $\bar y_n\in\,]py_n]$ so that $\bar x_n\to\bar x_\omega$ and $\bar y_n\to\bar y_\omega$ as $n\to\omega$. Clearly

$$\tilde\measuredangle^{\kappa}\left(p_n\,{}^{\bar x_n}_{\bar y_n}\right)\to\tilde\measuredangle^{\kappa}\left(p_\omega\,{}^{\bar x_\omega}_{\bar y_\omega}\right)$$

as $n\to\omega$.

From the hinge comparison theorem (see 8.14(c)), $\measuredangle\left[p_n\,{}^{x_n}_{y_n}\right]\geqslant\tilde\measuredangle^{\kappa}\left(p_n\,{}^{\bar x_n}_{\bar y_n}\right)$. Hence the result. □

8.41. Angle continuity. *Let $\mathcal{L}_1,\mathcal{L}_2,\dots$ be a sequence of complete length* CBB(κ) *spaces, and let $\mathcal{L}_n\to\mathcal{L}_\omega$ as $n\to\omega$. Assume that sequences of points p_n, x_n, y_n in $\mathcal{L}_n$ converge to points p_ω, x_ω, y_ω in $\mathcal{L}_\omega$ as $n\to\omega$, and the following two conditions hold:*

(a) $p_\omega\in\operatorname{Str}(x_\omega)$,

(b) $p_\omega\in\operatorname{Str}(y_\omega)$ *or* $y_\omega\in\operatorname{Str}(p_\omega)$.

Then

$$\measuredangle[p_\omega\,{}^{x_\omega}_{y_\omega}] = \lim_{n\to\omega} \measuredangle[p_n\,{}^{x_n}_{y_n}].$$

Proof. By Corollary 8.33, we may assume that $\kappa \leqslant 0$.

By Plaut's Theorem 8.11, the hinge $[p_\omega\,{}^{x_\omega}_{y_\omega}]$ is uniquely defined. Therefore the hinges $[p_n\,{}^{x_n}_{y_n}]$ converge to $[p_\omega\,{}^{x_\omega}_{y_\omega}]$ as $n \to \omega$. Hence by the angle semicontinuity, see 8.40, we have

$$\measuredangle[p_\omega\,{}^{x_\omega}_{y_\omega}] \leqslant \lim_{n\to\omega} \measuredangle[p_n\,{}^{x_n}_{y_n}].$$

It remains to show that

$$\measuredangle[p_\omega\,{}^{x_\omega}_{y_\omega}] \geqslant \lim_{n\to\omega} \measuredangle[p_n\,{}^{x_n}_{y_n}]. \tag{1}$$

Fix $\varepsilon > 0$. Since $p_\omega \in \operatorname{Str}(x_\omega)$, there is a point $q_\omega \in \mathcal{L}_\omega$ such that

$$\tilde\measuredangle^\kappa(p_\omega\,{}^{x_\omega}_{q_\omega}) > \pi - \varepsilon.$$

The hinge comparison theorem (see 8.14(c)) implies that

$$\measuredangle[p_\omega\,{}^{x_\omega}_{q_\omega}] > \pi - \varepsilon. \tag{2}$$

By the triangle inequality for angles (see 6.5),

$$\begin{aligned}\measuredangle[p_\omega\,{}^{x_\omega}_{y_\omega}] &\geqslant \measuredangle[p_\omega\,{}^{x_\omega}_{q_\omega}] - \measuredangle[p_\omega\,{}^{y_\omega}_{q_\omega}] > \\ &> \pi - \varepsilon - \measuredangle[p_\omega\,{}^{y_\omega}_{q_\omega}].\end{aligned} \tag{3}$$

Note that we can assume in addition that $q_\omega \in \operatorname{Str}(p_\omega)$. Choose $q_n \in \mathcal{L}_n$ such that $q_n \to q_\omega$ as $n \to \omega$. Note that by angle semicontinuity we again have

$$\begin{aligned}\measuredangle[p_\omega\,{}^{x_\omega}_{q_\omega}] &\leqslant \lim_{n\to\omega} \measuredangle[p_n\,{}^{x_n}_{q_n}], \\ \measuredangle[p_\omega\,{}^{y_\omega}_{q_\omega}] &\leqslant \lim_{n\to\omega} \measuredangle[p_n\,{}^{y_n}_{q_n}].\end{aligned} \tag{4}$$

By the CBB(κ) comparison Definition 8.2 and 8.17(b),

$$\measuredangle[p_n\,{}^{x_n}_{y_n}] + \measuredangle[p_n\,{}^{y_n}_{q_n}] + \measuredangle[p_n\,{}^{x_n}_{q_n}] \leqslant 2{\cdot}\pi$$

for all n. Together with (4), (2), and (3), this implies

$$\begin{aligned}\lim_{n\to\omega} \measuredangle[p_n\,{}^{x_n}_{y_n}] &\leqslant 2{\cdot}\pi - \lim_{n\to\omega} \measuredangle[p_n\,{}^{x_n}_{q_n}] - \lim_{n\to\omega} \measuredangle[p_n\,{}^{y_n}_{q_n}] \leqslant \\ &\leqslant 2{\cdot}\pi - \measuredangle[p_\omega\,{}^{x_\omega}_{q_\omega}] - \measuredangle[p_\omega\,{}^{y_\omega}_{q_\omega}] < \\ &< \measuredangle[p_\omega\,{}^{x_\omega}_{y_\omega}] + 2{\cdot}\varepsilon.\end{aligned}$$

Since $\varepsilon > 0$ is arbitrary, (1) follows. ☐

8.42. First variation formula. *Let $\mathcal{L}$ be a complete length* CBB *space. For any point q and any geodesic $[px]$ in $\mathcal{L}$ with $p \neq q$, we have*

$$|q - \mathrm{geod}_{[px]}(t)| = |q - p| - t \cdot \cos\phi + o(t), \tag{5}$$

where ϕ is the infimum of angles between $[px]$ and all geodesics from p to q in the ultrapower $\mathcal{L}^\omega$.

Remark. If $\mathcal{L}$ is a proper space, then $\mathcal{L}^\omega = \mathcal{L}$; see Section 4.B. Therefore the infimum ϕ is achieved on a particular geodesic from p to q.

As a corollary we obtain the following classical result:

8.43. Strong angle lemma. *Let $\mathcal{L}$ be a complete length* CBB *space, and let $p \neq q \in \mathcal{L}$ be such that there is unique geodesic from p to q in the ultrapower $\mathcal{L}^\omega$. Then for any hinge $[p\,{}^q_x]$ we have*

$$\measuredangle[p\,{}^q_x] = \lim_{\substack{\bar{x} \to p \\ \bar{x} \in\,]px]}} \tilde{\measuredangle}^\kappa(p\,{}^q_{\bar{x}}) \tag{6}$$

for any $\kappa \in \mathbb{R}$ such that $|p - q| < \varpi^\kappa$.

In particular, (6) *holds if $p \in \mathrm{Str}(q)$ as well as if $q \in \mathrm{Str}(p)$.*

Remark.

- The above lemma is essentially due to Alexandrov. The right-hand side in (6) is called the *strong angle* of the hinge $[p\,{}^q_x]$. Note that in a general metric space the angle and the strong angle of the same hinge might differ.
- As follows from Corollary 4.12, if there is a unique geodesic $[pq]$ in the ultrapower $\mathcal{L}^\omega$, then $[pq]$ lies in $\mathcal{L}$.

Proof of 8.43. The first statement follows directly from the first variation formula (see 8.42) and the definition of model angle (see Section 6.C).

The second statement follows from Plaut's Theorem 8.11 applied to $\mathcal{L}^\omega$. (Note that according to Proposition 8.4, $\mathcal{L}^\omega$ is a complete length CBB space.) □

Proof of 8.42. By Corollary 8.33, we can assume that $\kappa \leqslant 0$. The inequality

$$|q - \mathrm{geod}_{[px]}(t)| \leqslant |q - p| - t \cdot \cos\phi + o(t)$$

follows from the first variation inequality (see 6.7). Thus, it is sufficient to show that

$$|q - \mathrm{geod}_{[px]}(t)| \geqslant |q - p| - t \cdot \cos\phi + o(t).$$

Assume the contrary. Then there is $\varepsilon > 0$ such that $\phi+\varepsilon < \pi$, and for a sequence $t_n \to 0+$ we have

$$|q - \mathrm{geod}_{[px]}(t_n)| < |q - p| - t_n \cdot \cos(\phi - \varepsilon). \tag{7}$$

Let $x_n = \mathrm{geod}_{[px]}(t_n)$. Clearly

$$\tilde\measuredangle^\kappa\left(x_n\,{}^p_q\right) > \pi - \phi + \tfrac{\varepsilon}{2}$$

for all large n.

Assume $\mathcal{L}$ is geodesic. Choose a sequence of geodesics $[x_n q]$. Let $[x_n q] \to$ $\to [pq]_{\mathcal{L}^\omega}$ as $n \to \omega$ (in general $[pq]$ might lie in $\mathcal{L}^\omega$). Applying both parts of the hinge comparison theorem (see 8.14(c)), we have $\measuredangle\left[x_n\,{}^q_x\right] < \phi - \frac{\varepsilon}{2}$ for all large n. According to 8.40, the angle between $[pq]$ and $[px]$ is at most $\phi - \frac{\varepsilon}{2}$, a contradiction.

Finally, if $\mathcal{L}$ is not geodesic, choose a sequence $q_n \in \mathrm{Str}(x_n)$, such that $q_n \to q$ and the inequality

$$\tilde\measuredangle^\kappa\left(x_n\,{}^{p}_{q_n}\right) > \pi - \phi + \tfrac{\varepsilon}{2}$$

still holds. Then the same argument as above shows that $[x_n q_n]$ ω-converges to a geodesic $[pq]_{\mathcal{L}^\omega}$ from p to q having angle at most $\phi - \frac{\varepsilon}{2}$ with $[px]$. □

H. On positive lower bound

In this section we consider CBB(κ) spaces for $\kappa > 0$. Applying rescaling we can assume that $\kappa = 1$.

The following theorem states that if one ignores a few exceptional spaces, then the diameter of a space with positive lower curvature bound is bounded. Many authors (but not us) exclude these spaces in the definition of Alexandrov space with positive lower curvature bound.

8.44. On diameter of a space. *Let $\mathcal{L}$ be a complete length* CBB(1) *space. Then either*

(a) $\mathrm{diam}\,\mathcal{L} \leqslant \pi$; *or*

(b) *$\mathcal{L}$ is isometric to one of the following exceptional spaces:*
  (a) *real line* $\mathbb{R}$,
  (b) *a half-line* $\mathbb{R}_{\geqslant 0}$,
  (c) *a closed interval* $[0, a] \in \mathbb{R}$, $a > \pi$,
  (d) *a circle* $\mathbb{S}^1_a$ *of length* $a > 2\cdot\pi$.

Proof. Assume that $\mathcal{L}$ is a geodesic space and $\mathrm{diam}\,\mathcal{L} > \pi$. Choose $x, y \in \mathcal{L}$ so that $|x - y| = \pi + \varepsilon$, $0 < \varepsilon < \frac{\pi}{4}$. By moving y slightly, we can also assume that the geodesic $[xy]$ is unique; to prove this, use either Plaut's Theorem 8.11

or the fact that geodesics do not split (see 8.37). Let z be the midpoint of the geodesic $[xy]$.

Consider the function $f = \mathrm{dist}_x + \mathrm{dist}_y$. As follows from Lemma 8.45, f is concave in $\mathrm{B}(z, \frac{\varepsilon}{4})$. Let $p \in \mathrm{B}(z, \frac{\varepsilon}{4})$. Choose a geodesic $[zp]$, and let $h(t) = f \circ \mathrm{geod}_{[zp]}(t)$ and $\ell = |z-p|$. Clearly h is concave. From the adjacent angle lemma (Lemma 8.39), we have $h^+(0) = 0$. Therefore h is nonincreasing which means that

$$|x-p| + |y-p| = h(\ell) \leqslant h(0) = |x-y|.$$

Since the geodesic $[xy]$ is unique, this means that $p \in [xy]$, and hence $\mathrm{B}(z, \frac{\varepsilon}{4})$ only contains points of $[xy]$.

Since in CBB spaces, geodesics do not bifurcate (see 8.37(a)), it follows that all of $\mathcal{L}$ coincides with the maximal extension of $[xy]$ as a local geodesic γ (which might not be minimizing). In other words, $\mathcal{L}$ is isometric to a one-dimensional Riemannian manifold with possibly nonempty boundary. From this, it is easy to see that $\mathcal{L}$ falls into one of the exceptional spaces described in the theorem.

Lastly, if $\mathcal{L}$ is not geodesic and $\operatorname{diam} \mathcal{L} > \pi$, then the above argument applied to $\mathcal{L}^\omega$ yields that each metric component of $\mathcal{L}^\omega$ is isometric to one of the exceptional spaces. As all of those spaces are proper, $\mathcal{L}$ is a metric component in $\mathcal{L}^\omega$. □

8.45. Lemma. *Let $\mathcal{L}$ be a complete length* CBB(1) *space, and let $p \in \mathcal{L}$. Then $\mathrm{dist}_p : \mathcal{L} \to \mathbb{R}$ is concave in $\mathrm{B}(p, \pi) \setminus \mathrm{B}(p, \frac{\pi}{2})$.*

In particular, if $\operatorname{diam} \mathcal{L} \leqslant \pi$, then the complements $\mathcal{L} \setminus \mathrm{B}(p, r)$ and $\mathcal{L} \setminus \overline{\mathrm{B}}[p, r]$ are convex for any $r > \frac{\pi}{2}$.

Proof. This is a consequence of Theorem 8.23(b). □

8.46. Exercise. Let $\mathcal{L}$ be a complete length CBB(1) space that is not exceptional (that is, $\operatorname{diam} \mathcal{L} \leqslant \pi$). Assume that a group G acts on $\mathcal{L}$ by isometries, has closed orbits, and

$$\operatorname{diam}(\mathcal{L}/G) > \tfrac{\pi}{2}.$$

Show that the action of G has a fixed point in $\mathcal{L}$.

8.47. Advanced exercise. Let $\mathcal{L}$ be a complete length CBB(1) space Show that $\mathcal{L}$ contains at most three points with space of directions $\leqslant \frac{1}{2} \cdot \mathbb{S}^n$ (see Definition 5.9).

8.48. On perimeter of a triple. *Suppose $\mathcal{L}$ is a complete length* CBB(1) *space and* $\operatorname{diam} \mathcal{L} \leqslant \pi$. *Then the perimeter of any triple of points $p, q, r \in \mathcal{L}$ is at most $2 \cdot \pi$.*

Proof. Arguing by contradiction, suppose

(1) $$|p-q|+|q-r|+|r-p|>2\cdot\pi$$

for $p, q, r \in \mathcal{L}$. Rescaling the space slightly, we can assume that $\operatorname{diam}\mathcal{L} < \pi$, but inequality (1) still holds. By Corollary 8.33, after rescaling $\mathcal{L}$ is still CBB(1).

Since $\mathcal{L}$ is G-delta geodesic (see 8.11), it is sufficient to consider the case when there is a geodesic $[qr]$.

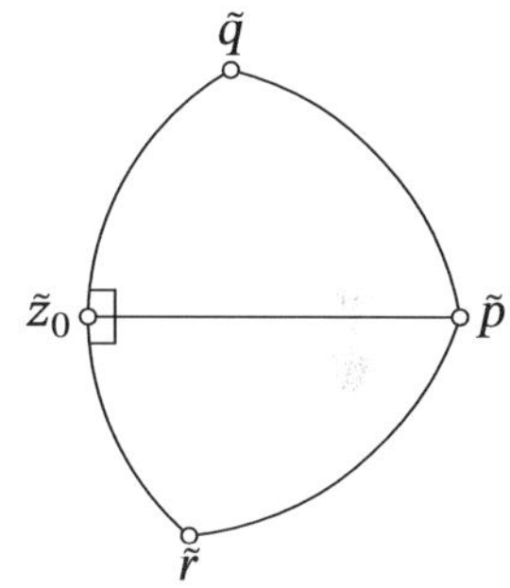

First note that since $\operatorname{diam}\mathcal{L} < \pi$, by Theorem 8.23(b) the function

$$y(t)=\operatorname{md}^1|p-\operatorname{geod}_{[qr]}(t)|$$

satisfies the differential inequality $y'' \leqslant 1-y$.

Take $z_0 \in [qr]$ so that the restriction $\operatorname{dist}_p|_{[qr]}$ attains its maximum at z_0, and set $t_0 = |q-z_0|$ so $z_0 = \operatorname{geod}_{[qr]}(t_0)$. Consider the following model configuration: two geodesics $[\tilde p\tilde z_0]$, $[\tilde q\tilde r]$ in $\mathbb{S}^2$ such that

$$|\tilde p-\tilde z_0|=|p-z_0|,\quad |\tilde q-\tilde r|=|q-r|,$$
$$|\tilde z_0-\tilde q|=|z_0-q|,\quad |\tilde z_0-\tilde r|=|z_0-q|,$$

and

$$\measuredangle\left[\tilde z_0\,{}^{\tilde q}_{\tilde p}\right]=\measuredangle\left[\tilde z_0\,{}^{\tilde r}_{\tilde p}\right]=\tfrac{\pi}{2}.$$

Clearly, $\bar y(t)=\operatorname{md}^1|\tilde p-\operatorname{geod}_{[\tilde q\tilde r]}(t)|$ satisfies $\bar y''=1-\bar y$ and $\bar y'(t_0)=0$, $\bar y(t_0)= = y(t_0)$. Since z_0 is a maximum point, $y(t)\leqslant y(t_0)+o(t-t_0)$; thus, $\bar y(t)$ is a barrier for $y(t)=\operatorname{md}^1|p-\operatorname{geod}_{[qr]}(t)|$ at t_0 by Theorem 3.14(d). From the barrier inequality in 3.14(d), we get

$$|\tilde p-\operatorname{geod}_{[\tilde q\tilde r]}(t)|\geqslant|p-\operatorname{geod}_{[qr]}(t)|,$$

and hence $|\tilde p-\tilde q|\geqslant|p-q|$ and $|\tilde p-\tilde r|\geqslant|p-r|$.

Therefore $|p-q|+|q-r|+|r-p|$ cannot exceed the perimeter of the spherical triangle $[\tilde p\tilde q\tilde r]$. In particular,

$$|p-q|+|q-r|+|r-p|\leqslant 2\cdot\pi,$$

a contradiction. □

Let $\kappa>0$. Consider the following extension $\tilde\measuredangle^{\kappa+}(*\,{}^*_*)$ of the model angle function $\tilde\measuredangle^{\kappa}(*\,{}^*_*)$. This definition works well for CBB spaces; for CAT spaces there is a similar definition. Some authors define the comparison angle to be $\tilde\measuredangle^{\kappa+}(*\,{}^*_*)$.

8.49. Definition of extended angle. *Suppose p, q, r are points in a metric space, and $p\neq q$, $p\neq r$. Let*

$$\tilde\measuredangle^{\kappa+}(p\,{}^q_r)=\sup\{\,\tilde\measuredangle^{\mathrm{K}}(p\,{}^q_r):\mathrm{K}\leqslant\kappa\,\}.$$

The value $\tilde\measuredangle^{\kappa+}(p\,{}^q_r)$ is called the extended model angle *of the triple p, q, r.*

8.50. Extended angle comparison. *Let $\kappa > 0$ and $\mathcal{L}$ be a complete length* CBB(κ) *space. Then for any hinge $[p\,{}^q_r]$ we have $\measuredangle[p\,{}^q_r] \geqslant \tilde{\measuredangle}^{\kappa+}(p\,{}^q_r)$.*

Moreover, the extended model angle $\tilde{\measuredangle}^{\kappa+}(p\,{}^q_r)$ can be calculated using the following rule:

(a) $\tilde{\measuredangle}^{\kappa+}(p\,{}^q_r) = \tilde{\measuredangle}^{\kappa}(p\,{}^q_r)$ *if* $\tilde{\measuredangle}^{\kappa}(p\,{}^q_r)$ *is defined;*

(b) $\tilde{\measuredangle}^{\kappa+}(p\,{}^q_r) = \tilde{\measuredangle}^{\kappa+}(p\,{}^r_q) = 0$ *if* $|p-q| + |q-r| = |p-r|$;

(c) $\tilde{\measuredangle}^{\kappa+}(p\,{}^q_r) = \pi$ *if none of the above is applicable.*

Proof. From Corollary 8.33, $\mathrm{K} < \kappa$ implies that any complete length CBB(K) space is CBB(κ); thus the extended angle comparison follows from the definition.

The rule for calculating extended angle is an easy consequence of its definition. □

I. Remarks

The question of whether the first part of Theorem 8.14(c) suffices to conclude that $\mathcal{L}$ is CBB(κ) is a long-standing open problem (possibly dating back to Alexandrov); it was first stated in [**37**, footnote in 4.1.5].

8.51. Open question. Let $\mathcal{L}$ be a complete geodesic space (you can also assume that $\mathcal{L}$ is homeomorphic to $\mathbb{S}^2$ or $\mathbb{R}^2$) such that for any hinge $[x\,{}^p_y]$ in $\mathcal{L}$, the angle $\measuredangle[x\,{}^p_y]$ is defined and

$$\measuredangle[x\,{}^p_y] \geqslant \tilde{\measuredangle}^0(x\,{}^p_y).$$

Is it true that $\mathcal{L}$ is CBB(0)?

Examples and constructions. Let us list important sources of examples of CBB spaces. We do not provide all the proofs, and some proofs are deferred to later chapters.

Complete Riemannian manifolds with sectional curvature at least κ, their *Gromov–Hausdorff limits*, as well as their *ultralimits*, are CBB(κ). This statement follows from Proposition 8.4, Theorem 5.16 and the Toponogov comparison which is a partial case of Globalization Theorem 8.31. For example, if M is a Riemannian manifold of nonnegative sectional curvature, then the limit of its rescalings $\frac{1}{n}\cdot M$ as $n \to \infty$ is CBB(0); this is the so-called *asymptotic cone* of M.

Most of applications to Riemannian geometry are based on the described sources and the following corollary of Gromov's Selection Theorem 5.6 and the Bishop–Gromov inequality.

8.52. Gromov compactness theorem. *Let M_n be a sequence of Riemannian manifolds with sectional curvature at least κ. Then, for any choice of marked points $p_n \in M_n$, a subsequence of M_n admits a Gromov–Hausdorff convergence such that the corresponding subsequence of p_n converges.*

By Corollary 8.34, the *target of submetry* from CBB(κ) is CBB(κ). In particular, if M is a Riemannian manifold with sectional curvature at least κ and G is a closed subgroup of isometries on M, then the *quotient space* M/G is CBB(κ).

Yet another source is given by *convex hypersurfaces*. Namely, suppose M is a complete Riemannian manifold of sectional curvature at least κ. Let N be a closed (as a subset) hypersurface in M. Then N with the induced inner metric is CBB(κ). In particular, any closed convex hypersurface in $\mathbb{R}^n$ is CBB(0).

The smooth case of this statement follows from the Gauss formula. This has been generalized by Sergei Buyalo [**46**] to the nonsmooth case and sharpened by the authors [**9**].

Let us mention that an analogous statement about convex hypersurfaces in CBB(κ) space is completely open. Namely, it is unknown whether *the boundary of a complete finite-dimensional* CBB(κ) *length space has to be* CBB(κ).

Further, CBB spaces behave nicely with respect to several natural constructions. For example, the product of CBB(0) spaces is a CBB(0). Also, the Euclidean cone over CBB(1) space is a CBB(0). These are the first examples of the so-called *warped products* that are discussed in Chapter 11; a general statement is given in Theorem 11.10. Perelman's doubling theorem can be considered as a partial case; it states that if $\mathcal{L}$ is a finite-dimensional CBB(κ) length space with nonempty boundary, then its *doubling across the boundary* is CBB(κ) as well.

More conceptually, *Wasserstein space* of order 2 over CBB(0) space is CBB(0). Also, there is a natural metric on the *space of metric-measure spaces* that makes it CBB(0) space; it was constructed by Karl-Theodor Sturm [**154**].

Among less important examples, let us mention *polyhedral spaces*: An if-and-only-if condition is given in Theorem 12.5. Also, in addition to Perelman's doubling theorem, there are several versions of gluing theorems [**67, 98, 133**]; they give conditions that guarantee that gluing of two (or more) spaces is CBB(κ).

Chapter 9

Fundamentals of curvature bounded above

A. Four-point comparison

9.1. Four-point comparison. *A quadruple of points p^1, p^2, x^1, x^2 in a metric space satisfies* CAT(κ) comparison *if*

(a) $\tilde\measuredangle^\kappa\left(p^1\,{}^{x^1}_{x^2}\right) \leqslant \tilde\measuredangle^\kappa\left(p^1\,{}^{p^2}_{x^1}\right) + \tilde\measuredangle^\kappa\left(p^1\,{}^{p^2}_{x^2}\right)$, *or*

(b) $\tilde\measuredangle^\kappa\left(p^2\,{}^{x^1}_{x^2}\right) \leqslant \tilde\measuredangle^\kappa\left(p^2\,{}^{p^1}_{x^1}\right) + \tilde\measuredangle^\kappa\left(p^2\,{}^{p^1}_{x^2}\right)$, *or*

(c) *one of the six model angles*

$$\tilde\measuredangle^\kappa\left(p^1\,{}^{x^1}_{x^2}\right),\quad \tilde\measuredangle^\kappa\left(p^1\,{}^{p^2}_{x^1}\right),\quad \tilde\measuredangle^\kappa\left(p^1\,{}^{p^2}_{x^2}\right),$$
$$\tilde\measuredangle^\kappa\left(p^2\,{}^{x^1}_{x^2}\right),\quad \tilde\measuredangle^\kappa\left(p^2\,{}^{p^1}_{x^1}\right),\quad \tilde\measuredangle^\kappa\left(p^2\,{}^{p^1}_{x^2}\right)$$

is undefined.

Here is a more intuitive formulation.

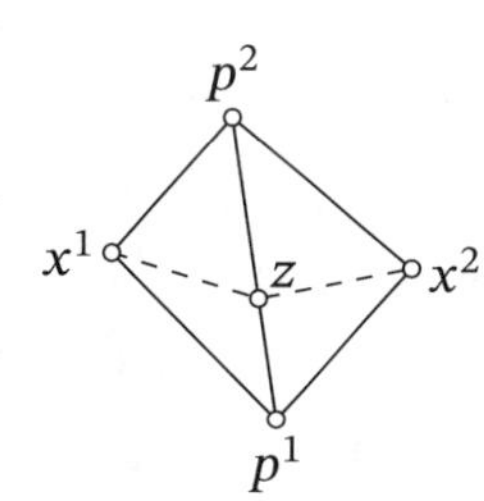

9.2. Reformulation. *Let $\mathcal{X}$ be a metric space. A quadruple $p^1, p^2, x^1, x^2 \in \mathcal{X}$ satisfies* CAT(κ) *comparison if one of the following holds:*

(a) *One of the triples (p^1, p^2, x^1) or (p^1, p^2, x^2) has perimeter $> 2\cdot\varpi^\kappa$.*

(b) *If* $[\tilde{p}^1\tilde{p}^2\tilde{x}^1] = \tilde{\Delta}^{\kappa}(p^1p^2x^1)$ *and* $[\tilde{p}\tilde{p}^2\tilde{x}^2] = \tilde{\Delta}^{\kappa}p^1p^2x^2$, *then*

$$|\tilde{x}^1 - \tilde{z}| + |\tilde{z} - \tilde{x}^2| \geqslant |x^1 - x^2|,$$

for any $\tilde{z} \in [\tilde{p}^1\tilde{p}^2]$.

9.3. Definition. Let $\mathcal{U}$ be a metric space.

(a) $\mathcal{U}$ is CAT(κ) if any quadruple in $\mathcal{X}$ satisfies CAT(κ) comparison.

(b) $\mathcal{U}$ is *locally* CAT(κ) if any point $q \in \mathcal{U}$ admits a neighborhood $\Omega \ni q$ such that any quadruple in Ω satisfies CAT(κ) comparison.

(c) $\mathcal{U}$ is a CAT space if $\mathcal{U}$ is CAT(κ) for some $\kappa \in \mathbb{R}$.

The condition $\mathcal{U}$ is CAT(κ) should be understood as "$\mathcal{U}$ has global curvature $\leqslant \kappa$". In Proposition 9.18, it will be shown that this formulation makes sense; in particular, if $\kappa \leqslant$ K, then any CAT(κ) space is CAT(K).

This terminology was introduced by Michael Gromov: CAT stands for Élie Cartan, Alexandr Alexandrov, and Victor Toponogov. Originally these spaces were called $\mathfrak{R}_\kappa$ *domains*; this is Alexandrov's terminology and is still in use.

9.4. Exercise. Let $\mathcal{U}$ be a metric space. Show that $\mathcal{U}$ is CAT(κ) if and only if every quadruple of points in $\mathcal{U}$ admits a labeling by (p, x^1, x^2, x^3) such that the three angles $\tilde{\measuredangle}^{\kappa}\left(p\,{}^{x^1}_{x^2}\right)$, $\tilde{\measuredangle}^{\kappa}\left(p\,{}^{x^2}_{x^3}\right)$, and $\tilde{\measuredangle}^{\kappa}\left(p\,{}^{x^1}_{x^3}\right)$ satisfy all three triangle inequalities or one of these angles is undefined.

9.5. Exercise. Show that $\mathcal{U}$ is CAT(κ) if and only if for any quadruple of points p^1, p^2, x^1, x^2 in $\mathcal{U}$ such that $|p^1 - p^2|, |x^1 - x^2| \leqslant \varpi^{\kappa}$, there is a quadruple q^1, q^2, y^1, y^2 in $\mathbb{M}^m(\kappa)$ such that

$$|q^1 - q^2| = |p^1 - p^2|, \qquad |y^1 - y^2| = |x^1 - x^2|, \qquad |q^i - y^j| \leqslant |p^i - x^j|$$

for any i and j.

9.6. Advanced exercise. Let $\mathcal{U}$ be a complete length space such that for any quadruple $p, q, x, y \in \mathcal{L}$ the following inequality holds

$$|p - q|^2 + |x - y|^2 \leqslant |p - x|^2 + |p - y|^2 + |q - x|^2 + |q - y|^2. \tag{1}$$

Prove that $\mathcal{U}$ is CAT(0).

Construct a four-point metric space $\mathcal{X}$ that satisfies inequality (1) for any relabeling of its points by p, q, x, y, and such that $\mathcal{X}$ is not CAT(0).

The next proposition follows directly from Definition 9.3 and the definitions of ultralimit and ultrapower; see Section 4.B for the related definitions. Recall that ω denotes a fixed selective ultrafilter on $\mathbb{N}$.

9.7. Proposition. *Let* $\mathcal{U}_n$ *be a* CAT(κ_n) *space for each* $n \in \mathbb{N}$. *Assume* $\mathcal{U}_n \to \mathcal{U}_\omega$ *and* $\kappa_n \to \kappa_\omega$ *as* $n \to \omega$. *Then* $\mathcal{U}_\omega$ *is* CAT(κ_ω).

Moreover, a metric space $\mathcal{U}$ *is* CAT(κ) *if and only if so is its ultrapower* $\mathcal{U}^\omega$.

B. Geodesics

9.8. Uniqueness of geodesics. *In a complete length* CAT(κ) *space, pairs of points at distance* $< \varpi^\kappa$ *are joined by unique geodesics, and these geodesics depend continuously on their endpoint pairs.*

Proof. Fix a complete length CAT(κ) space $\mathcal{U}$. Fix two points $p^1, p^2 \in \mathcal{U}$ such that

$$|p^1 - p^2|_{\mathcal{U}} < \varpi^\kappa.$$

Choose a sequence of approximate midpoints z_n between p^1 and p^2; that is,

$$(1) \qquad |p^1 - z_n|, |p^2 - z_n| \to \tfrac{1}{2}\cdot|p^1 - p^2| \quad \text{as} \quad n \to \infty.$$

By the law of cosines, $\tilde{\measuredangle}^\kappa\left(p^1\,{}^{z_n}_{p^2}\right)$ and $\tilde{\measuredangle}^\kappa\left(p^2\,{}^{z_n}_{p^1}\right)$ are arbitrarily small when n is sufficiently large.

Let us apply CAT(κ) comparison (see 9.1) to the quadruple p^1, p^2, z_n, z_k with large n and k. We conclude that $\tilde{\measuredangle}^\kappa\left(p\,{}^{z_n}_{z_k}\right)$ is arbitrarily small when n, k are sufficiently large and p is either p^1 or p^2. By (1) and the law of cosines, the sequence z_n converges.

Since $\mathcal{U}$ is complete, the sequence z_n converges to a midpoint between p^1 and p^2. By Lemma 2.13 we obtain the existence of a geodesic $[p^1 p^2]$.

Now suppose $p^1_n \to p^1$, $p^2_n \to p^2$ as $n \to \infty$. Let z_n be the midpoint of a geodesic $[p^1_n p^2_n]$, and let z be the midpoint of a geodesic $[p^1 p^2]$.

It suffices to show that

$$(2) \qquad |z_n - z| \to 0 \quad \text{as} \quad n \to \infty.$$

By the triangle inequality, the z_n are approximate midpoints between p^1 and p^2. Apply the CAT(κ) comparison (see 9.1) to the quadruple p^1, p^2, z_n, z. For $p = p^1$ or $p = p^2$, we see that $\tilde{\measuredangle}^\kappa\left(p\,{}^{z_n}_{z}\right)$ is arbitrarily small when n is sufficiently large. By the law of cosines, (2) follows. □

9.9. Exercise. Let $\mathcal{U}$ be a complete length CAT space. Assume $\mathcal{U}$ is a topological manifold. Show that any geodesic in $\mathcal{U}$ can be extended as a two-side infinite local geodesic.

Moreover the same holds for any locally geodesic locally CAT space $\mathcal{U}$ with nontrivial local homology groups at any point; the latter holds in particular if $\mathcal{U}$ is a homological manifold.

9.10. Exercise. Assume $\mathcal{U}$ is a locally compact geodesic CAT space with extendable geodesics; that is, any geodesic in $\mathcal{U}$ can be extended to a both-sided infinite local geodesic.

Show that the space of geodesic directions Σ'_p is complete for any $p \in \mathcal{U}$.

By the uniqueness of geodesics (see 9.8), we have the following.

9.11. Corollary. *Any complete length* CAT(κ) *space is* ϖ^κ*-geodesic.*

9.12. Proposition. *The completion* $\bar{\mathcal{U}}$ *of any geodesic* CAT(κ) *space* $\mathcal{U}$ *is a complete length* CAT(κ) *space.*

Moreover, $\mathcal{U}$ *is a geodesic* CAT(κ) *space if and only if there is a complete length* CAT(κ) *space* $\bar{\mathcal{U}}$ *that contains a* ϖ^κ*-convex dense set isometric to* $\mathcal{U}$*.*

Proof. By Theorem 9.8, in order to show that $\bar{\mathcal{U}}$ is CAT(κ), it is sufficient to verify that the completion of a length space is a length space; this is straightforward.

For the second part, note that the completion $\bar{\mathcal{U}}$ contains the original space $\mathcal{U}$ as a dense ϖ^κ-convex subset, and the metric on $\mathcal{U}$ coincides with the induced length metric from $\bar{\mathcal{U}}$. □

Here is a corollary from Proposition 9.12 and Theorem 9.8.

9.13. Corollary. *Let* $\mathcal{U}$ *be a* ϖ^κ*-geodesic* CAT(κ) *space. Then pairs of points in* $\mathcal{U}$ *at distance less than* ϖ^κ *are joined by unique geodesics, and these geodesics depend continuously on their endpoint pairs.*

Moreover for any pair of points $p, q \in \mathcal{U}$ *and any value*

$$\ell > \sup\left\{ \frac{\operatorname{sn}^\kappa r}{\operatorname{sn}^\kappa |p-q|} : 0 \leqslant r \leqslant |p-q| \right\}$$

there are neighborhoods $\Omega_p \ni p$ *and* $\Omega_q \ni q$ *such that the map*

$$(x, y, t) \mapsto \operatorname{path}_{[xy]}(t)$$

is ℓ*-Lipschitz in* $\Omega_p \times \Omega_q \times [0,1]$.

Proof. By Proposition 9.12, any geodesic CAT(κ) space is isometric to a convex dense subset of a complete length CAT(κ) space. It remains to apply Theorem 9.8. □

C. More comparisons

Here we give a few reformulations of Definition 9.3.

9.14. Theorem. *If* $\mathcal{U}$ *is a* CAT(κ) *space, then the following conditions hold for all triples* $p, x, y \in \mathcal{U}$ *of perimeter* $< 2\cdot\varpi^\kappa$*:*

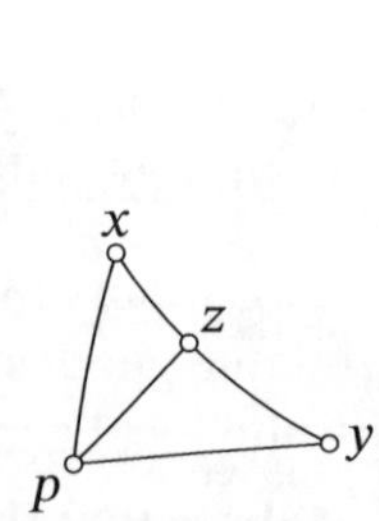

(a) Adjacent angle comparison. *For any geodesic* $[xy]$ *and* $z \in \,]xy[$*, we have*

$$\tilde\measuredangle^\kappa(z\,{}^p_x) + \tilde\measuredangle^\kappa(z\,{}^p_y) \geqslant \pi.$$

(b) Point-on-side comparison. *For any geodesic* $[xy]$ *and* $z \in]xy[$, *we have*

$$\tilde{\measuredangle}^\kappa(x\,{}^p_y) \geqslant \tilde{\measuredangle}^\kappa(x\,{}^p_z),$$

or equivalently,

$$|\tilde{p} - \tilde{z}| \geqslant |p - z|,$$

where $[\tilde{p}\tilde{x}\tilde{y}] = \tilde{\triangle}^\kappa(pxy)$, $\tilde{z} \in]\tilde{x}\tilde{y}[$, $|\tilde{x} - \tilde{z}| = |x - z|$.

(c) Hinge comparison. *For any hinge* $[x\,{}^p_y]$, *the angle* $\measuredangle[x\,{}^p_y]$ *exists and*

$$\measuredangle[x\,{}^p_y] \leqslant \tilde{\measuredangle}^\kappa(x\,{}^p_y),$$

or equivalently,

$$\tilde{\gamma}^\kappa[x\,{}^p_y] \leqslant |p - y|.$$

Moreover, if $\mathcal{U}$ *is* ϖ^κ*-geodesic, then the converse holds in each case.*

Remark. In the following proof, the part (c) $\Rightarrow$ (a) only requires that the CAT(κ) comparison (see 9.1) hold for any quadruple, and does not require the existence of geodesics at distance $< \varpi^\kappa$. The same is true of the parts (a)$\Leftrightarrow$(b) and (b) $\Rightarrow$ (c). Thus the conditions (a), (b), and (c) are valid for any metric space (not necessarily a length space) that satisfies CAT(κ) comparison (see 9.1). The converse does not hold; for example, all these conditions are vacuously true in a totally disconnected space, while CAT(κ) comparison is not.

Proof.

(a). Since the perimeter of p, x, y is $< 2\cdot\varpi^\kappa$, so is the perimeter of any subtriple of p, z, x, y by the triangle inequality. By Alexandrov's Lemma 6.3,

$$\tilde{\measuredangle}^\kappa(p\,{}^z_x) + \tilde{\measuredangle}^\kappa(p\,{}^z_y) < \tilde{\measuredangle}^\kappa(p\,{}^x_y) \quad \text{or} \quad \tilde{\measuredangle}^\kappa(z\,{}^p_x) + \tilde{\measuredangle}^\kappa(z\,{}^p_y) = \pi.$$

In the former case, the CAT(κ) comparison (see 9.1) applied to the quadruple p, z, x, y implies

$$\tilde{\measuredangle}^\kappa(z\,{}^p_x) + \tilde{\measuredangle}^\kappa(z\,{}^p_y) \geqslant \tilde{\measuredangle}^\kappa(z\,{}^x_y) = \pi.$$

(a)$\Leftrightarrow$(b). Follows from Alexandrov's Lemma 6.3.

(b) $\Rightarrow$ (c). By (b), for $\bar{p} \in]xp]$ and $\bar{y} \in]xy]$ the function $(|x - \bar{p}|, |x - \bar{y}|) \mapsto \tilde{\measuredangle}^\kappa(x\,{}^{\bar{p}}_{\bar{y}})$ is nondecreasing in each argument. In particular, $\measuredangle[x\,{}^p_y] = \inf \tilde{\measuredangle}^\kappa(x\,{}^{\bar{p}}_{\bar{y}})$. Thus $\measuredangle[x\,{}^p_y]$ exists and is at most $\tilde{\measuredangle}^\kappa(x\,{}^p_y)$.

Converse. Assume $\mathcal{U}$ is ϖ^κ-geodesic. Let us first show that in this case (c) $\Rightarrow$ (a).

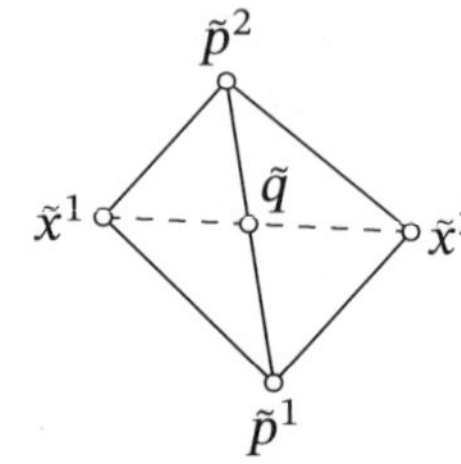

Indeed, by (c) and the triangle inequality for angles (see 6.5),

$$\tilde\measuredangle^\kappa\left(z\,{}^{p}_{x}\right) + \tilde\measuredangle^\kappa\left(z\,{}^{p}_{y}\right) \geqslant \measuredangle\left[z\,{}^{p}_{x}\right] + \measuredangle\left[z\,{}^{p}_{y}\right] \geqslant \pi.$$

It remains to prove the converse for (b).

Given a quadruple p^1, p^2, x^1, x^2 whose subtriples have perimeter $< 2\cdot\varpi^\kappa$, we must verify the CAT(κ) comparison (see 9.1). In $\mathbb{M}^2(\kappa)$, construct the model triangles $[\tilde p^1\tilde p^2\tilde x^1] = \tilde\triangle^\kappa(p^1p^2x^1)$ and $[\tilde p^1\tilde p^2\tilde x^2] = \tilde\triangle^\kappa(p^1p^2x^2)$, lying on either side of a common segment $[\tilde p^1\tilde p^2]$. We may suppose

$$\tilde\measuredangle^\kappa\left(p^1\,{}^{p^2}_{x^1}\right) + \tilde\measuredangle^\kappa\left(p^1\,{}^{p^2}_{x^2}\right) \leqslant \pi \quad\text{and}\quad \tilde\measuredangle^\kappa\left(p^2\,{}^{p^1}_{x^1}\right) + \tilde\measuredangle^\kappa\left(p^2\,{}^{p^1}_{x^2}\right) \leqslant \pi,$$

since otherwise CAT(κ) comparison holds trivially. Then $[\tilde p^1\tilde p^2]$ and $[\tilde x^1\tilde x^2]$ intersect, say at $\tilde q$.

By assumption, there is a geodesic $[p^1p^2]$. Choose $q \in [p^1p^2]$ corresponding to $\tilde q$; that is, $|p^1 - q| = |\tilde p^1 - \tilde q|$. Then

$$|x^1 - x^2| \leqslant |x^1 - q| + |q - x^2| \leqslant |\tilde x^1 - \tilde q| + |\tilde q - \tilde x^2| = |\tilde x^1 - \tilde x^2|,$$

where the second inequality follows from (b). By monotonicity of the function $a \mapsto \tilde\measuredangle^\kappa\{a; b, c\}$ (see 1.1(c)),

$$\tilde\measuredangle^\kappa\left(p^1\,{}^{x^1}_{x^2}\right) \leqslant \measuredangle\left[\tilde p^1\,{}^{\tilde x^1}_{\tilde x^2}\right] = \tilde\measuredangle^\kappa\left(p^1\,{}^{p^2}_{x^1}\right) + \tilde\measuredangle^\kappa\left(p^1\,{}^{p^2}_{x^2}\right). \qquad \square$$

Let us display a corollary of the proof of Theorem 9.14, namely, monotonicity of the model angle with respect to adjacent sidelengths.

9.15. Angle-sidelength monotonicity. *Suppose $\mathcal{U}$ is a ϖ^κ-geodesic* CAT(κ) *space, and p, x, $y \in \mathcal{U}$ have perimeter $< 2\cdot\varpi^\kappa$. Then for $\bar y \in\ ]xy]$, the function*

$$|x - \bar y| \mapsto \tilde\measuredangle^\kappa\left(x\,{}^{p}_{\bar y}\right)$$

is nondecreasing.

In particular, if $\bar p \in\ ]xp]$, then

(a) *the function*

$$(|x - \bar y|, |x - \bar p|) \mapsto \tilde\measuredangle^\kappa\left(x\,{}^{\bar p}_{\bar y}\right)$$

is nondecreasing in each argument;

(b) $\measuredangle\left[x\,{}^{p}_{y}\right] = \inf\left\{\,\tilde\measuredangle^\kappa\left(x\,{}^{\bar p}_{\bar y}\right) : \bar p \in\ ]xp],\ \bar y \in\ ]xy]\,\right\}.$

9.16. Exercise. Let $\mathcal{U}$ be $\mathbb{R}^m$ with the metric defined by a norm. Show that $\mathcal{U}$ is a complete length CAT space if and only if $\mathcal{U} \overset{\text{iso}}{=\!=} \mathbb{E}^m$.

9.17. Exercise. Assume $\mathcal{U}$ is a geodesic CAT(0) space. Show that for any two geodesic paths $\gamma, \sigma\colon [0,1] \to \mathcal{U}$ the function

$$t \mapsto |\gamma(t) - \sigma(t)|$$

is convex.

9.18. Proposition. *Assume $\kappa < \mathrm{K}$. Then any complete length* CAT(κ) *space is* CAT(K).

Moreover a space $\mathcal{U}$ is CAT(κ) *if $\mathcal{U}$ is* CAT(K) *for all* $\mathrm{K} > \kappa$.

Proof. The first statement follows from Corollary 9.11, the adjacent angles comparison (see 9.14(a)) and the monotonicity of the function $\kappa \mapsto \tilde{\measuredangle}^\kappa \left(x\,{}^y_z\right)$ (see 1.1(d)).

The second statement follows since the function $\kappa \mapsto \tilde{\measuredangle}^\kappa \left(x\,{}^y_z\right)$ is continuous. □

D. Thin triangles

In this section we define thin triangles and use them to characterize CAT spaces. Inheritance for thin triangles with respect to decomposition is the main result of this section. It will lead to two fundamental constructions: Alexandrov's patchwork globalization (see 9.30) and Reshetnyak gluing (see 9.39).

9.19. Definition of κ-thin triangles. *Let $[x^1x^2x^3]$ be a triangle of perimeter $< 2\cdot\varpi^\kappa$ in a metric space and $[\tilde{x}^1\tilde{x}^2\tilde{x}^3] = \tilde{\triangle}^\kappa(x^1x^2x^3)$. Consider the* natural map *$[\tilde{x}^1\tilde{x}^2\tilde{x}^3] \to [x^1x^2x^3]$ that sends a point $\tilde{z} \in [\tilde{x}^i\tilde{x}^j]$ to the corresponding point $z \in [x^ix^j]$ (that is, such that $|\tilde{x}^i - \tilde{z}| = |x^i - z|$ and therefore $|\tilde{x}^j - \tilde{z}| = |x^j - z|$).*

We say the triangle $[x^1x^2x^3]$ is κ-thin *if the natural map $[\tilde{x}^1\tilde{x}^2\tilde{x}^3] \to [x^1x^2x^3]$ is short.*

9.20. Exercise. Let $\mathcal{U}$ be a ϖ^κ-geodesic CAT(κ) space. Let $[xyz]$ be a triangle in $\mathcal{U}$, and let $[\tilde{x}\tilde{y}\tilde{z}]$ be its model triangle in $\mathbb{M}^2(\kappa)$. Prove that the natural map $f\colon [\tilde{x}\tilde{y}\tilde{z}] \to [xyz]$ is distance-preserving if and only if one of the following conditions hold:

(a) $\measuredangle\left[x\,{}^y_z\right] = \tilde{\measuredangle}^\kappa\left(x\,{}^y_z\right)$,

(b) $|x - w| = |\tilde{x} - \tilde{w}|$ for some $\tilde{w} \in]\tilde{y}\tilde{z}[$ and $w = f(\tilde{w})$,

(c) $|v - w| = |\tilde{v} - \tilde{w}|$ for some $\tilde{v} \in]\tilde{x}\tilde{y}[$, $\tilde{w} \in]\tilde{x}\tilde{z}[$ and $v = f(\tilde{v})$, $w = f(\tilde{w})$.

9.21. Proposition. *Let $\mathcal{U}$ be a ϖ^κ-geodesic space. Then $\mathcal{U}$ is* CAT(κ) *if and only if every triangle of perimeter $< 2\cdot\varpi^\kappa$ in $\mathcal{U}$ is κ-thin.*

Proof. The if part follows from the point-on-side comparison (see 9.14(b)). The only-if part follows from angle-sidelength monotonicity (see 9.15(a)). □

9.22. Corollary. *Suppose $\mathcal{U}$ is a ϖ^κ-geodesic* CAT(κ) *space. Then any local geodesic in $\mathcal{U}$ of length $< \varpi^\kappa$ is length-minimizing.*

Proof. Suppose $\gamma\colon [0,\ell] \to \mathcal{U}$ is a local geodesic that is not minimizing, with $\ell < \varpi^\kappa$. Choose a to be the maximal value such that γ is minimizing on $[0,a]$. Further choose $b > a$ so that γ is minimizing on $[a,b]$.

Since triangle $[\gamma(0)\gamma(a)\gamma(b)]$ is κ-thin, we have

$$|\gamma(a-\varepsilon) - \gamma(a+\varepsilon)| < 2\cdot\varepsilon$$

for all small $\varepsilon > 0$, a contradiction. □

Now let us formulate the main result of this section. The inheritance lemma states that in any metric space, a triangle is κ-thin if it decomposes into κ-thin triangles. In contrast, κ-thickness of triangles (Exercise 8.21) is not inherited in this way.

9.23. Inheritance lemma. *In a metric space, consider a triangle $[pxy]$ that* decomposes *into two triangles $[pxz]$ and $[pyz]$; that is, $[pxz]$ and $[pyz]$ have common side $[pz]$, and the sides $[xz]$ and $[zy]$ together form the side $[xy]$ of $[pxy]$.*

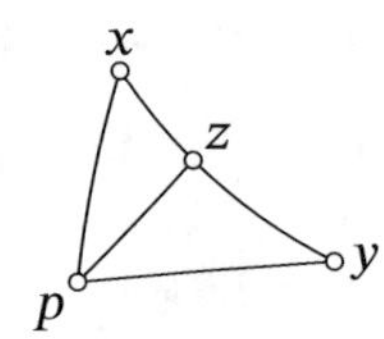

If the triangle $[pxy]$ has perimeter $< 2\cdot\varpi^\kappa$ and both triangles $[pxz]$ and $[pyz]$ are κ-thin, then triangle $[pxy]$ is κ-thin.

The following model-space lemma is extracted from Lemma 2 in [**142**].

9.24. Lemma. *Let $[\tilde p\tilde x\tilde y]$ be a triangle in $\mathbb{M}^2(\kappa)$, and let $\tilde z \in [\tilde x\tilde y]$. Consider the solid triangle $\tilde D =$ Conv$[\tilde p\tilde x\tilde y]$. Construct points $\dot p, \dot x, \dot z, \dot y \in \mathbb{M}^2(\kappa)$ such that*

$$\begin{aligned} &|\dot p - \dot x| = |\tilde p - \tilde x|, && |\dot p - \dot y| = |\tilde p - \tilde y|, && |\dot p - \dot z| \leqslant |\tilde p - \tilde z|,\\ &|\dot x - \dot z| = |\tilde x - \tilde z|, && |\dot y - \dot z| = |\tilde y - \tilde z|, \end{aligned}$$

where points $\dot x$ and $\dot y$ lie on either side of $[\dot p\dot z]$. Set

$$\dot D = \mathrm{Conv}[\dot p\dot x\dot z] \cup \mathrm{Conv}[\dot p\dot y\dot z].$$

Then there is a short map $F\colon \tilde D \to \dot D$ that maps $\tilde p$, $\tilde x$, $\tilde y$ and $\tilde z$ to $\dot p$, $\dot x$, $\dot y$, and $\dot z$, respectively.

Proof. By Alexandrov's Lemma 6.3, there are nonoverlapping triangles $[\tilde p\tilde x\tilde z_y] \overset{\mathrm{iso}}{=\!=} [\dot p\dot x\dot z]$ and $[\tilde p\tilde y\tilde z_x] \overset{\mathrm{iso}}{=\!=} [\dot p\dot y\dot z]$ inside triangle $[\tilde p\tilde x\tilde y]$.

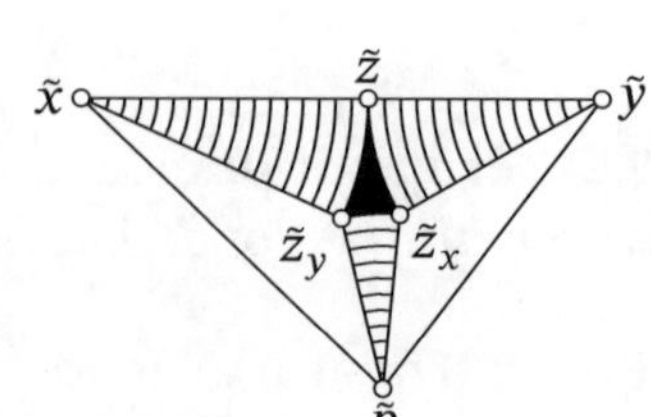

Connect points in each pair $(\tilde z, \tilde z_x)$, $(\tilde z_x, \tilde z_y)$ and $(\tilde z_y, \tilde z)$ with arcs of circles centered at $\tilde y$, $\tilde p$, and $\tilde x$ respectively. Define F as follows.

- Map $\operatorname{Conv}[\tilde{p}\tilde{x}\tilde{z}_y]$ isometrically onto $\operatorname{Conv}[\dot{p}\dot{x}\dot{y}]$; similarly map $\operatorname{Conv}[\tilde{p}\tilde{y}\tilde{z}_x]$ onto $\operatorname{Conv}[\dot{p}\dot{y}\dot{z}]$.
- If w is in one of the three circular sectors, say at distance r from the center of the circle, let $F(w)$ be the point on $[\dot{p}\dot{z}]$, $[\dot{x}\dot{z}]$, or $[\dot{y}\dot{z}]$ whose distance from the left-hand endpoint of the segment is r.
- Finally, if w lies in the remaining curvilinear triangle $\tilde{z}\tilde{z}_x\tilde{z}_y$, set $F(w) = \dot{z}$.

By construction, F meets the conditions of the lemma. □

Proof of Inheritance Lemma 9.23. Construct model triangles $[\dot{p}\dot{x}\dot{z}] = \tilde{\triangle}^\kappa(pxz)$ and $[\dot{p}\dot{y}\dot{z}] = \tilde{\triangle}^\kappa(pyz)$ so that $\dot{x}$ and $\dot{y}$ lie on opposite sides of $[\dot{p}\dot{z}]$.

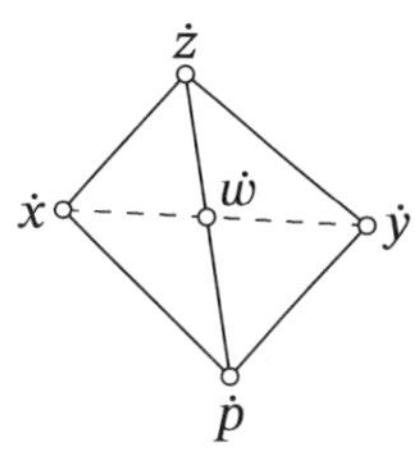

Suppose

$$\tilde{\measuredangle}^\kappa\left(z\,{}^p_x\right) + \tilde{\measuredangle}^\kappa\left(z\,{}^p_y\right) < \pi.$$

Then for some point $\dot{w} \in [\dot{p}\dot{z}]$, we have

$$|\dot{x} - \dot{w}| + |\dot{w} - \dot{y}| < |\dot{x} - \dot{z}| + |\dot{z} - \dot{y}| = |x - y|.$$

Let $w \in [pz]$ correspond to $\dot{w}$; that is, $|z - w| = |\dot{z} - \dot{w}|$. Since $[pxz]$ and $[pyz]$ are κ-thin, we have

$$|x - w| + |w - y| < |x - y|,$$

contradicting the triangle inequality.

Thus

$$\tilde{\measuredangle}^\kappa\left(z\,{}^p_x\right) + \tilde{\measuredangle}^\kappa\left(z\,{}^p_y\right) \geqslant \pi.$$

By Alexandrov's Lemma 6.3, this is equivalent to

$$(1) \qquad \tilde{\measuredangle}^\kappa\left(x\,{}^p_z\right) \leqslant \tilde{\measuredangle}^\kappa\left(x\,{}^p_y\right).$$

Let $[\tilde{p}\tilde{x}\tilde{y}] = \tilde{\triangle}^\kappa(pxy)$ and $\tilde{z} \in [\tilde{x}\tilde{y}]$ correspond to z; that is, $|x - z| = |\tilde{x} - \tilde{z}|$. Inequality (1) is equivalent to $|p - z| \leqslant |\tilde{p} - \tilde{z}|$. Hence Lemma 9.24 applies. Therefore there is a short map F that sends $[\tilde{p}\tilde{x}\tilde{y}]$ to $\dot{D} = \operatorname{Conv}[\dot{p}\dot{x}\dot{z}] \cup \operatorname{Conv}[\dot{p}\dot{y}\dot{z}]$ in such a way that $\tilde{p} \mapsto \dot{p}$, $\tilde{x} \mapsto \dot{x}$, $\tilde{z} \mapsto \dot{z}$ and $\tilde{y} \mapsto \dot{y}$.

By assumption, the natural maps $[\dot{p}\dot{x}\dot{z}] \to [pxz]$ and $[\dot{p}\dot{y}\dot{z}] \to [pyz]$ are short. By composition, the natural map from $[\tilde{p}\tilde{x}\tilde{y}]$ to $[pyz]$ is short, as claimed. □

E. Function comparison

In this section we give analytic and geometric ways of viewing the point-on-side comparison (see 9.14(b)) as a convexity condition.

First we obtain a corresponding differential inequality for the distance function in $\mathcal{U}$; see Section 3.F for the definition.

9.25. Theorem. *Suppose $\mathcal{U}$ is a ϖ^κ-geodesic space. Then the following are equivalent:*

(a) *$\mathcal{U}$ is* CAT(κ),

(b) *for any $p \in \mathcal{U}$, the function $f = \operatorname{md}^\kappa \circ \operatorname{dist}_p$ satisfies*

$$f'' + \kappa \cdot f \geqslant 1$$

in $\operatorname{B}(p, \varpi^\kappa)$.

9.26. Corollary. *A geodesic space $\mathcal{U}$ is* CAT(0) *if and only if for any $p \in \mathcal{U}$, the function* $\operatorname{dist}_p^2 : \mathcal{U} \to \mathbb{R}$ *is 2-convex.*

Proof of 9.25. Fix a sufficiently short geodesic $[xy]$ in $\operatorname{B}(p, \varpi^\kappa)$. We can assume that the model triangle $[\tilde{p}\tilde{x}\tilde{y}] = \tilde{\triangle}^\kappa(pxy)$ is defined. Let

$$\tilde{r}(t) = |\tilde{p} - \operatorname{geod}_{[\tilde{x}\tilde{y}]}(t)|, \qquad r(t) = |p - \operatorname{geod}_{[xy]}(t)|.$$

Let $\tilde{f} = \operatorname{md}^\kappa \circ \tilde{r}$ and $f = \operatorname{md}^\kappa \circ r$. By 1.1(a), we have $\tilde{f}'' = 1 - \kappa \cdot \tilde{f}$. Clearly $\tilde{f}(t)$ and $f(t)$ agree at $t = 0$ and $t = |x - y|$. The point-on-side comparison (see 9.14(b)) is the condition $r(t) \leqslant \tilde{r}(t)$ for all $t \in [0, |x - y|]$. Since md^κ is increasing on $[0, \varpi^\kappa)$, then $r \leqslant \tilde{r}$ and $f \leqslant \tilde{f}$ are equivalent. Thus the claim follows by Jensen's inequality (see 3.14(c)). □

9.27. Corollary. *Suppose $\mathcal{U}$ is a ϖ^κ-geodesic* CAT(κ) *space. Then any ball (closed or open) of radius $R < \frac{\varpi^\kappa}{2}$ in $\mathcal{U}$ is convex.*

Moreover, any open ball of radius $\frac{\varpi^\kappa}{2}$ is convex and any closed ball of radius $\frac{\varpi^\kappa}{2}$ is ϖ^κ-convex.

Proof. Suppose $p \in \mathcal{U}$, $R \leqslant \varpi^\kappa/2$, and two points x and y lie in $\overline{\operatorname{B}}[p, R]$ or $\operatorname{B}(p, R)$. By the triangle inequality, if $|x - y| < \varpi^\kappa$, then any geodesic $[xy]$ lies in $\operatorname{B}(p, \varpi^\kappa)$.

By the function comparison in Theorem 9.25, the geodesic $[xy]$ lies in $\overline{\operatorname{B}}[p, R]$ or $\operatorname{B}(p, R)$, respectively.

Thus any ball (closed or open) of radius $R < \frac{\varpi^\kappa}{2}$ is ϖ^κ-convex. This implies convexity unless there is a pair of points in the ball at distance at least ϖ^κ. By the triangle inequality, the latter is possible only for the closed ball of radius $\frac{\varpi^\kappa}{2}$. □

Recall that Busemann functions are defined in Proposition 6.1. The following exercise is analogous to Exercise 8.25.

9.28. Exercise. Let $\mathcal{U}$ be a complete length CAT(κ) space, and let $\operatorname{bus}_\gamma : \mathcal{U} \to$ $\to \mathbb{R}$ be the Busemann function for a half-line $\gamma : [0, \infty) \to \mathcal{L}$.

(a) If $\kappa = 0$, then the Busemann function $\operatorname{bus}_\gamma$ is convex.

(b) If $\kappa = -1$, then the function $f = \exp\circ\operatorname{bus}_\gamma$ satisfies

$$f'' - f \geqslant 0.$$

F. Development

Geometrically, the development construction (Lemma-definition 8.26) translates distance comparison into a local convexity statement for subsets of $\mathbb{M}^2(\kappa)$. Recall that a curve in $\mathbb{M}^2(\kappa)$ is (*locally*) *concave* with respect to p if (locally) its supergraph with respect to p is a convex subset of $\mathbb{M}^2(\kappa)$; see Definition 8.27.

9.29. Development criterion. *For a ϖ^κ-geodesic space $\mathcal{U}$, the following statements hold:*

(a) *For any $p \in \mathcal{U}$ and any geodesic $\gamma\colon [0,T] \to \mathrm{B}(p, \varpi^\kappa)$, suppose the κ-development $\tilde{\gamma}$ in $\mathbb{M}^2(\kappa)$ of γ with respect to p is locally concave. Then $\mathcal{U}$ is* CAT(κ).

(b) *If $\mathcal{U}$ is* CAT(κ), *then for any geodesic $\gamma\colon [0,T] \to \mathcal{U}$ and $p \in \mathcal{U}$ such that the triangle $[p\gamma(0)\gamma(T)]$ has perimeter $< 2\cdot\varpi^\kappa$, the κ-development $\tilde{\gamma}$ in $\mathbb{M}^2(\kappa)$ of γ with respect to p is concave.*

Proof.

(a). Let $\gamma = \operatorname{geod}_{[xy]}$ and $T = |x - y|$. Let $\tilde{\gamma}\colon [0,T] \to \mathbb{M}^2(\kappa)$ be the concave κ-development based at $\tilde{p}$ of γ with respect to p. Let us show that the function

(1) $$t \mapsto \tilde{\measuredangle}^\kappa\left(x\,{}^{p}_{\gamma(t)}\right)$$

is nondecreasing.

For a partition $0 = t^0 < t^1 < \dots < t^n = T$, let

$$\tilde{y}^i = \tilde{\gamma}(t^i) \quad\text{and}\quad \tau^i = |\tilde{y}^0 - \tilde{y}^1| + |\tilde{y}^1 - \tilde{y}^2| + \dots + |\tilde{y}^{i-1} - \tilde{y}^i|.$$

Since $\tilde{\gamma}$ is locally concave, for a sufficiently fine partition the polygonal line $\tilde{y}^0\tilde{y}^1\cdots\tilde{y}^n$ is locally concave with respect to $\tilde{p}$. Alexandrov's Lemma 6.3, applied inductively to pairs of triangles $\tilde{\triangle}^\kappa\{\tau^{i-1}, |p - \tilde{y}^0|, |p - \tilde{y}^{i-1}|\}$ and $\tilde{\triangle}^\kappa\{|\tilde{y}^{i-1} - \tilde{y}^i|, |p - \tilde{y}^{i-1}|, |p - \tilde{y}^i|\}$, shows that the sequence $\tilde{\measuredangle}^\kappa\{|\tilde{p} - \tilde{y}^i|;\ |\tilde{p} - \tilde{y}^0|, \tau^i\}$ is nondecreasing.

Taking finer partitions and passing to the limit, we get

$$\max_i\{|\tau^i - t^i|\} \to 0.$$

Therefore (1) and the point-on-side comparison in Theorem 9.14(b) follows.

(b). Consider a partition $0 = t^0 < t^1 < \dots < t^n = T$, and let $x^i = \gamma(t^i)$. Construct a chain of model triangles $[\tilde p\tilde x^{i-1}\tilde x^i] = \tilde\triangle^\kappa(px^{i-1}x^i)$ with the direction of $[\tilde p\tilde x^i]$ turning counterclockwise as i grows. By the angle comparison in Theorem 9.14(c),

$$\measuredangle\left[\tilde x^i\,{}^{\tilde x^{i-1}}_{\tilde p}\right] + \measuredangle\left[\tilde x^i\,{}^{\tilde x^{i+1}}_{\tilde p}\right] \geqslant \pi. \tag{2}$$

Since γ is a geodesic,

$$\operatorname{length}\gamma = \sum_{i=1}^{n}|x^{i-1} - x^i| \leqslant |p - x^0| + |p - x^n|. \tag{3}$$

By repeated application of Alexandrov's Lemma 6.3, and inequality (3),

$$\sum_{i=1}^{n}\measuredangle\left[\tilde p\,{}^{\tilde x^{i-1}}_{\tilde x^i}\right] \leqslant \tilde\measuredangle^\kappa\left(p\,{}^{x^0}_{x^n}\right) \leqslant \pi.$$

Then by (2), the polygonal line $\tilde p\tilde x^0\tilde x^1\dots\tilde x^n$ are concave with respect to $\tilde p$.

Note that under finer partitions, the polygonal line $\tilde x^0\tilde x^1\dots\tilde x^n$ approach the development of γ with respect to p. Since the polygonal lines are convex, their lengths converge to the length of γ. Hence the result. □

G. Patchwork globalization

If $\mathcal U$ is a CAT(κ) space, then it is locally CAT(κ). The converse does not hold even for complete length space. For example, $\mathbb S^1$ is locally isometric to $\mathbb R$, and so is locally CAT(0), but it is easy to find a quadruple of points in $\mathbb S^1$ that violates CAT(0) comparison.

The following theorem was essentially proved by Alexandr Alexandrov [**17**, Satz 9]; it gives a global condition on geodesics that is necessary and sufficient for a locally CAT(κ) space to be globally CAT(κ). The proof uses thin-triangle decompositions and the inheritance lemma (Lemma 9.23).

9.30. Patchwork globalization theorem. *For any complete length space $\mathcal U$, the following two statements are equivalent:*

(a) *$\mathcal U$ is* CAT(κ).

(b) *$\mathcal U$ is locally* CAT(κ)*; moreover, pairs of points in $\mathcal U$ at distance $< \varpi^\kappa$ are joined by unique geodesics, and these geodesics depend continuously on their endpoint pairs.*

Note that the implication (a) ⇒ (b) follows from Theorem 9.8.

9.31. Corollary. *Let Ω be an open locally* CAT(κ) *subset in a complete length space $\mathcal{U}$. Then for any point $p \in \Omega$ there is $R > 0$ such that $\overline{\mathrm{B}}[p, R]$ is a convex subset of $\mathcal{U}$ and $\overline{\mathrm{B}}[p, R]$ is* CAT(κ).

Proof. Fix $R > 0$ such that CAT(κ) comparison holds in $\mathrm{B}(p, R)$.

We may assume that $\mathrm{B}(p, R) \subset \Omega$ and $R < \varpi^{\mathrm{K}}$. The same argument as in the proof of the theorem on uniqueness of geodesics in 9.8 shows that any two points in $\overline{\mathrm{B}}[p, \frac{R}{2}]$ can be joined by a unique geodesic that depends continuously on the endpoints.

The same argument as in the proof of Corollary 9.27 shows that $\overline{\mathrm{B}}[p, \frac{R}{2}]$ is a convex set. Then (b) $\Rightarrow$ (a) of the patchwork globalization theorem implies that $\overline{\mathrm{B}}[p, \frac{R}{2}]$ is CAT(κ). □

The proof of patchwork globalization uses the following construction:

9.32. Definition (Line-of-sight map). *Let p be a point, and let α be a curve of finite length in a length space $\mathcal{U}$. Let $\bar{\alpha} : [0, 1] \to \mathcal{U}$ be the constant-speed parametrization of α. If $\gamma_t : [0, 1] \to \mathcal{U}$ is a geodesic path from p to $\bar{\alpha}(t)$, we say that the map $[0, 1] \times [0, 1] \to \mathcal{U}$ defined by*

$$(t, s) \mapsto \gamma_t(s)$$

is a line-of-sight map *for α with respect to p.*

Note that a line-of-sight map is closely related to geodesic homotopy (Section 9.M).

Proof of 9.30. It only remains to prove (b) $\Rightarrow$ (a).

Let $[pxy]$ be a triangle of perimeter $< 2 \cdot \varpi^\kappa$ in $\mathcal{U}$. According to Propositions 9.21 and 9.18, it is sufficient to show the triangle $[pxy]$ is κ-thin.

Since pairs of points at distance $< \varpi^\kappa$ are joined by unique geodesics and these geodesics depend continuously on their endpoint pairs, there is a unique and continuous line-of-sight map for $[xy]$ with respect to p.

For a partition

$$0 = t^0 < t^1 < \cdots < t^N = 1,$$

let $x^{i,j} = \gamma_{t^i}(t^j)$. Since the line-of-sight map is continuous, we may assume each triangle $[x^{i,j} x^{i,j+1} x^{i+1,j+1}]$ and $[x^{i,j} x^{i+1,j} x^{i+1,j+1}]$ is κ-thin (see Proposition 9.21).

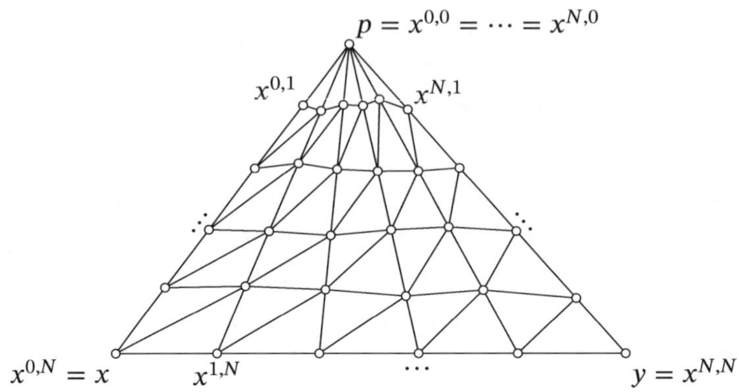

Now we show that the κ-thin property propagates to $[pxy]$, by repeated application of the inheritance lemma (Lemma 9.23):

- First, for fixed i, sequentially applying the lemma shows that the triangles $[xx^{i,1}x^{i+1,2}]$, $[xx^{i,2}x^{i+1,2}]$, $[xx^{i,2}x^{i+1,3}]$, and so on are κ-thin.

In particular, for each i, the long triangle $[xx^{i,N}x^{i+1,N}]$ is κ-thin.

- Applying the lemma again shows that the triangles $[xx^{0,N}x^{2,N}]$, $[xx^{0,N}x^{3,N}]$, and so on are κ-thin.

In particular, $[pxy] = [px^{0,N}x^{N,N}]$ is κ-thin. □

The following exercise implies that if the space is proper, then one can drop the condition on continuous dependence of geodesics in the formulation of patchwork globalization.

9.33. Exercise.

(a) Suppose pairs of points in a geodesic space $\mathcal{U}$ are joined by unique geodesics. Show that if $\mathcal{U}$ is proper, then these geodesics depend continuously on their endpoint pairs.

(b) Construct an example of a complete geodesic space $\mathcal{U}$ such that pairs of points in $\mathcal{U}$ are joined by unique geodesics, but these geodesics do not depend continuously on their endpoint pairs.

H. Angles

Recall that ω denotes a selective nonprincipal ultrafilter on $\mathbb{N}$, see Section 4.B.

9.34. Angle semicontinuity. *Suppose $\mathcal{U}_1, \mathcal{U}_2, \dots$ is a sequence of ϖ^κ-geodesic* CAT(κ) *spaces and $\mathcal{U}_n \to \mathcal{U}_\omega$ as $n \to \omega$. Assume that a sequence of hinges $[p_n\,{}^{x_n}_{y_n}]$ in $\mathcal{U}_n$ converges to a hinge $[p_\omega\,{}^{x_\omega}_{y_\omega}]$ in $\mathcal{U}_\omega$ as $n \to \omega$. Then*

$$\measuredangle[p_\omega\,{}^{x_\omega}_{y_\omega}] \geqslant \lim_{n\to\omega} \measuredangle[p_n\,{}^{x_n}_{y_n}].$$

Proof. By the angle-sidelength monotonicity (see 9.15),

$$\measuredangle\left[p_\omega\,{}^{x_\omega}_{y_\omega}\right]=\inf\left\{\tilde\measuredangle^\kappa\left(p_\omega\,{}^{\bar x_\omega}_{\bar y_\omega}\right) : \bar x_\omega\in\,]p_\omega x_\omega],\ \bar y_\omega\in\,]p_\omega y_\omega]\right\}.$$

For fixed $\bar x_\omega \in\,]p_\omega x_\omega]$ and $\bar y_\omega \in\,]p_\omega x_\omega]$, choose $\bar x_n \in\,]px_n]$ and $\bar y_n \in$ $\in\,]py_n]$ so that $\bar x_n \to \bar x_\omega$ and $\bar y_n \to \bar y_\omega$ as $n\to\omega$. Clearly

$$\tilde\measuredangle^\kappa\left(p_n\,{}^{\bar x_n}_{\bar y_n}\right)\to\tilde\measuredangle^\kappa\left(p_\omega\,{}^{\bar x_\omega}_{\bar y_\omega}\right)$$

as $n\to\omega$.

By the angle comparison (see Theorem 9.14(c)), $\measuredangle\left[p_n\,{}^{x_n}_{y_n}\right]\leqslant\tilde\measuredangle^\kappa\left(p_n\,{}^{\bar x_n}_{\bar y_n}\right)$. Hence the result. □

Now we verify that the first variation formula holds in the CAT setting. Compare it to the first variation inequality (see 6.7) which holds for general metric spaces and to the strong angle lemma (see 8.42) for CBB spaces.

9.35. Strong angle lemma. *Let $\mathcal U$ be a ϖ^κ-geodesic* CAT(κ) *space. Then for any hinge $\left[p\,{}^q_y\right]$ in $\mathcal U$, we have*

(1) $$\measuredangle\left[p\,{}^q_y\right]=\lim_{\bar y\to p}\left\{\tilde\measuredangle^\kappa\left(p\,{}^q_{\bar y}\right) : \bar y\in\,]py]\right\}$$

for any $\kappa\in\mathbb R$ such that $|p-q|<\varpi^\kappa$.

Proof. By angle-sidelength monotonicity (see 9.15), the right-hand side is defined and is bigger than or equal to the left-hand side.

By Lemma 6.4, we may take $\kappa = 0$ in (1). By the cosine law and the first variation inequality (see 6.7), the right-hand side is less than or equal to the left-hand side. □

9.36. First variation. *Let $\mathcal U$ be a ϖ^κ-geodesic* CAT(κ) *space. For any nontrivial geodesic $[py]$ in $\mathcal U$ and point $q\neq p$ such that $|p-q|<\varpi^\kappa$, we have*

$$|q-\mathrm{geod}_{[py]}(t)|=|q-p|-t\cdot\cos\measuredangle\left[p\,{}^q_y\right]+o(t).$$

Proof. The first variation equation is equivalent to the strong angle lemma (Lemma 9.35), as follows from the cosine law. □

9.37. Both-endpoints first variation. *Let $\mathcal U$ be a ϖ^κ-geodesic* CAT(κ) *space. Then for any nontrivial geodesics $[py]$ and $[qz]$ in $\mathcal U$ such that $p\neq q$ and $|p-q|<$ $<\varpi^\kappa$, we have*

$$|\mathrm{geod}_{[py]}(t)-\mathrm{geod}_{[qz]}(\tau)|=|q-p|-t\cdot\cos\measuredangle\left[p\,{}^q_y\right]-\tau\cdot\cos\measuredangle\left[q\,{}^p_z\right]+o(t+\tau).$$

Proof. By Theorem 9.14(c),

$$\begin{aligned}
&|\mathrm{geod}_{[py]}(t) - \mathrm{geod}_{[qz]}(\tau)| \\
&\quad\geqslant |q - \mathrm{geod}_{[py]}(t)| - \tau\cdot\cos\measuredangle\left[q\,{}^{\mathrm{geod}_{[py]}(t)}_{z}\right] + o(\tau) \\
&\quad\geqslant |q-p| - t\cdot\cos\measuredangle\left[p\,{}^{q}_{y}\right] + o(t) - \tau\cdot\cos\measuredangle\left[q\,{}^{\mathrm{geod}_{[py]}(t)}_{z}\right] + o(\tau) \\
&\quad= |q-p| - t\cdot\cos\measuredangle\left[p\,{}^{q}_{y}\right] - \tau\cdot\cos\measuredangle\left[q\,{}^{p}_{z}\right] + o(t+\tau).
\end{aligned}$$

Here the final equality follows from

$$\lim_{t\to 0}\measuredangle\left[q\,{}^{\mathrm{geod}_{[py]}(t)}_{z}\right] = \measuredangle\left[q\,{}^{p}_{z}\right]. \tag{2}$$

The angle semicontinuity (see 9.34) implies "$\leqslant$" in (2), and "$\geqslant$" holds by the triangle inequality for angles, since the angle comparison (in Theorem 9.14(c)) gives

$$\lim_{t\to 0}\measuredangle\left[q\,{}^{p}_{\mathrm{geod}_{[py]}(t)}\right] = 0.$$

The opposite inequality follows from 9.36 and the triangle inequality

$$|\mathrm{geod}_{[py]}(t) - \mathrm{geod}_{[qz]}(\tau)| \leqslant |\mathrm{geod}_{[py]}(t) - m| + |m - \mathrm{geod}_{[qz]}(\tau)|,$$

where m is the midpoint of $[pq]$. □

We have given elementary proofs of the first-variation statements 9.35, 9.36, and 9.37. Note however that the no-conjugate-point theorem (see 9.46) not only provides proofs of these statements but also extends the statements from geodesics in CAT(κ) spaces to local geodesics in locally CAT(κ) spaces as follows:

9.38. First variation for local geodesics. *Let $\gamma_t\colon [0,1]\to\mathcal{U}$ be a continuous family of local geodesics in a locally* CAT(κ). *Set $\alpha(t)=\gamma_t(0)$ and $\beta(t)=\gamma_t(1)$. Suppose that γ_0 is unit-speed and $\alpha^+(0)$ and $\beta^+(0)$ are defined. Then*

$$\operatorname{length}\gamma_t = \operatorname{length}\gamma_0 - (\langle\alpha^+(0),\gamma_0^+(0)\rangle + \langle\beta^+(0),\gamma_0^-(1)\rangle)\cdot t + o(t).$$

I. Reshetnyak gluing theorem

The following theorem was proved by Yuriy Reshetnyak [**142**], assuming $\mathcal{U}^1$, $\mathcal{U}^2$ are proper and complete. In the following form, the theorem appears in the book of Martin Bridson and André Haefliger [**34**].

9.39. Reshetnyak gluing theorem. *Suppose $\mathcal{U}^1$, $\mathcal{U}^2$ are ϖ^κ-geodesic spaces with isometric complete ϖ^κ-convex sets $A^i\subset\mathcal{U}^i$. Let $\iota\colon A^1\to A^2$ be an isometry. Let $\mathcal{W}=\mathcal{U}^1\sqcup_\iota\mathcal{U}^2$; that is, $\mathcal{W}$ is the gluing of $\mathcal{U}^1$ and $\mathcal{U}^2$ along ι (see Section 2.E).*

Then:

(a) *Both canonical mappings* $\jmath_i : \mathcal{U}^i \to \mathcal{W}$ *are distance-preserving and the images* $\jmath_i(\mathcal{U}^i)$ *are* ϖ^κ*-convex subsets in* $\mathcal{W}$.

(b) *If* $\mathcal{U}^1, \mathcal{U}^2$ *are* CAT(κ)*, then so is* $\mathcal{W}$.

Proof. Part (a) follows directly from ϖ^κ-convexity of the A^i.

(b). According to (a), we can identify $\mathcal{U}^i$ with its image $\jmath_i(\mathcal{U}^i)$ in $\mathcal{W}$; in this way, the subsets $A^i \subset \mathcal{U}^i$ will be identified and denoted further by A. Thus $A = \mathcal{U}^1 \cap \mathcal{U}^2 \subset \mathcal{W}$, and A is ϖ^κ-convex in $\mathcal{W}$.

Part (b) can be reformulated as follows:

9.40. Reformulation of Theorem 9.39(b). *Let* $\mathcal{W}$ *be a length space having two* ϖ^κ*-convex subsets* $\mathcal{U}^1, \mathcal{U}^2 \subset \mathcal{W}$ *such that* $\mathcal{W} = \mathcal{U}^1 \cup \mathcal{U}^2$. *Assume the subset* $A = \mathcal{U}^1 \cap \mathcal{U}^2$ *is complete and* ϖ^κ*-convex in* $\mathcal{W}$*, and* $\mathcal{U}^1$*,* $\mathcal{U}^2$ *are* CAT(κ) *spaces. Then* $\mathcal{W}$ *is a* CAT(κ) *space.*

(1) *If* $\mathcal{W}$ *is* ϖ^κ*-geodesic, then* $\mathcal{W}$ *is* CAT(κ).

Indeed, according to Proposition 9.21, it is sufficient to show that any triangle $[x^0x^1x^2]$ of perimeter $< 2\cdot\varpi^\kappa$ in $\mathcal{W}$ is κ-thin. This is obviously true if all three points x^0, x^1, x^2 lie in a single $\mathcal{U}^i$. Thus, without loss of generality, we may assume that $x^0 \in \mathcal{U}^1$ and $x^1, x^2 \in \mathcal{U}^2$.

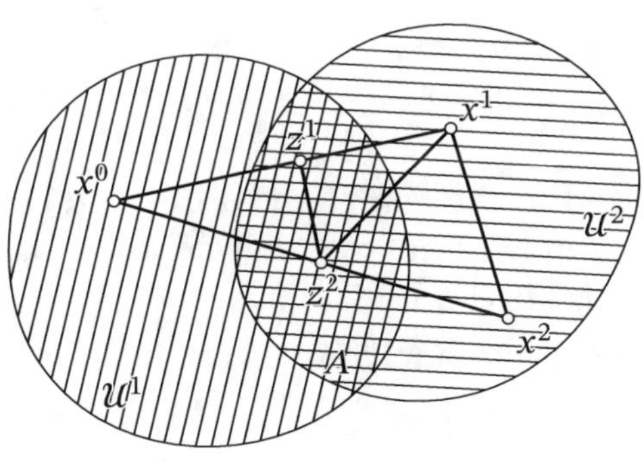

Choose points $z^1, z^2 \in A = \mathcal{U}^1 \cap \mathcal{U}^2$ lying respectively on the sides $[x^0x^1]$, $[x^0x^2]$. Note that all distances between any pair of points from x^0, x^1, x^2, z^1, z^2 are less than ϖ^κ. Therefore,

- triangle $[x^0z^1z^2]$ lies in $\mathcal{U}^1$,
- both triangles $[x^1z^1z^2]$ and $[x^1z^2x^2]$ lie in $\mathcal{U}^2$.

In particular, each triangle $[x^0z^1z^2]$, $[x^1z^1z^2]$, $[x^1z^2x^2]$ is κ-thin.

Applying the inheritance lemma (Lemma 9.23) for thin triangles twice, we get that $[x^0x^1z^2]$ and consequently $[x^0x^1x^2]$ is κ-thin. △

(2) *$\mathcal{W}$ is* CAT(κ) *if* $\kappa \leqslant 0$.

By (1) it suffices to prove that $\mathcal{W}$ is geodesic.

For $p^1 \in \mathcal{U}^1$, $p^2 \in \mathcal{U}^2$, we may choose a sequence $z_n \in A$ such that $|p^1 - z_n| + |p^2 - z_n|$ converges to $|p^1 - p^2|$, and $|p^1 - z_n|$ and $|p^2 - z_n|$ converge. Since A is complete, it suffices to show z_n is a Cauchy sequence. In that case, the limit point z of z_n satisfies $|p^1 - z| + |p^2 - z| = |p^1 - p^2|$, so the geodesics $[p^1z]$ in $\mathcal{U}^1$ and $[p^2z]$ in $\mathcal{U}^2$ together give a geodesic $[p^1p^2]$ in $\mathcal{U}$.

Suppose z_n is not a Cauchy sequence. Then there are subsequences x_n and y_n of z_n satisfying $\lim |x_n - y_n| > 0$. Let m_n be the midpoint of $[x_ny_n]$. Since $|p^1 - m_n| + |p^2 - m_n| \geqslant |p^1 - p^2|$, and $|p^1 - x_n| + |p^2 - x_n|$ and $|p^1 - y_n| + |p^2 - y_n|$ converge to $|p^1 - p^2|$, then for any $\epsilon > 0$, we may assume (taking subsequences and possibly relabeling p^1 and p^2)

$$|p^1 - m_n| \geqslant |p^1 - x_n| - \epsilon, \qquad |p^1 - m_n| \geqslant |p^1 - y_n| - \epsilon.$$

Since triangle $[p^1x_ny_n]$ is thin, the analogous inequalities hold for the Euclidean model triangle $[\tilde{p}^1\tilde{x}_n\tilde{y}_n]$. Then there is a nondegenerate limit triangle $[pxy]$ in the Euclidean plane satisfying $|p - x| = |p - y| \leqslant |p - m|$ where m is the midpoint of $[xy]$. This contradiction proves the claim. △

Finally suppose $\kappa > 0$; by rescaling, take $\kappa = 1$. Consider the Euclidean cones Cone $\mathcal{U}^i$ (see Section 6.E). By Theorem 11.7(a), Cone $\mathcal{U}^i$ is a CAT(0) space for $i = 1, 2$.

Geodesics contained in the complement of the tip of Cone $\mathcal{U}^i$ project to geodesics of length $< \pi$ in $\mathcal{U}^i$. It follows that Cone A is convex in Cone $\mathcal{U}^1$ and Cone $\mathcal{U}^2$. By the cone distance formula, Cone A is complete since A is complete.

Gluing along Cone A and applying (1) and (2) for $\kappa = 0$, we find that Cone $\mathcal{W}$ is a CAT(0) space. By Theorem 11.7(a), $\mathcal{W}$ is a CAT(1) space. □

9.41. Exercise. Let Q be the nonconvex subset of the plane bounded by two half-lines γ_1 and γ_2 with a common starting point and angle α between them. Assume $\mathcal{U}$ is a complete length CAT(0) space and γ_1', γ_2' are two half-lines in $\mathcal{U}$ with a common starting point and angle α between them. Show that the space glued from Q and $\mathcal{U}$ along the corresponding half-lines is a CAT(0) space.

9.42. Exercise. Suppose $\mathcal{U}$ is a complete length CAT(0) space and $A \subset \mathcal{U}$ is a closed subset. Assume that the doubling of $\mathcal{U}$ in A is CAT(0). Show that A is a convex set of $\mathcal{U}$.

9.43. Exercise. Let $\mathcal{U}$ be a complete length CAT(1) space, and let $K \subset \mathcal{U}$ be a closed π-convex set. Assume $K \subset \overline{\mathrm{B}}[p, \frac{\pi}{2}]$ for $p \in K$. Show that there is a decreasing continuous one-parameter family of closed convex sets K_t for $t \in$ $\in [0,1]$ such that $K_0 = \overline{\mathrm{B}}[p, \frac{\pi}{2}]$ and $K_1 = K$.

(Decreasing means with respect to inclusion; that is $K_{t_0} \supset K_{t_1}$ if $t_0 \leqslant t_1$. Continuous means with respect to Hausdorff distance; that is $K_t \underset{\mathrm{H}}{\longrightarrow} K_{t_0}$ as $t \to t_0$.)

9.44. Exercise. Let A and B be two closed convex sets in a complete length CAT(0) space. Assume $A \cap B \neq \varnothing$. Show that the union $A \cup B$ equipped with induced length metric is CAT(0).

J. Space of geodesics

In this section we prove a no-conjugate-point theorem for spaces with upper curvature bounds and derive from it a number of statements about local geodesics. These statements will be used to prove Hadamard–Cartan Theorem 9.65 and Lifting Globalization Theorem 9.50, in much the same way as the exponential map is used in Riemannian geometry.

9.45. Proposition. *Let $\mathcal{U}$ be a locally* CAT(κ) *space. Let $\gamma_n : [0,1] \to \mathcal{U}$ be a sequence of local geodesic paths converging to a path $\gamma_\infty : [0,1] \to \mathcal{U}$. Then γ_∞ is a local geodesic path. Moreover*

$$\operatorname{length}\gamma_n \to \operatorname{length}\gamma_\infty$$

as $n \to \infty$.

Proof. Fix $t \in [0,1]$. By Corollary 9.31, we may choose R satisfying $0 < R <$ $< \varpi^\kappa$, and such that the ball $\mathcal{B} = \mathrm{B}(\gamma_\infty(t), R)$ is a convex subset of $\mathcal{U}$ and forms a CAT(κ) space.

A local geodesic segment with length less than $R/2$ that intersects $\mathrm{B}(\gamma_\infty(t), R/2)$ cannot leave $\mathcal{B}$, and hence is minimizing by Corollary 9.22. In particular, for all sufficiently large n, if a subsegment of γ_n has length less than $R/2$ and contains $\gamma_n(t)$, then it is a geodesic.

Since $\mathcal{B}$ is CAT(κ), geodesic segments in $\mathcal{B}$ depend uniquely and continuously on their endpoint pairs by Theorem 9.8. Thus there is a subinterval $\mathbb{I}$ of $[0,1]$ that contains a neighborhood of t in $[0,1]$ and such that $\gamma_n|_{\mathbb{I}}$ is minimizing for all large n. It follows that the restriction $\gamma_\infty|_{\mathbb{I}}$ is a geodesic, and therefore γ_∞ is a local geodesic. □

The following theorem was proved by the first author and Richard Bishop [5]. In analogy with Riemannian geometry, the main statement of the following theorem could be restated as: *In a space of curvature $\leqslant \kappa$, two points cannot be conjugate along a local geodesic of length $< \varpi^\kappa$.*

9.46. No-conjugate-point theorem. *Suppose $\mathcal{U}$ is a locally complete, length, locally* CAT(κ) *space. Let $\gamma\colon [0,1] \to \mathcal{U}$ be a local geodesic path with length $< \varpi^\kappa$. Then for some neighborhoods $\Omega^0 \ni \gamma(0)$ and $\Omega^1 \ni \gamma(1)$, there is a unique continuous map from the direct product $\Omega^0 \times \Omega^1 \times [0,1]$ to $\mathcal{U}$,*

$$(x,y,t) \mapsto \gamma_{xy}(t),$$

such that $\gamma_{xy}\colon [0,1] \to \mathcal{U}$ is a local geodesic path with $\gamma_{xy}(0) = x$ and $\gamma_{xy}(1) = y$ for each $(x,y) \in \Omega^0 \times \Omega^1$, and the family γ_{xy} contains γ. Moreover, we can assume that the map

$$(x,y,t) \mapsto \gamma_{xy}(t)\colon \Omega^0 \times \Omega^1 \times [0,1] \to \mathcal{U}$$

is ℓ-Lipschitz for any $\ell > \max\left\{ \frac{\operatorname{sn}^\kappa r}{\operatorname{sn}^\kappa \ell} : 0 \leqslant r \leqslant \ell \right\}$.

9.47. Patchwork along a geodesic. *Let $\mathcal{U}$ be a locally complete, length, locally* CAT(κ) *space, and $\alpha\colon [a,b] \to \mathcal{U}$ be a local geodesic.*

Then there is a complete length CAT(κ) *space $\mathcal{N}$ with an open set $\hat{\Omega} \subset \mathcal{N}$, a local geodesic $\hat{\alpha}\colon [a,b] \to \hat{\Omega}$, and an open locally distance-preserving map $\Phi\colon \hat{\Omega} \nrightarrow \mathcal{U}$ such that $\Phi \circ \hat{\alpha} = \alpha$.*

Moreover if α is simple, then one can assume in addition that Φ is an open embedding; thus $\hat{\Omega}$ is locally isometric to a neighborhood of $\Omega = \Phi(\hat{\Omega})$ of α.

This lemma and its proof were suggested by Alexander Lytchak. The proof proceeds by piecing together CAT(κ) neighborhoods of points on a curve to construct a new CAT(κ) space. Exercise 9.81 is inspired by the original idea of the proof of the no-conjugate-point theorem (see Theorem 9.46), which was given in [**5**].

Proof. According to Corollary 9.31, we can choose $r > 0$ such that for any $t \in [a,b]$ the closed ball $\overline{\mathrm{B}}[\alpha(t), r]$ is a convex set that forms a complete length CAT(κ) space.

Choose balls $\mathcal{B}_i = \overline{\mathrm{B}}[\alpha(t_i), r]$ for some partition $a = t_0 < t_1 < \cdots < t_n = b$ in such a way that

$$\operatorname{Interior} \mathcal{B}_i \supset \alpha([t_{i-1}, t_i])$$

for all $i > 0$. We can assume in addition that $\mathcal{B}_{i-1} \cap \mathcal{B}_{i+1} \subset \mathcal{B}_i$ if $0 < i < n$.

Consider the disjoint union $\bigsqcup_i \mathcal{B}_i = \{(i,x) : x \in \mathcal{B}_i\}$ with the minimal equivalence relation $\sim$ such that $(i,x) \sim (i-1,x)$ for all $i > 0$. Let $\mathcal{N}$ be the space obtained by gluing the $\mathcal{B}_i$ along $\sim$. Note that $A_i = \mathcal{B}_i \cap \mathcal{B}_{i-1}$ is convex in $\mathcal{B}_i$ and in $\mathcal{B}_{i-1}$. Applying Reshetnyak Gluing Theorem 9.39 several times, we conclude that $\mathcal{N}$ is a complete length CAT(κ) space.

For $t \in [t_{i-1}, t_i]$, let $\hat{\alpha}(t)$ be the equivalence class of $(i, \alpha(t))$ in $\mathcal{N}$. Let $\hat{\Omega}$ be the ε-neighborhood of $\hat{\alpha}$ in $\mathcal{N}$, where $\varepsilon > 0$ is chosen so that $\mathrm{B}(\alpha(t), \varepsilon) \subset \mathcal{B}_i$ for all $t \in [t_{i-1}, t_i]$.

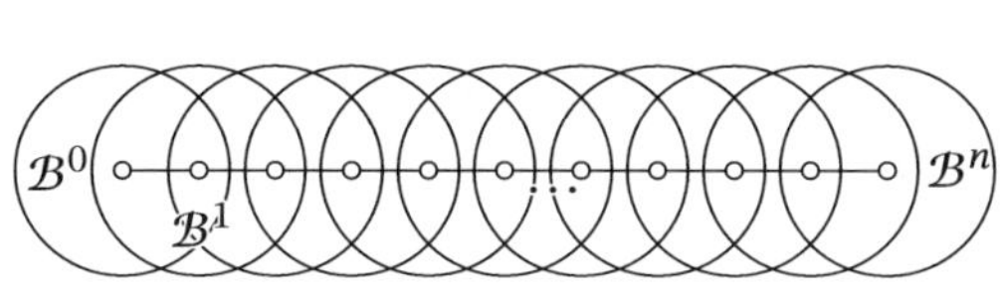

Define $\Phi\colon \hat\Omega \to \mathcal{U}$ by sending the equivalence class of (i, x) to x. It is straightforward to check that $\Phi\colon \mathcal{N} \to \mathcal{U}$, $\hat\alpha\colon [a, b] \to \mathcal{N}$ and $\hat\Omega \subset \mathcal{N}$ satisfy the conclusion of the main part of the lemma.

To prove the final statement in the lemma, we only have to choose $\varepsilon > 0$ so that in addition, $|\alpha(\tau) - \alpha(\tau')| > 2\cdot\varepsilon$ if $\tau \leqslant t_{i-1}$ and $t_i \leqslant \tau'$ for some i. □

Proof of Theorem 9.46. Apply patchwork along γ (see 9.47). □

No-Conjugate-Point Theorem 9.46 allows us to move a local geodesic so that its endpoints follow given trajectories. The following corollary describes how this process might terminate.

9.48. Corollary. *Let $\mathcal{U}$ be a locally complete, length, locally* CAT(κ) *space. Suppose $\gamma\colon [0, 1] \to \mathcal{U}$ is a local geodesic with length $< \varpi^\kappa$. Let $\alpha^i\colon [0, 1] \to \mathcal{U}$, for $i = 0, 1$, be curves starting at $\gamma(0)$ and $\gamma(1)$, respectively.*

Then there is a uniquely determined pair consisting of an interval $\mathbb{I}$ satisfying $0 \in \mathbb{I} \subset [0, 1]$, and a continuous family of local geodesics $\gamma_t\colon [0, 1] \to \mathcal{U}$ for $t \in \mathbb{I}$, such that

(a) *$\gamma_0 = \gamma$, $\gamma_t(0) = \alpha^0(t)$, $\gamma_t(1) = \alpha^1(t)$, and γ_t has length $< \varpi^\kappa$;*
(b) *if $\mathbb{I} \neq [0, 1]$, then $\mathbb{I} = [0, a)$, where either γ_t converges uniformly to a local geodesic γ_a of length ϖ^κ, or for some fixed $s \in [0, 1]$ the curve $\gamma_t(s)\colon [0, a) \to \mathcal{U}$ is a Lipschitz curve with no limit as $t \to a-$.*

Proof. Uniqueness follows from Theorem 9.46.

Let $\mathbb{I}$ be the maximal interval for which there is a family γ_t satisfying condition (a). By Theorem 9.46, such an interval exists and is open in $[0, 1]$. Suppose $\mathbb{I} \neq [0, 1]$. Then $\mathbb{I} = [0, a)$ for $0 < a \leqslant 1$.

For each fixed $s \in [0, 1]$, define the curve $\alpha_s\colon [0, a) \to \mathcal{U}$ by $\alpha_s(t) = \gamma_t(s)$. By Theorem 9.46, each α_s is locally Lipschitz.

If α_s for some value of s does not converge as $t \to a-$, then condition (b) holds. If each α_s converges as $t \to a-$, then γ_t converges as $t \to a-$, say to γ_a. By Proposition 9.45, γ_a is a local geodesic and

$$\operatorname{length}\gamma_t \to \operatorname{length}\gamma_a \leqslant \varpi^\kappa.$$

By maximality of $\mathbb{I}$, length $\gamma_a = \varpi^\kappa$ and so condition (b) again holds. □

9.49. Corollary. *Let $\mathcal{U}$ be a complete locally* CAT(κ) *length space, and* $\alpha\colon[0,1]\to$ $\to\mathcal{U}$ *be a path of length* $<\varpi^\kappa$ *that starts at* p *and ends at* q. *Then:*

(a) *there is a unique homotopy of local geodesic paths* $\gamma_t : [0,1]\to\mathcal{U}$ *such that* $\gamma_0(t)=\gamma_t(0)=p$ *and* $\gamma_t(1)=\alpha(t)$ *for any* t.

(b) *for any* $t\in[0,1]$,

$$\operatorname{length}\gamma_t\leqslant\operatorname{length}(\alpha|_{[0,t]}),$$

and equality holds for given t *if and only if the restriction* $\alpha|_{[0,t]}$ *is a reparametrization of* γ_t.

Moreover, instead of completeness of $\mathcal{U}$, *one can assume that the subspace*

$$W=\{\,x\in\mathcal{U} : |x-p|+|x-q|\leqslant\ell\,\}$$

is complete.

Proof. By Corollary 9.48, taking $\alpha^0(t)=p$ and $\alpha^1(t)=\alpha(t)$ for all $t\in[0,1]$, there is an interval $\mathbb{I}$ such that (a) holds for all $t\in\mathbb{I}$, and either $\mathbb{I}=[0,1]$ or $\mathbb{I}=[0,a)$ for $a\leqslant1$.

By patchwork along a curve (see 9.47), the values of t for which condition (b) holds form an open subset of $\mathbb{I}$ containing 0; clearly this subset is also closed in $\mathbb{I}$. Therefore (b) holds on all of $\mathbb{I}$.

Corollary 9.48 implies that $\mathbb{I}=[0,1]$. Indeed if $\mathbb{I}=[0,a)$, then either $\operatorname{length}\gamma_t\to\varpi^\kappa$ as $t\to a-$, or for some fixed $s\in[0,1]$ the Lipschitz curve $\gamma_t(s)$ has no limit as $t\to a-$. Since $\operatorname{length}\alpha<\varpi^\kappa$, Corollary 9.48 implies that neither of these is possible. □

K. Lifting globalization

Hadamard–Cartan Theorem 9.65 states that the universal metric cover of a complete locally CAT(0) space is CAT(0). The lifting globalization theorem gives an appropriate generalization of the above statement to arbitrary curvature bounds; it could be also described as a global version of Gauss's lemma.

9.50. Lifting globalization theorem. *Suppose* $\mathcal{U}$ *is a complete length locally* CAT(κ) *space and* $p\in\mathcal{U}$. *Then there is a complete* CAT(κ) *length space* $\mathcal{B}$, *with a point* $\hat p$ *such that there is a locally distance-preserving map* $\Phi:\mathcal{B}\to\mathcal{U}$ *such that* $\Phi(\hat p)=p$ *and the following lifting property holds: for any path* $\alpha:[0,1]\to\mathcal{U}$ *with* $\alpha(0)=p$ *and* $\operatorname{length}\alpha<\varpi^\kappa/2$, *there is a unique path* $\hat\alpha:[0,1]\to\mathcal{B}$ *such that* $\hat\alpha(0)=\hat p$ *and* $\Phi\circ\hat\alpha\equiv\alpha$.

Note that the lifting property implies that $\Phi(\mathcal{B})\supset\mathrm{B}(p,\varpi^\kappa/2)$ and by completeness $\Phi(\mathcal{B})\supset\overline{\mathrm{B}}[p,\varpi^\kappa/2]$. Also, since $\mathcal{B}$ is CAT(κ), the closed ball $\overline{\mathrm{B}}[\hat p,\tfrac{\varpi^\kappa}{2}]_{\mathcal{B}}$ is a weakly convex set in $\mathcal{B}$ (see Corollary 9.27); in particular $\overline{\mathrm{B}}[\hat p,\tfrac{\varpi^\kappa}{2}]_{\mathcal{B}}$ is a complete length CAT(κ) space. Therefore we can assume in addition that $|\hat{p}-\hat{x}| \leqslant \varpi^\kappa/2$ for any $\hat{x} \in \mathcal{B}$; or equivalently

$$\overline{\mathrm{B}}[\hat{p}, \tfrac{\varpi^\kappa}{2}]_{\mathcal{B}} = \mathcal{B}.$$

Before proving the theorem, we state and prove its corollary.

9.51. Corollary. *Suppose $\mathcal{U}$ is a complete length locally* CAT(κ) *space. Then for any $p \in \mathcal{U}$ there is $\rho_p > 0$ such that $\overline{\mathrm{B}}[p, \rho_p]$ is a complete length* CAT(κ) *space.*

Moreover, we can assume that $\rho_p < \frac{\varpi^\kappa}{2}$ for any p and the function $p \mapsto \rho_p$ is 1-Lipschitz.

Proof. Assume $\Phi\colon \mathcal{B} \to \mathcal{U}$ and $\hat{p} \in \mathcal{B}$ are provided by Lifting Globalization Theorem 9.50.

Since Φ is local isometry, we can choose $r > 0$ so that the restriction of Φ to $\overline{\mathrm{B}}[\hat{p}, r]$ is distance-preserving. By the lifting globalization, the image $\Phi(\overline{\mathrm{B}}[\hat{p}, r])$ coincides with the ball $\overline{\mathrm{B}}[p, r]$. This proves the first part of the theorem.

To prove the second part, let us choose ρ_p to be the maximal value $\leqslant \frac{\varpi^\kappa}{2}$ such that $\overline{\mathrm{B}}[p, \rho_p]$ is a complete length CAT(κ) space. By Corollary 9.27, the ball

$$\overline{\mathrm{B}}[q, \rho_p - |p-q|]$$

is weakly convex in $\overline{\mathrm{B}}[p, \rho_p]$. Therefore

$$\overline{\mathrm{B}}[q, \rho_p - |p-q|]$$

is a complete length CAT(κ) space for any $q \in \mathrm{B}(p, \rho_p)$. In particular, $\rho_q \geqslant \rho_p - |p-q|$ for any $p, q \in \mathcal{U}$. Hence the second statement follows. ☐

The proof of the lifting globalization theorem relies heavily on the properties of the space of local geodesic paths discussed in Section 9.J. The following lemma is a key step in the proof; it was proved by the first author and Richard Bishop [3].

9.52. Radial lemma. *Let $\mathcal{U}$ be a length locally* CAT(κ) *space, and suppose $p \in \mathcal{U}$, $R \leqslant \varpi^\kappa$. Assume the ball $\overline{\mathrm{B}}[p, \bar{R}]$ is complete for any $\bar{R} < R$, and there is a unique geodesic path,* $\mathrm{path}_{[px]}$, *from p to any point $x \in \mathrm{B}(p, R)$ that depends continuously on x. Then $\mathrm{B}(p, \frac{R}{2})$ is a ϖ^κ-geodesic* CAT(κ) *space.*

Proof. Without loss of generality, we may assume $\mathcal{U} = \mathrm{B}(p, R)$.

Set $f = \mathrm{md}^\kappa \circ \mathrm{dist}_p$. Let us show that

$$f'' + \kappa \cdot f \geqslant 1. \qquad (1)$$

Fix $z \in \mathcal{U}$. We will apply No-Conjugate-Point Theorem 9.46 for the unique geodesic path γ from p to z. The notations $\Omega^0, \Omega^1, \gamma_{xy}, \mathcal{N}, \hat{x}, \hat{y}$ will be as in the formulation of Lifting Globalization Theorem 9.50; in particular, $z \in \Omega^1$.

By assumption, $\gamma_{py} = \text{path}_{[py]}$ for any $y \in \Omega^1$. Consequently, $f(y) = \text{md}^\kappa |\hat{p} - \hat{y}|_{\mathcal{N}}$. Applying the function comparison (see 9.25) in $\mathcal{N}$, we have that $f'' + \kappa \cdot f \geqslant 1$ in Ω^1; whence (1) follows. □

Fix $r < \frac{R}{2}$. Proving the following claim takes most of the remaining proof:

(2) $\overline{\text{B}}[p, r]$ *is a convex set in* $\mathcal{U}$.

Choose arbitrary $x, z \in \overline{\text{B}}[p, r]$. First note that (1) implies the following claim.

(3) *If* $\gamma\colon [0,1] \to \mathcal{U}$ *is a local geodesic path from* x *to* z *and* $\text{length}\,\gamma < \varpi^\kappa$, *then* $\text{length}\,\gamma \leqslant 2\cdot r$ *and* γ *lies completely in* $\overline{\text{B}}[p, r]$.

Note that $|x - z| < \varpi^\kappa$. Thus to prove claim (2), it is sufficient to show that there is a geodesic path from x to z. Note that by assumption $\overline{\text{B}}[p, 2\cdot r]$ is complete. Therefore Corollary 9.49 implies the following:

(4) *Given a path* $\alpha\colon [0,1] \to \mathcal{U}$ *from* x *to* z *with* $\text{length}\,\alpha < 2\cdot r$, *there is a local geodesic path* γ *from* x *to* z *such that*

$$\text{length}\,\gamma \leqslant \text{length}\,\alpha.$$

Further, let us prove the following:

(5) *There is a unique local geodesic path* γ_{xz} *in* $\overline{\text{B}}[p, r]$ *from* x *to* z.

Denote by Δ_{xz} the set of all local geodesic paths in $\overline{\text{B}}[p, r]$ from x to z. By Corollary 9.48, there is a bijection $\Delta_{xz} \to \Delta_{pp}$. According to (1), Δ_{pp} contains only the constant path. Claim 5 follows.

Note that claims (3), (4), and (5) imply that γ_{xz} is minimizing; hence claim (2).

Further, claim (3) and No-Conjugate-Point Theorem 9.46 together imply that the map $(x, z) \mapsto \gamma_{xz}$ is continuous.

By Patchwork Globalization Theorem 9.30, $\overline{\text{B}}[p, r]$ is a ϖ^κ-geodesic CAT(κ) space.

Since

$$\text{B}(p, R) = \bigcup_{r<R} \overline{\text{B}}[p, r],$$

then $\text{B}(p, R)$ is convex in $\mathcal{U}$ and CAT(κ) comparison holds for any quadruple in $\text{B}(p, R)$. Therefore $\text{B}(p, \varpi^\kappa/2)$ is CAT(κ). △

In the following proof, we construct a space $\mathfrak{G}_p$ of local geodesic paths that start at p. The space $\mathfrak{G}_p$ comes with a marked point $\hat{p}$ and the endpoint map $\Phi\colon \mathfrak{G}_p \to \mathcal{U}$. One can think of the map Φ as an analogue of the exponential map $\exp_p$ in Riemannian geometry; in this case, the space $\mathfrak{G}_p$ corresponds to the ball of radius ϖ^κ in the tangent space at p, equipped with the metric pulled back by $\exp_p$.

We are going to set $\mathcal{B} = \mathrm{B}(\hat{p}, \varpi^\kappa/2) \subset \mathfrak{G}_p$, and use Radial Lemma 9.52 to prove that $\mathcal{B}$ is a ϖ^κ-geodesic CAT(κ) space.

Proof of Theorem 9.50. Suppose $\hat{\gamma}$ is a homotopy of local geodesic paths that start at p. Thus the map

$$\hat{\gamma}\colon (t,\tau) \mapsto \hat{\gamma}_t(\tau)\colon [0,1]\times[0,1] \to \mathcal{U}$$

is continuous, and the following holds for each t:

- $\hat{\gamma}_t(0) = p$,
- $\hat{\gamma}_t\colon [0,1] \to \mathcal{U}$ is a local geodesic path in $\mathcal{U}$.

Denote by $\theta(\hat{\gamma})$ the length traced by the ends of $\hat{\gamma}_t$; that is, $\theta(\hat{\gamma})$ is the length of the path $t \mapsto \hat{\gamma}_t(1)$.

Let $\mathfrak{G}_p$ be the set of all local geodesic paths with length $< \varpi^\kappa$ in $\mathcal{U}$ that start at p. Denote by $\hat{p} \in \mathfrak{G}_p$ the constant path $\hat{p}(t) \equiv p$. Given $\alpha, \beta \in \mathfrak{G}_p$, define

$$|\alpha - \beta|_{\mathfrak{G}_p} = \inf_{\hat{\gamma}}\{\theta(\hat{\gamma})\},$$

with the exact lower bound taken along all homotopies $\hat{\gamma}\colon [0,1]\times[0,1] \to \mathcal{U}$ such that $\hat{\gamma}_0 = \alpha$, $\hat{\gamma}_1 = \beta$ and $\hat{\gamma}_t \in \mathfrak{G}_p$ for all $t \in [0,1]$.

By No-Conjugate-Point Theorem 9.46, we have $|\alpha-\beta|_{\mathfrak{G}_p} > 0$ for distinct α and β; that is,

(6) $|{*}-{*}|_{\mathfrak{G}_p}$ *is a metric on* $\mathfrak{G}_p$.

Further, again from No-Conjugate-Point Theorem 9.46, we have

(7) *The map*

$$\Phi\colon \xi \mapsto \xi(1)\colon \mathfrak{G}_p \to \mathcal{U}$$

is a local isometry. In particular, $\mathfrak{G}_p$ *is locally* CAT(κ).

Let $\alpha\colon [0,1] \to \mathcal{U}$ be a path, length $\alpha < \varpi^\kappa$, and $\alpha(0) = p$. The homotopy constructed in Corollary 9.49 can be regarded as a path in $\mathfrak{G}_p$, say $\hat{\alpha}\colon [0,1] \to \mathfrak{G}_p$, such that $\hat{\alpha}(0) = \hat{p}$ and $\Phi\circ\hat{\alpha} = \alpha$; in particular $\hat{\alpha}_t(1) \equiv \alpha(t)$ for any t. By (7),

$$\mathrm{length}(\hat{\alpha})_{\mathfrak{G}_p} = \mathrm{length}(\alpha)_{\mathcal{U}}.$$

Moreover, it follows that α is a local geodesic path of $\mathcal{U}$ if and only if $\hat{\alpha}$ is a local geodesic path of $\mathfrak{G}_p$.

Further, from Corollary 9.49, for any $\xi \in \mathfrak{G}_p$ and path $\hat{\alpha}\colon [0,1] \to \mathfrak{G}_p$ from $\hat{p}$ to ξ, we have

$$\begin{aligned}\operatorname{length}\hat{\alpha} &= \operatorname{length}\Phi\circ\hat{\alpha} \geqslant \\ &\geqslant \operatorname{length}\xi = \\ &= \operatorname{length}\hat{\xi},\end{aligned}$$

where equality holds only if $\hat{\alpha}$ is a reparametrization of $\hat{\xi}$. In particular,

(8) $$|\hat{p} - \xi|_{\mathfrak{G}_p} = \operatorname{length}\xi,$$

and $\hat{\xi}\colon [0,1] \to \mathfrak{G}_p$ is the unique geodesic path from $\hat{p}$ to ξ. Clearly, the map $\xi \mapsto \hat{\xi}$ is continuous.

By (8) and Proposition 9.45,

(9) *For any $\bar{R} < \varpi^\kappa$, the closed ball $\overline{\mathrm{B}}[\hat{p}, \bar{R}]$ in $\mathfrak{G}_p$ is complete.*

Take $\mathrm{B}(\hat{p}, \varpi^\kappa/2)$ and Φ constructed above. According to Radial Lemma 9.52, $\mathrm{B}(\hat{p}, \varpi^\kappa/2)$ is a ϖ^κ-geodesic CAT(κ) space. The map Φ extends to its completion $\mathcal{B} = \overline{\mathrm{B}}[\hat{p}, \varpi^\kappa/2]$. All the remaining statements are already proved. □

L. Reshetnyak majorization

9.53. Definition. Let $\mathcal{X}$ be a metric space, let $\tilde{\alpha}$ be a simple closed curve of finite length in $\mathbb{M}^2(\kappa)$, and let $D \subset \mathbb{M}^2(\kappa)$ be a closed region bounded by $\tilde{\alpha}$. A length-nonincreasing map $F\colon D \to \mathcal{X}$ is called *majorizing* if it is length-preserving on $\tilde{\alpha}$.

In this case, we say that D *majorizes* the curve $\alpha = F \circ \tilde{\alpha}$ under the map F.

The following proposition is a consequence of the definition.

9.54. Proposition. *Let α be a closed curve in a metric space $\mathcal{X}$. Suppose $D \subset \mathbb{M}^2(\kappa)$ majorizes α under $F\colon D \to \mathcal{X}$. Then any geodesic subarc of α is the image under F of a subarc of $\partial_{\mathbb{M}^2(\kappa)}D$ that is geodesic in the length metric of D.*

In particular, if D is convex, then the corresponding subarc is a geodesic in $\mathbb{M}^2(\kappa)$.

Proof. For a geodesic subarc $\gamma\colon [a,b] \to \mathcal{X}$ of $\alpha = F \circ \tilde{\alpha}$, set

$$\tilde{r} = |\tilde{\gamma}(a) - \tilde{\gamma}(b)|_D, \qquad \tilde{\gamma} = (F|_{\partial D})^{-1} \circ \gamma,$$
$$s = \operatorname{length}\gamma, \qquad \tilde{s} = \operatorname{length}\tilde{\gamma}.$$

Then

$$\tilde{r} \geqslant r = s = \tilde{s} \geqslant \tilde{r}.$$

Therefore $\tilde{s} = \tilde{r}$. □

9.55. Corollary. *Let $[pxy]$ be a triangle of perimeter $< 2\cdot\varpi^\kappa$ in a metric space $\mathcal{X}$. Assume a convex region $D \subset \mathbb{M}^2(\kappa)$ majorizes $[pxy]$. Then $D = \operatorname{Conv}[\tilde{p}\tilde{x}\tilde{y}]$ for a model triangle $[\tilde{p}\tilde{x}\tilde{y}] = \tilde{\triangle}^\kappa(pxy)$, and the majorizing map sends $\tilde{p}$, $\tilde{x}$, and $\tilde{y}$, respectively, to p, x, and y.*

Now we come to the main theorem of this section.

9.56. Majorization theorem. *Any closed curve α with length smaller than $2\cdot\varpi^\kappa$ in a ϖ^κ-geodesic* CAT(κ) *space is majorized by a convex region in $\mathbb{M}^2(\kappa)$.*

This theorem was proved by Yuriy Reshetnyak [**142**]; our proof uses a trick that we learned from the lectures of Werner Ballmann [**21**]. Another proof can be built on Kirszbraun's Theorem 10.14, but it works only for complete spaces.

The case when α is a triangle, say $[pxy]$, is the base and is nontrivial. In this case, by Corollary 9.55, the majorizing convex region has to be isometric to $\operatorname{Conv}[\tilde{p}\tilde{x}\tilde{y}]$, where $[\tilde{p}\tilde{x}\tilde{y}] = \tilde{\triangle}^\kappa(pxy)$. There is a majorizing map for $[pxy]$ whose image W is the image of the line-of-sight map (Definition 9.32) for $[xy]$ from p, but as one can see from the following example, the line-of-sight map is not majorizing in general.

Example. Let Ω be a solid quadrangle $[pxzy]$ in $\mathbb{E}^2$ formed by two congruent triangles, which is nonconvex at z (as in the picture). Equip Ω with the length metric. Then Ω is CAT(0) by Reshetnyak Gluing Theorem 9.39. For triangle $[pxy]_\Omega$ in Ω and its model triangle $[\tilde{p}\tilde{x}\tilde{y}]$ in $\mathbb{E}^2$, we have

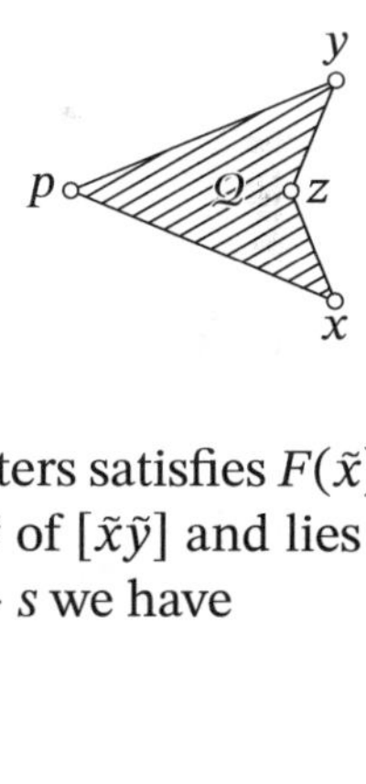

$$|\tilde{x} - \tilde{y}| = |x - y|_\Omega = |x - z| + |z - y|.$$

Then the map F defined by matching line-of-sight parameters satisfies $F(\tilde{x}) = x$ and $|x - F(\tilde{w})| > |\tilde{x} - \tilde{w}|$ if $\tilde{w}$ is near the midpoint $\tilde{z}$ of $[\tilde{x}\tilde{y}]$ and lies on $[\tilde{p}\tilde{z}]$. Indeed, by First Variation Formula (8.42), for $\varepsilon = 1 - s$ we have

$$|\tilde{x} - \tilde{w}| = |\tilde{x} - \tilde{\gamma}_{\frac{1}{2}}(s)| = |x - z| + o(\varepsilon)$$

and

$$|x - F(\tilde{w})| = |x - \gamma_{\frac{1}{2}}(s)| = |x - z| - \varepsilon\cdot\cos\measuredangle\left[z\,{}^{p}_{x}\right] + o(\varepsilon).$$

Thus F is not majorizing.

In the following proofs, $x^1 \cdots x^n$ ($n \geqslant 3$) denotes a polygonal line $x^1, \dots, x^n$, and $[x^1 \cdots x^n]$ denotes the corresponding (closed) polygon. For a subset R of the ambient metric space, we denote by $[x^1 \cdots x^n]_R$ a polygon in the length metric of R.

Our first lemma gives a model space construction based on repeated application of Lemma 9.24 from the proof of the inheritance. Recall that convex and concave curves with respect to a point are defined in Definition 8.27.

9.57. Lemma. *In $\mathbb{M}^2(\kappa)$, let β be a curve from x to y that is concave with respect to p. Let D be the subgraph of β with respect to p. Assume*

$$\operatorname{length}\beta + |p-x| + |p-y| < 2\cdot\varpi^\kappa.$$

(a) *Then β forms a geodesic $[xy]_D$ in D and therefore β, $[px]$ and $[py]$ form a triangle $[pxy]_D$ in the length metric of D.*

(b) *Let $[\tilde p\tilde x\tilde y]$ be the model triangle for $[pxy]_D$. Then there is a short map*

$$G\colon \operatorname{Conv}[\tilde p\tilde x\tilde y] \to D$$

such that $\tilde p \mapsto p$, $\tilde x \mapsto x$, $\tilde y \mapsto y$, and G is length-preserving on each side of $[\tilde p\tilde x\tilde y]$. In particular, Conv$[\tilde p\tilde x\tilde y]$ *majorizes triangle $[pxy]_D$ in D under G.*

Proof. We prove the lemma for a polygonal line β; the general case then follows by approximation. Namely, since β is concave it can be approximated by polygonal lines that are concave with respect to p, with their lengths converging to length β. Passing to a partial limit, we will obtain the needed map G.

Suppose $\beta = x^0x^1\cdots x^n$ is a polygonal line with $x^0 = x$ and $x^n = y$. Consider a sequence of polygonal lines $\beta_i = x^0x^1\cdots x^{i-1}y_i$ such that $|p-y_i| = |p-y|$ and β_i has same length as β; that is,

$$|x^{i-1}-y_i| = |x^{i-1}-x^i| + |x^i - x^{i+1}| + \dots + |x^{n-1}-x^n|.$$

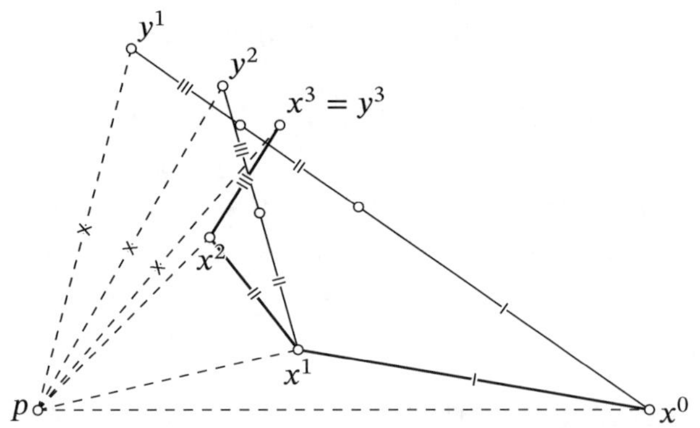

Clearly $\beta_n = \beta$. Sequentially applying Alexandrov's Lemma 6.3 shows that each of the polygonal lines $\beta_{n-1}, \beta_{n-2}, \dots, \beta_1$ is concave with respect to p. Let D_i be the subgraph of β_i with respect to p. Applying Lemma 9.24 gives a short map $G_i\colon D_i \to D_{i+1}$ that maps $y_i \mapsto y_{i+1}$ and does not move p and x (in fact, G_i is the identity everywhere except on Conv$[px^{i-1}y_i]$). Thus the composition

$$G_{n-1}\circ\dots\circ G_1\colon D_1 \to D_n$$

is short. The result follows since $D_1 \overset{\text{iso}}{=\!=} \operatorname{Conv}[\tilde p\tilde x\tilde y]$. □

9.58. Lemma. *Let $[pxy]$ be a triangle of perimeter $< 2\cdot\varpi^\kappa$ in a ϖ^κ-geodesic* $\mathrm{CAT}(\kappa)$ *space $\mathcal{U}$. In $\mathbb{M}^2(\kappa)$, let $\tilde\gamma$ be the κ-development of $[xy]$ with respect to p, where $\tilde\gamma$ has basepoint $\tilde p$ and subgraph D. Consider the map $H : D \to \mathcal{U}$ that sends the point with parameter (t, s) under the line-of-sight map for $\tilde\gamma$ with respect to $\tilde p$, to the point with the same parameter under the line-of-sight map f for $[xy]$ with respect to p. Then H is length-nonincreasing. In particular, D majorizes triangle $[pxy]$.*

Proof. Let $\gamma = \mathrm{geod}_{[xy]}$ and $T = |x - y|$. As in the proof of Development Criterion 9.29, take a partition

$$0 = t^0 < t^1 < \dots < t^n = T,$$

and set $x^i{=}\gamma(t^i)$. Construct a chain of model triangles $[\tilde p\tilde x^{i-1}\tilde x^i]{=}\tilde\Delta^\kappa(px^{i-1}x^i)$, with $\tilde x^0 = \tilde x$ and the direction of $[\tilde p\tilde x^i]$ turning counterclockwise as i grows. Let D_n be the subgraph with respect to $\tilde p$ of the polygonal line $\tilde x^0 \cdots \tilde x^n$.

Let δ_n be the maximum radius of a circle inscribed in any of the triangles $[\tilde p\tilde x^{i-1}\tilde x^i]$.

Now we construct a map $H_n : D_n \to \mathcal{U}$ that increases distances by at most $2\cdot\delta_n$. Suppose $w \in D_n$. Then w lies on or inside some triangle $[\tilde p\tilde x^{i-1}\tilde x^i]$. Define $H_n(w)$ by first mapping w to a nearest point on $[\tilde p\tilde x^{i-1}\tilde x^i]$ (choosing one if there are several), followed by the natural map to the triangle $[px^{i-1}x^i]$.

Since triangles in $\mathcal{U}$ are κ-thin (see Proposition 9.21), the restriction of H_n to each triangle $[\tilde p\tilde x^{i-1}\tilde x^i]$ is short. Then the triangle inequality implies that the restriction of H_n to

$$U_n = \bigcup_{1\leqslant i\leqslant n} [\tilde p\tilde x^{i-1}\tilde x^i]$$

is short with respect to the length metric on D_n. Since nearest-point projection from D_n to U_n increases the D_n-distance between two points by at most $2\cdot\delta_n$, the map H_n also increases the D_n-distance by at most $2\cdot\delta_n$.

Consider converging sequences $v_n \to v$ and $w_n \to w$ such that $v_n, w_n \in D_n$ and therefore $v, w \in D$. Note that

$$|H_n(v_n) - H_n(w_n)| \leqslant |v_n - w_n|_{D_n} + 2\cdot\delta_n, \tag{1}$$

for each n. Since $\delta_n \to 0$ and geodesics in $\mathcal{U}$ vary continuously with their endpoints (see 9.30), we have $H_n(v_n) \to H(v)$ and $H_n(w_n) \to H(w)$. Therefore the left-hand side in (1) converges to $|H(v) - H(w)|$ and the right-hand side converges to $|v - w|_D$; it follows that H is short. □

Proof of Theorem 9.56. We begin by proving the theorem in case α is polygonal.

First suppose α is a triangle, say $[pxy]$. By assumption, the perimeter of $[pxy]$ is less than $2\cdot\varpi^\kappa$. This is the base case for the induction.

Let $\tilde{\gamma}$ be the κ-development of $[xy]$ with respect to p, where $\tilde{\gamma}$ has basepoint $\tilde{p}$ and subgraph D. By Development Criterion 9.29, $\tilde{\gamma}$ is concave. By Lemma 9.57, there is a short map $G\colon \operatorname{Conv}\tilde{\triangle}^{\kappa}(pxy) \to D$. Further, by Lemma 9.58, D majorizes $[pxy]$ under a majorizing map $H\colon D \to \mathcal{U}$. Clearly $H \circ G$ is a majorizing map for $[pxy]$.

Now we claim that any closed n-gon $[x^1x^2\cdots x^n]$ of perimeter less than $2\cdot\varpi^{\kappa}$ in a CAT(κ) space is majorized by a convex polygonal region

$$R_n = \operatorname{Conv}[\tilde{x}^1\tilde{x}^2\cdots\tilde{x}^n]$$

under a map F_n such that $F_n\colon \tilde{x}^i \mapsto x^i$ for each i.

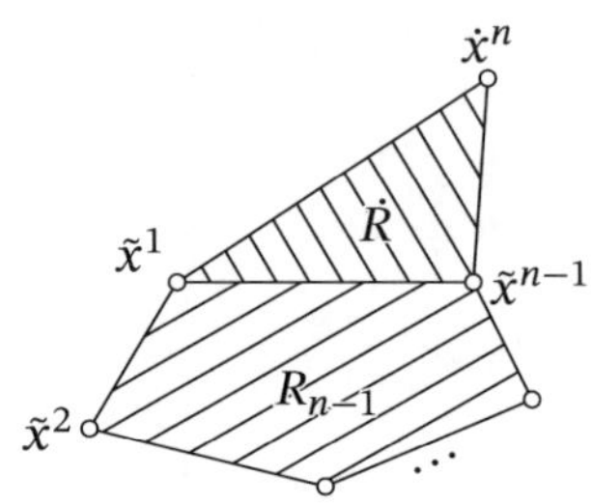

Assume the statement is true for $(n-1)$-gons, $n \geqslant 4$. Then $[x^1x^2\cdots x^{n-1}]$ is majorized by a convex polygonal region

$$R_{n-1} = \operatorname{Conv}[\tilde{x}^1\tilde{x}^2\cdots\tilde{x}^{n-1}],$$

in $\mathbb{M}^2(\kappa)$ under a map F_{n-1} satisfying $F_{n-1}(\tilde{x}^i) = x^i$ for all i. Take $\dot{x}^n \in \mathbb{M}^2(\kappa)$ such that $[\tilde{x}^1\tilde{x}^{n-1}\dot{x}^n] = \tilde{\triangle}^{\kappa}(x^1x^{n-1}x^n)$ and this triangle lies on the other side of $[\tilde{x}^1\tilde{x}^{n-1}]$ from R_{n-1}. Let $\dot{R} = \operatorname{Conv}[\tilde{x}^1\tilde{x}^{n-1}\dot{x}^n]$, and let $\dot{F}\colon \dot{R} \to \mathcal{U}$ be a majorizing map for $[x^1x^{n-1}x^n]$ as provided above.

Set $R = R_{n-1} \cup \dot{R}$, where R carries its length metric. Since F_n and F agree on $[\tilde{x}^1\tilde{x}^{n-1}]$, we may define $F\colon R \to \mathcal{U}$ by

$$F(x) = \begin{cases} F_{n-1}(x), & x \in R_{n-1}, \\ \dot{F}(x), & x \in \dot{R}. \end{cases}$$

Then F is length-nonincreasing and is a majorizing map for $[x^1x^2\cdots x^n]$ (as in Definition 9.53).

If R is a convex subset of $\mathbb{M}^2(\kappa)$, we are done.

If R is not convex, the total internal angle of R at $\tilde{x}^1$ or $\tilde{x}^{n-1}$ or both is $> \pi$. By relabeling, we may suppose this holds for $\tilde{x}^{n-1}$.

The region R is obtained by gluing R_{n-1} to $\dot{R}$ by $[x^1x^{n-1}]$. Thus, by Reshetnyak Gluing Theorem 9.39, R carrying its length metric is a CAT(κ)-space. Moreover $[\tilde{x}^{n-2}\tilde{x}^{n-1}] \cup [\tilde{x}^{n-1}\dot{x}^n]$ is a geodesic of R. Thus $[\tilde{x}^1\tilde{x}^2\cdots\tilde{x}^{n-2}\dot{x}^n]_R$ is a closed $(n-1)$-gon in R, to which the induction hypothesis applies. The resulting short map from a convex region in $\mathbb{M}^2(\kappa)$ to R, followed by F, is the desired majorizing map.

Note that in fact we have proved the following:

(2) *Let F_{n-1} be a majorizing map for the polygon $[x^1x^2\cdots x^{n-1}]$, and let $\dot F$ be a majorizing map for the triangle $[x^1x^{n-1}x^n]$. Then there is a majorizing map F_n for the polygon $[x^1x^2\cdots x^n]$ such that*

$$\operatorname{Im} F_{n+1} = \operatorname{Im} F_n \cup \operatorname{Im} \dot F.$$

We now use this claim to prove the theorem for general curves.

Assume $\alpha\colon [0,\ell] \to \mathcal{U}$ is an arbitrary closed curve with natural parameter. Choose a sequence of partitions $0 = t_n^0 < t_n^2 < \dots < t_n^n = \ell$ so that:

- the set $\{t_{n+1}^i\}_{i=0}^{n+1}$ is obtained from the set $\{t_n^i\}_{i=0}^n$ by adding one element.
- for a sequence $\varepsilon_n \to 0+$, we have $t_n^i - t_n^{i-1} < \varepsilon_n$ for all i.

Inscribe in α a sequence of polygons P_n with vertexes $\alpha(t_n^i)$. Apply the claim above, to get a sequence of majorizing maps $F_n\colon R_n \to \mathcal{U}$. Note that for all $m > n$, we have

- $\operatorname{Im} F_m$ lies in an ε_n-neighborhood of $\operatorname{Im} F_n$,
- $\operatorname{Im} F_m \setminus \operatorname{Im} F_n$ lies in an ε_n-neighborhood of α.

It follows that the set

$$K = \alpha \cup \left(\bigcup_n \operatorname{Im} F_n\right)$$

is compact. Therefore the sequence (F_n) has a partial limit as $n \to \infty$; say F. Clearly F is a majorizing map for α. □

If $p_1 \cdots p_n$ is a polygon, then values $\theta_i = \pi - \measuredangle\left[p_i\,{}^{p_{i-1}}_{p_{i+1}}\right]$ for all $i \pmod n$ are called *external angles* of the polygon. The following exercise is a generalization of Fenchel's theorem.

9.59. Exercise. Show that the sum of external angles of any polygon in a complete length CAT(0) space cannot be smaller than $2\cdot\pi$.

9.60. Very advanced exercise. *Suppose that a simple polygon β in a complete length* CAT(0) *space does not bound an embedded disc. Show that the sum of external angles of β cannot be smaller than $4\cdot\pi$.*

Give an example of such a polygon β with the sum of external angles exactly $4\cdot\pi$.

The following exercise is the rigidity case of the majorization theorem.

9.61. Exercise. Let $\mathcal{U}$ be a ϖ^κ-geodesic CAT(κ) space, and let $\alpha\colon [0,\ell] \to \mathcal{U}$ be a closed curve with arclength parametrization. Assume that $\ell < 2\cdot\varpi^\kappa$ and there is a closed convex curve $\tilde{\alpha}\colon [0,\ell] \to \mathbb{M}^2(\kappa)$ such that

$$|\alpha(t_0) - \alpha(t_1)|_{\mathcal{U}} = |\tilde{\alpha}(t_0) - \tilde{\alpha}(t_1)|_{\mathbb{M}^2(\kappa)}$$

for any t_0 and t_1. Then there is a distance-preserving map $F\colon \operatorname{Conv}\tilde{\alpha} \to \mathcal{U}$ such that $F\colon \tilde{\alpha}(t) \mapsto \alpha(t)$ for any t.

9.62. Exercise. Two majorizations $F\colon D \to \mathcal{U}$ and $F'\colon D' \to \mathcal{U}$ will be called *equivalent* if $F' = F \circ \iota$ for an isometry $\iota\colon D \to D'$.

Show that a closed rectifiable curve in a CAT(0) space has an isometric majorization map if and only if the majorization map is unique up to equivalence.

The following lemma states, in particular, that in a CAT(κ) space, a sharp triangle comparison implies the presence of an isometric copy of the convex hull of the model triangle. The latter statement was proved by Alexandr Alexandrov [**17**].

9.63. Arm lemma. *Let $\mathcal{U}$ be a ϖ^κ-geodesic* CAT(κ) *space, and let $P = [x^0x^1\cdots x^{n+1}]$ be a polygon of length $< 2\cdot\varpi^\kappa$ in $\mathcal{U}$. Suppose $\tilde{P} = [\tilde{x}^0\tilde{x}^1\cdots\tilde{x}^{n+1}]$ is a convex polygon in $\mathbb{M}^2(\kappa)$ such that*

$$(3)\qquad |\tilde{x}^i - \tilde{x}^{i-1}|_{\mathbb{M}^2(\kappa)} = |x^i - x^{i-1}|_{\mathcal{U}} \quad \text{and} \quad \measuredangle\left[x^i\,{}^{x^{i-1}}_{x^{i+1}}\right] \geqslant \measuredangle\left[\tilde{x}^i\,{}^{\tilde{x}^{i-1}}_{\tilde{x}^{i+1}}\right]$$

for all i. Then

(a) $|\tilde{x}^0 - \tilde{x}^{n+1}|_{\mathbb{M}^2(\kappa)} \leqslant |x^0 - x^{n+1}|_{\mathcal{U}}$.

(b) *Equality holds in* (a) *if and only if the map $\tilde{x}^i \mapsto x^i$ can be extended to a distance-preserving map of* $\operatorname{Conv}(\tilde{x}^0, \tilde{x}^1 \dots \tilde{x}^{n+1})$ *onto* $\operatorname{Conv}(x^0, x^1 \dots x^{n+1})$.

Proof.

(a). By Majorization Theorem 9.56, P is majorized by a convex region $\tilde{D}$ in $\mathbb{M}^2(\kappa)$. By Proposition 9.54 and the definition of angle, $\tilde{D}$ is bounded by a convex polygon $\tilde{P}_R = [\tilde{y}^0\tilde{y}^1\cdots\tilde{y}^{n+1}]$ that satisfies

$$|\tilde{y}^i - \tilde{y}^{i\pm1}|_{\mathbb{M}^2(\kappa)} = |x^i - x^{i\pm1}|_{\mathcal{U}}, \qquad |\tilde{y}^0 - \tilde{y}^{n+1}|_{\mathbb{M}^2(\kappa)} = |x^0 - x^{n+1}|_{\mathcal{U}},$$

$$\measuredangle\left[\tilde{y}^i\,{}^{\tilde{y}^{i-1}}_{\tilde{y}^{i+1}}\right] \geqslant \measuredangle\left[x^i\,{}^{x^{i-1}}_{x^{i+1}}\right] \geqslant \measuredangle\left[\tilde{x}^i\,{}^{\tilde{x}^{i-1}}_{\tilde{x}^{i+1}}\right]$$

for $1 \leqslant i \leqslant n$; the last inequality follows from (3).

The classical arm lemma [**145**] gives $|\tilde{x}^0 - \tilde{x}^{n+1}| \leqslant |\tilde{y}^0 - \tilde{y}^{n+1}|$. Since $|\tilde{y}^0 - \tilde{y}^{n+1}| = |x^0 - x^{n+1}|$, part (a) follows.

(b). Suppose equality holds in (a). Then angles at the jth vertex of $\tilde{P}$, P, and $\tilde{P}_R$ are equal for $1 \leqslant j \leqslant n$, and we may take $\tilde{P} = \tilde{P}_R$.

Let $F\colon \tilde{D} \to \mathcal{U}$ be the majorizing map for P, where $\tilde{D}$ is the convex region bounded by $\tilde{P}$, and $F|_{\tilde{P}}$ is length-preserving.

(4) *Let $\tilde{x}, \tilde{y}, \tilde{z}$ be three vertexes of $\tilde{P}$, and let x, y, z be the corresponding vertexes of P. If $|\tilde{x} - \tilde{y}| = |x - y|$, $|\tilde{y} - \tilde{z}| = |x - z|$ and $\measuredangle\left[\tilde{y}\,{}^{\tilde{x}}_{\tilde{z}}\right] = \measuredangle\left[y\,{}^{x}_{z}\right]$. Then $F|_{\mathrm{Conv}(\tilde{x},\tilde{y},\tilde{z})}$ is distance-preserving.*

Indeed, since F is majorizing, F restricts to distance-preserving maps from $[\tilde{x}\tilde{y}]$ to $[xy]$ and $[\tilde{y}\tilde{z}]$ to $[yz]$. Suppose $\tilde{p} \in [\tilde{x}\tilde{y}]$ and $\tilde{q} \in [\tilde{y}\tilde{z}]$. Then

(5) $$|\tilde{p} - \tilde{q}|_{\mathbb{M}^2(\kappa)} = |F(\tilde{p}) - F(\tilde{q})|_{\mathcal{U}}.$$

This inequality holds in one direction by majorization, and in the other direction by the angle comparison (see Theorem 9.14(c)). By the first variation formula (see 9.37), it follows that each pair of corresponding angles of triangles $[\tilde{x}\tilde{y}\tilde{z}]$ and $[xyz]$ are equal. But then (5) holds for p, q on any two sides of these triangles, so F is distance-preserving on every geodesic of $\mathrm{Conv}(\tilde{p}, \tilde{x}, \tilde{y})$. Hence the claim. △

(6) *Suppose $F|_{\mathrm{Conv}(\tilde{x}^0,\tilde{x}^1,\dots,\tilde{x}^k)}$ is distance-preserving for $2 \leqslant k \leqslant n - 1$. Then the restriction $F|_{\mathrm{Conv}(\tilde{x}^0,\tilde{x}^1,\dots,\tilde{x}^{k+1})}$ is distance-preserving.*

To verify this claim, let

$$\tilde{p} = [\tilde{x}^{k-1}\tilde{x}^{k+1}] \cap [\tilde{x}^k\tilde{x}^0] \quad \text{and} \quad p = F(p).$$

Note that the following maps are distance-preserving:

(i) $F|_{\mathrm{Conv}(\tilde{x}^{k-1},\tilde{x}^k,\tilde{x}^{k+1})}$,

(ii) $F|_{\mathrm{Conv}(\tilde{x}^{k+1},\tilde{x}^{k-1},\tilde{x}^0)}$,

(iii) $F|_{\mathrm{Conv}(\tilde{x}^0,\tilde{x}^k,\tilde{x}^{k+1})}$.

Indeed, (i) follows from (4). Therefore $|\tilde{x}^{k-1} - \tilde{x}^{k+1}| = |x^{k-1} - x^{k+1}|$, and so F restricts to a distance-preserving map from $[\tilde{x}^{k-1}\tilde{x}^{k+1}]$ onto $[x^{k-1}x^{k+1}]$. With the induction hypothesis in (6), it follows that $p = [x^{k-1}x^{k+1}] \cap [x^k x^0]$, hence

(7) $$\measuredangle\left[\tilde{x}^{k-1}\,{}^{\tilde{x}^{k+1}}_{\tilde{x}^0}\right] = \measuredangle\left[x^{k-1}\,{}^{x^{k+1}}_{x^0}\right].$$

Then (ii) follows from (7) and (4). Since $|\tilde{x}^k - \tilde{x}^0| = |x^k - x^0|$, (iii) follows from (7) and (i).

Let $\tilde{\gamma}$ be a geodesic of $\mathrm{Conv}(\tilde{x}^0, \tilde{x}^0, \tilde{x}^1 \cdots \tilde{x}^{k+1})$. Then $\operatorname{length}\tilde{\gamma} < \varpi^\kappa$. If $\tilde{\gamma}$ does not contain the point $\tilde{p}$, then by the induction hypothesis in (6) and (i) + (ii) + (iii), we get that $\gamma = F \circ \tilde{\gamma}$ is a local geodesic of length $< \varpi^\kappa$. By Corollary 9.22, γ is a geodesic.

By continuity, $F \circ \tilde{\gamma}$ is a geodesic for all $\tilde{\gamma}$; so (6) follows.

The base of the induction is provided by (4). It finishes the proof of part (b). □

9.64. Exercise. Let $\mathcal{U}$ be a complete length CAT(0) space and suppose for four points $x^1, x^2, x^3, x^4 \in \mathcal{U}$ there is a convex quadrangle $[\tilde{x}^1\tilde{x}^2\tilde{x}^3\tilde{x}^4]$ in $\mathbb{E}^2$ such that

$$|x^i - x^j|_{\mathcal{U}} = |\tilde{x}^i - \tilde{x}^j|_{\mathbb{E}^2}$$

for all i and j. Show that $\mathcal{U}$ contains an isometric copy of *solid quadrangle* $[\tilde{x}^1\tilde{x}^2\tilde{x}^3\tilde{x}^4]$; that is, the convex hull of $\tilde{x}^1, \tilde{x}^2, \tilde{x}^3, \tilde{x}^4$ in $\mathbb{E}^2$.

M. Hadamard–Cartan theorem

The development of Alexandrov geometry was greatly influenced by the Hadamard–Cartan theorem. Its original formulation states that if M is a complete Riemannian manifold with nonpositive sectional curvature, then the exponential map at any point $p \in M$ is a covering; in particular it implies that the universal cover of M is diffeomorphic to the Euclidean space of the same dimension.

In this generality, the theorem appeared in the lectures of Élie Cartan [**47**]. For surfaces in the Euclidean space, the theorem was proved by Hans von Mangoldt [**113**], and a few years later independently by Jacques Hadamard [**79**].

Formulations for metric spaces of different generality were proved by Herbert Busemann [**45**], Willi Rinow [**143**], and Michael Gromov [**76**, p. 119]. A detailed proof of Gromov's statement when $\mathcal{U}$ is proper was given by Werner Ballmann [**20**], using Birkhoff's curve-shortening. A proof in the nonproper geodesic case was given by the first author and Richard Bishop [**5**]. This proof applies more generally, to *convex spaces* (see Exercise 9.81). It was pointed out by Bruce Kleiner [**21**] and independently by Martin Bridson and André Haefliger [**34**] that this proof extends to length spaces, as well as geodesic spaces, giving the following statement:

9.65. Hadamard–Cartan theorem. *Let $\kappa \leqslant 0$, and $\mathcal{U}$ be a complete, simply connected length locally* CAT(κ) *space. Then $\mathcal{U}$ is* CAT(κ).

Proof. Since $\varpi^\kappa = \infty$, Theorem 9.50 implies that there is a CAT(κ) space $\mathcal{B}$ and a *metric covering* $\Phi\colon \mathcal{B} \to \mathcal{U}$; that is, Φ is a length-preserving covering map.

Since $\mathcal{U}$ is simply connected, $\Phi\colon \mathcal{B} \to \mathcal{U}$ is an isometry—hence the result. □

To formulate the generalized Hadamard–Cartan theorem, we need the following definition.

9.66. Definition. Given $\ell \in (0, \infty]$, a metric space $\mathcal{X}$ is called ℓ-simply connected if it is connected and any closed curve of length $< \ell$ is null-homotopic in the class of curves of length $< \ell$ in $\mathcal{X}$.

Note that there is a subtle difference between simply connected and ∞-simply connected spaces; the first states that any closed curve is null-homotopic while the second means that any rectifiable curve is null-homotopic in the class of rectifiable curves. However, as follows from Proposition 9.69, for locally CAT(κ) spaces these two definitions are equivalent. This fact makes it possible to deduce the Hadamard–Cartan theorem directly from the generalized Hadamard–Cartan theorem.

9.67. Generalized Hadamard–Cartan theorem. *A complete length space $\mathcal{U}$ is* CAT(κ) *if and only if $\mathcal{U}$ is $2\cdot\varpi^\kappa$-simply connected and $\mathcal{U}$ is locally* CAT(κ).

For proper spaces, the generalized Hadamard–Cartan theorem was proved by Brian Bowditch [**31**]. In the proof we need the following lemma.

9.68. Lemma. *Let $\mathcal{U}$ be a complete length locally* CAT(κ) *space, $\varepsilon > 0$, and $\gamma_1, \gamma_2 : \mathbb{S}^1 \to \mathcal{U}$ be two closed curves. Assume*

(a) length $\gamma_i < 2\cdot\varpi^\kappa - 4\cdot\varepsilon$ *for $i = 1, 2$;*

(b) $|\gamma_1(x) - \gamma_2(x)| < \varepsilon$ *for any $x \in \mathbb{S}^1$, and the geodesic $[\gamma_1(x)\gamma_2(x)]$ is uniquely defined and depends continuously on x;*

(c) *γ_1 is majorized by a convex region in $\mathbb{M}^2(\kappa)$.*

Then γ_2 is majorized by a convex region in $\mathbb{M}^2(\kappa)$.

Proof. Let D be a convex region in $\mathbb{M}^2(\kappa)$ that majorizes γ_1 under the map $F : D \to \mathcal{U}$ (see Definition 9.53). Denote by $\tilde{\gamma}_1$ the curve bounding D such that $F \circ \tilde{\gamma}_1 = \gamma_1$. Since

$$\begin{aligned}\operatorname{length}\tilde{\gamma}_1 &= \operatorname{length}\gamma_1\\ &< 2\cdot\varpi^\kappa - 4\cdot\varepsilon,\end{aligned}$$

there is a point $\tilde{p} \in D$ such that $|\tilde{p} - \tilde{\gamma}(x)|_{\mathbb{M}^2(\kappa)} < \frac{\varpi^\kappa}{2} - \varepsilon$ for any $x \in \mathbb{S}^1$. Denote by α_x the concatenation of the paths $F \circ \mathrm{path}_{[p\tilde{\gamma}_1(x)]_{\mathbb{M}^2(\kappa)}}$ and $\mathrm{path}_{[\gamma_1(x)\gamma_2(x)]}$ in $\mathcal{U}$. Note that α_x depends continuously on x, and

$$\operatorname{length}\alpha_x < \tfrac{\varpi^\kappa}{2} \quad\text{and}\quad \alpha_x(1) = \gamma_2(x)$$

hold for any x.

Let us apply Lifting Globalization Theorem 9.50 for $p = F(\tilde{p})$. We obtain a ϖ^κ-geodesic CAT(κ) space $\mathcal{B}$ and a locally distance-preserving map $\Phi : \mathcal{B} \to \mathcal{U}$ with $\Phi(\hat{p}) = p$ for some $\hat{p} \in \mathcal{B}$, and with the lifting property for the curves starting at p with length $< \varpi^\kappa/2$. Applying the lifting property for α_x, we get existence of a curve $\hat{\gamma}_2 : \mathbb{S}^1 \to \mathcal{B}$ such that

$$\gamma_2 = \Phi \circ \hat{\gamma}_2.$$

Since $\mathcal{B}$ is a geodesic CAT(κ) space, we can apply Majorization Theorem 9.56 for $\hat{\gamma}_2$. The composition of the obtained majorization with Φ is a majorization of γ_2. □

Proof of Theorem 9.67. The only-if part follows from Reshetnyak Majorization Theorem 9.56.

Let γ_t, $t \in [0,1]$ be a null-homotopy of curves in $\mathcal{U}$; that is, $\gamma_0(x) = p$ for some $p \in \mathcal{U}$ and any $x \in \mathbb{S}^1$. Assume further that length $\gamma_t < 2\cdot\varpi^\kappa$ for any t. To prove the if part, it is sufficient to show that γ_1 is majorized by a convex region in $\mathbb{M}^2(\kappa)$ if $\mathcal{U}$ is locally CAT(κ).

By semicontinuity of length (see 2.6), we can choose $\varepsilon > 0$ sufficiently small that

$$\operatorname{length}\gamma_t < 2\cdot\varpi^\kappa - 4\cdot\varepsilon$$

for all t.

By Corollary 9.51, we may assume in addition that $\mathrm{B}(\gamma_t(x), \varepsilon)$ is CAT(κ) for any t and x.

Choose a partition $0 = t_0 < t_1 < \dots < t_n = 1$ so that $|\gamma_{t_i}(x) - \gamma_{t_{i-1}}(x)| < \varepsilon$ for any i and x. According to 9.8, for any i, the geodesic $[\gamma_{t_i}(x)\gamma_{t_{i-1}}(x)]$ depends continuously on x.

Note that $\gamma_0 = \gamma_{t_0}$ is majorized by a convex region in $\mathbb{M}^2(\kappa)$. Applying the lemma n times, we see that the same holds for $\gamma_1 = \gamma_{t_n}$. □

9.69. Proposition. *Let $\mathcal{U}$ be a complete length locally* CAT(κ) *space. Then $\mathcal{U}$ is simply connected if and only if it is ∞-simply connected.*

Proof.

If part. It is sufficient to show that any closed curve in $\mathcal{U}$ is homotopic to a polygon.

Let γ_0 be a closed curve in $\mathcal{U}$. According to Corollary 9.51, there is $\varepsilon > 0$ such that $\mathrm{B}(\gamma(x), \varepsilon)$ is CAT(κ) for any x.

Choose a polygon γ_1 such that $|\gamma_0(x) - \gamma_1(x)| < \varepsilon$ for any x. By 9.8, $\mathrm{path}_{[\gamma_0(x)\gamma_1(x)]}$ is uniquely defined and depends continuously on x.

Hence $\gamma_t(x) = \mathrm{path}_{[\gamma_0(x)\gamma_1(x)]}(t)$ gives a homotopy from γ_0 to γ_1.

Only-if part. The proof is similar.

Assume γ_t is a homotopy between two rectifiable curves γ_0 and γ_1. Fix $\varepsilon > 0$ so that the ball $\mathrm{B}(\gamma_t(x), \varepsilon)$ is CAT(κ) for any t and x. Choose a partition $0 = t_0 < t_1 < \dots < t_n = 1$ so that

$$|\gamma_{t_{i-1}}(x) - \gamma_{t_i}(x)| < \tfrac{\varepsilon}{10}$$

for any i and x. Set $\hat{\gamma}_{t_0} = \gamma_0$, $\hat{\gamma}_{t_n} = \gamma_{t_n}$. For each $0 < i < n$, approximate γ_{t_i} by a polygon $\hat{\gamma}_i$.

Construct the *geodesic homotopy* from $\hat{\gamma}_{t_{i-1}}$ to $\hat{\gamma}_{t_i}$; that is, set

$$\hat{\gamma}_t = \mathrm{path}_{[\hat{\gamma}_{t_{i-1}}(x)\hat{\gamma}_{t_i}(x)]}(t)$$

for $t \in [t_{i-1}, t_i]$. Since ε is sufficiently small, by Corollary 9.13, we get that

$$\mathrm{length}\,\hat{\gamma}_t < 10\cdot(\mathrm{length}\,\hat{\gamma}_{t_{i-1}} + \mathrm{length}\,\hat{\gamma}_{t_i})$$

for any $t \in [t_{i-1}, t_i]$. In particular, $\hat{\gamma}_t$ is rectifiable for all t.

Joining the obtained homotopies for all i, we obtain a homotopy from γ_0 to γ_1 in the class of rectifiable curves. □

9.70. Exercise. Let $\mathcal{X}$ be a double cover of $\mathbb{E}^3$ that branches along two distinct lines ℓ and m. Show that $\mathcal{X}$ is CAT(0) if and only if ℓ intersects m at a right angle.

9.71. Exercise. Let $\mathcal{U}$ be a complete length CAT(0) space. Assume $\tilde{\mathcal{U}} \to \mathcal{U}$ is a metric covering branching along a geodesic. Show that $\tilde{\mathcal{U}}$ is CAT(0).

More generally, assume $A \subset \mathcal{U}$ is a closed convex subset and $f : \mathcal{X} \to \to \mathcal{U} \setminus A$ is a metric cover. Denote by $\bar{\mathcal{X}}$ the completion of $\mathcal{X}$, and $\bar{f} : \bar{\mathcal{X}} \to \mathcal{U}$ the continuous extension of f. Let $\tilde{\mathcal{U}}$ be the space glued from $\bar{\mathcal{X}}$ and A by identifying x and $\bar{f}(x)$ if $\bar{f}(x) \in A$. Show that $\tilde{\mathcal{U}}$ is CAT(0).

9.72. Exercise. Let $\mathcal{K}$ be a compact geodesic, locally CAT(κ) that is not CAT(κ). Prove that $\mathcal{K}$ contains a *geodesic circle* γ of length less that $2 \cdot \varpi^\kappa$; that is, γ is a simple closed curve such that for any two points $p, q \in \gamma$, one of the arcs of γ with endpoints p and q is a geodesic.

N. Convex sets

Recall that according to Corollary 9.27, any ball (closed or open) of radius $R < < \frac{\varpi^\kappa}{2}$ in a ϖ^κ-geodesic CAT(κ) space is convex. From the uniqueness of geodesics in CAT(κ) spaces (see 9.8) we get the following:

9.73. Observation. Any weakly ϖ^κ-convex set in a complete length CAT(κ) space is ϖ^κ-convex.

9.74. Closest-point projection lemma. *Let p be a point in a complete length* CAT(κ) *space $\mathcal{U}$, and let $K \subset \mathcal{U}$ be a closed ϖ^κ-convex set. If $\mathrm{dist}_K\, p < \frac{\varpi^\kappa}{2}$, then there is a unique point $p^* \in K$ that minimizes the distance to p; that is, $|p^* - p| = \mathrm{dist}_K\, p$.*

Proof. Fix r properly between $\mathrm{dist}_K\, p$ and $\frac{\varpi^\kappa}{2}$. By the function comparison (see 9.25), the function $f = \mathrm{md}^\kappa \circ \mathrm{dist}_p$ is strongly convex in $\overline{\mathrm{B}}[p, r]$.

The lemma follows from Lemma 14.4 applied to the subspace $K' = K \cap \cap \overline{\mathrm{B}}[p, r]$ and the restriction $f|_{K'}$. □

9.75. Exercise. Let $\mathcal{U}$ be a complete length CAT(0) space and $K \subset \mathcal{U}$ be a closed convex set. Show that the closest-point projection $\mathcal{U} \to K$ is short.

9.76. Advanced exercise. Let $\mathcal{U}$ be a complete length CAT(1) space, and let $K \subset \mathcal{U}$ be a closed π-convex set. Assume $K \subset \overline{\mathrm{B}}[p, \frac{\pi}{2}]$ for $p \in K$. Show that there is a short retraction of $\mathcal{U}$ to K.

9.77. Proposition. *Let $\mathcal{U}$ be a ϖ^κ-geodesic* CAT(κ) *space, and let $K \subset \mathcal{U}$ be a closed ϖ^κ-convex set. Let*

$$f = \operatorname{sn}^\kappa \circ \operatorname{dist}_K .$$

Then

$$f'' + \kappa \cdot f \geqslant 0$$

holds in $\mathrm{B}(K, \frac{\varpi^\kappa}{2})$.

Proof. It is sufficient to show that Jensen's inequality (see 3.14(c)) holds on a sufficiently short geodesic $[pq]$ in $\mathrm{B}(K, \frac{\varpi^\kappa}{2})$. We may assume that

(1) $$|p - q| + \operatorname{dist}_K p + \operatorname{dist}_K q < \varpi^\kappa.$$

For each $x \in [pq]$, we need to find a value $h(x) \in \mathbb{R}$ such that

$$h(p) = f(p), \qquad h(q) = f(q), \qquad h(x) \leqslant f(x)$$

for any $x \in [pq]$, and

(2) $$h'' + \kappa \cdot h \geqslant 0$$

along $[pq]$.

Denote by p^* and q^* the closest-point projections of p and q on K; they are provided by Closest Point Projection Lemma 9.74. From (1) and the triangle inequality, we have

$$|p^* - q^*| < \varpi^\kappa.$$

Since K is ϖ^κ-convex, $K \supset [p^*q^*]$; in particular

$$\operatorname{dist}_K x \leqslant \operatorname{dist}_{[p^*q^*]} x$$

for any $x \in \mathcal{U}$.

There is a majorizing map $F : D \to \mathcal{U}$ for quadrangle $[pp^*q^*q]$, as in Definition 9.53 and Reshetnyak Majorization Theorem 9.56. By Proposition 9.54, the figure D is a solid convex quadrangle $[\tilde{p}\tilde{p}^*\tilde{q}^*\tilde{q}]$ in $\mathbb{M}^2(\kappa)$ such that

$$|\tilde{p} - \tilde{p}^*|_{\mathbb{M}^2(\kappa)} = |p - p^*|_{\mathcal{U}}, \qquad |\tilde{p} - \tilde{q}|_{\mathbb{M}^2(\kappa)} = |p - q|_{\mathcal{U}},$$
$$|\tilde{q} - \tilde{q}^*|_{\mathbb{M}^2(\kappa)} = |q - q^*|_{\mathcal{U}}, \qquad |\tilde{p}^* - \tilde{q}^*|_{\mathbb{M}^2(\kappa)} = |p^* - q^*|_{\mathcal{U}}.$$

Given $x \in [pq]$, denote by $\tilde{x}$ the corresponding point on $[\tilde{p}\tilde{q}]$. Then

$$\operatorname{dist}_{[pq]} x \leqslant \operatorname{dist}_{[\tilde{p}\tilde{q}]} \tilde{x}.$$

Set

$$h(x) = \operatorname{sn}^\kappa \circ \operatorname{dist}_{[\tilde p \tilde q]} \tilde x.$$

By straightforward calculations, (2) holds and hence the statement follows. □

9.78. Corollary. *Let $\mathcal{U}$ be a complete length* CAT(κ) *space, and let $K \subset \mathcal{U}$ be a closed locally convex set. Then there is an open set $\Omega \supset K$ such that the function $f = \operatorname{sn}^\kappa \circ \operatorname{dist}_K$ satisfies*

$$f'' + \kappa \cdot f \geqslant 0$$

in Ω.

Proof. Fix $p \in K$. By Corollary 9.27, $\overline{\mathrm{B}}[p, r]$ is convex for all small $r > 0$.

Since K is locally convex, there is $r_p > 0$ such that the intersection $K' = K \cap \mathrm{B}(p, r_p)$ is convex.

Note that

$$\operatorname{dist}_K x = \operatorname{dist}_{K'} x$$

for any $x \in \mathrm{B}(p, \frac{r_p}{2})$. Therefore the statement holds for

$$\Omega = \bigcup_{p \in K} \mathrm{B}(p, \tfrac{r_p}{2}).$$

□

9.79. Theorem. *Assume $\mathcal{U}$ is a complete length* CAT(κ) *space, and $K \subset \mathcal{U}$ is a closed connected locally convex set. Assume $|x - y| < \varpi^\kappa$ for any $x, y \in K$. Then K is convex.*

In particular, if $\kappa \leqslant 0$, then any closed connected locally convex set in $\mathcal{U}$ is convex.

The following proof is due to Sergei Ivanov [**88**].

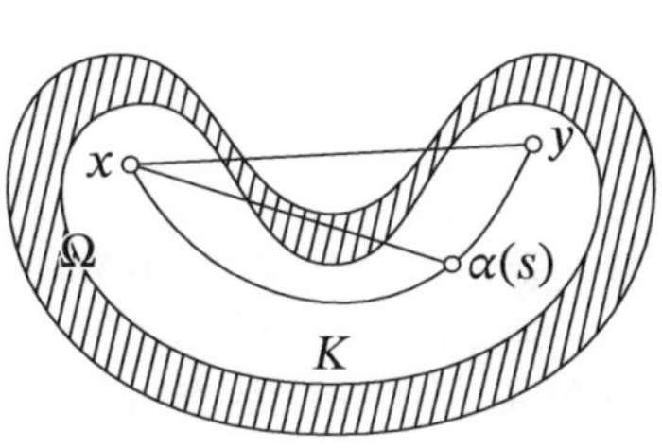

Proof. Since K is locally convex, it is locally path-connected. Since K is connected and locally path connected it is path-connected.

Fix two points $x, y \in K$. Let us connect x to y by a path $\alpha\colon [0,1] \to K$. Since $|x - \alpha(s)| < \varpi^\kappa$ for any s, Theorem 9.8 implies that the geodesic $[x\alpha(s)]$ is uniquely defined and depends continuously on s.

Let $\Omega \supset K$ be the open set provided by Corollary 9.78. If $[xy] = [x\alpha(1)]$ does not completely lie in K, then there is a value $s \in [0,1]$ such that $[x\alpha(s)]$ lies in Ω but does not completely lie in K. By Corollary 9.78, the function $f = \operatorname{sn}^\kappa \circ \operatorname{dist}_K \mathcal{U}$ satisfies the differential inequality

$$f'' + \kappa \cdot f \geqslant 0 \tag{3}$$

along $[x\alpha(s)]$.

Since

$$|x - \alpha(s)| < \varpi^\kappa, \qquad f(x) = f(\alpha(s)) = 0,$$

then the barrier inequality (see 3.14(b)) implies that $f(z) \leqslant 0$ for $z \in [x\alpha(s)]$; that is $[x\alpha(s)] \subset K$, a contradiction. □

O. Remarks

The following question was known in folklore in the 1980s, but it seems that in print it was first mentioned by Michael Gromov [**74**, 6.B$_1$(f)]. We do not see any reason why it should be true, but we also cannot construct a counterexample.

9.80. Open question. Let $\mathcal{U}$ be a complete length CAT(0) space, and let $K \subset \mathcal{U}$ be a compact set. Is it true that K lies in a convex compact set $\bar{K} \subset \mathcal{U}$?

The question can easily be reduced to the case when K is finite; so far it is not even known if any three points in a complete length CAT(0) space lie in a compact convex set.

We expect that 9.47 can be extended to all curves, not necessarily local geodesics, but the proof does not admit a straightforward generalization.

About convex spaces. A *convex space* $\mathcal{X}$ is a geodesic space such that the function $t \mapsto |\gamma(t) - \sigma(t)|$ is convex for any two geodesic paths $\gamma, \sigma : [0,1] \to \mathcal{X}$. A *locally convex space* is a length space in which every point has a neighborhood that is a convex space in the restricted metric.

9.81. Exercise. Assume $\mathcal{X}$ is a convex space such that the angle of any hinge is defined. Show that $\mathcal{X}$ is CAT(0).

The following exercise gives an analogue of the Hadamard–Cartan theorem for locally convex spaces; see also [**5**].

9.82. Exercise. Show that a complete, simply connected, locally convex space is a convex space.

Examples and constructions. Let us list important sources of examples of CAT spaces. This should be beneficial to the reader despite that we do not provide all the proofs and some proofs are deferred to later chapters.

Riemannian manifolds with sectional curvature at most κ are locally CAT(κ). This statement follows from the Rauch comparison, and it is one of the main motivations for CAT(κ) comparison. *Hilbert spaces* are another motivating example of CAT(0) spaces.

The question of when a Riemannian *manifold-with-boundary* is locally CAT(κ) has been completely answered by the first author, David Berg, and Richard Bishop [**1**]. If the sectional curvatures of the interior and the outward sectional curvatures of the boundary do not exceed κ, then it is locally CAT(κ)

(where an outward sectional curvature is one that corresponds to a tangent 2-plane all of whose normal curvature vectors point outward). In particular, if a Riemannian manifold and its boundary have sectional curvature at most κ, then it is locally CAT(κ).

Subsets of *positive reach* in Riemannian manifolds were studied in this context by the first author and Richard Bishop [**2**] and by Alexander Lytchak [**112**]. In particular, any compact subset of positive reach in a Riemannian manifold is CAT; as usual, we assume that it is equipped with the induced length metric.

It was shown by Alexander Lytchak and Stephan Stadler [**109**] that any *simply connected subset* of a contractible two-dimensional CAT(κ) length space (equipped with induced length metric) is CAT(κ). In higher dimensions things are more complicated, even for a Euclidean ambient space; see [**10**, Chapter 4] for related questions (open and solved).

By the Gauss formula smooth saddle surfaces in manifolds with sectional curvature at most κ are locally CAT(κ). The nonsmooth analogue of this statement is wide open; this is the so-called Shefel conjecture [**10**, Chapter 4]. However, the following weaker statement was proved by the third author and Stephan Stadler [**128**]: *metric-minimizing surfaces* in CAT(κ) space are locally CAT(κ); metric minimizing means that it is impossible to decrease its length metric by a small deformation.

Note that any *metric tree* (see Section 4.C) is CAT($-\infty$); that is, a metric tree is CAT(κ) for *any* $\kappa \in \mathbb{R}$. In particular, any tree (in the graph-theoretical sense) with a length metric such that every edge is isometric to a line segment is CAT($-\infty$).

Suppose $A^1, \dots, A^n$ is an array of convex closed sets in the Euclidean space $\mathbb{E}^m$. Let us prepare $n+1$ copy of $\mathbb{E}^m$ and glue successive pairs of spaces along $A^1, \dots, A^n$. The resulting space is called *puff pastry*; by the Reshetnyak gluing theorem it is CAT(0). (This observation was used in one of the most beautiful applications of the Reshetnyak Gluing Theorem given by Dmitri Burago, Serge Ferleger, and Alexey Kononenko [**38–41**]. They use it to study billiards; a short survey on the subject was written by Dmitri Burago [**36**]; see also our book [**10**].)

An if-and-only-if condition on *polyhedral spaces* is given in Theorem 12.2. It implies the so-called Gromov Flag Condition 12.10 which provides in particular a flexible way to construct CAT(0) *cube complexes*. (Several applications are mentioned in Section 12.D.)

By Proposition 9.7 and Theorem 5.16, the *ultralimits*, as well as *Gromov–Hausdorff limits* of CAT(κ) spaces are CAT(κ). In particular, if $\mathcal{U}$ is a CAT(0) space, then its *asymptotic cone* defined as the ultralimit of its rescalings $\frac{1}{n}\cdot\mathcal{U}$ as $n \to \omega$ is again CAT(0). Unlike in the case of CBB(0) spaces, the use of ultralimits is necessary even if $\mathcal{U}$ is a manifold due to the lack of compactness theorem. Asymptotic cones of CAT(0) spaces and their generalizations play an important role in geometric group theory.

Conformal deformations of CAT spaces were studied by Alexander Lytchak and Stephan Stadler [**108**]. In particular, if $\mathcal{U}$ is a CAT(0) space and $f : \mathcal{U} \to \mathbb{R}$ is continuous, convex and bounded below, then the conformally equivalent space with conformal factor e^f is CAT(0).

Further, CAT spaces behave nicely with respect to some natural constructions. For example, the product of CAT(0) spaces is again CAT(0). Also, the Euclidean cone over a CAT(1) space is CAT(0). These are the first examples of the so-called *warped products* that are discussed in Chapter 11; a general statement is given in Theorem 11.11. Also, as it was observed by Karl-Theodor Sturm [**155**, Prop. 3.10], the *space of L^2-maps* from a measure space to a complete CAT(0) length space is CAT(0).

Among more conceptual examples, let us mention a result of Brian Clarke [**54**]: the *space of Riemannian metrics* on a compact, orientable smooth manifold with respect to the L^2-distance is CAT(0); a shorter proof of this statement was given by Nicola Cavallucci [**48**]. By a result of Tamás Darvas [**57**], the *space of Kähler potentials* on a compact Kähler manifold is CAT(0). The *Teichmüller space* with the Weil–Petersson metric is CAT(0); the latter was shown by Sumio Yamada [**161**].

Chapter 10

Kirszbraun revisited

This chapter is based on our paper [**8**] and an earlier paper of Urs Lang and Viktor Schroeder [**99**].

A. Short map extension definitions

10.1. Theorem. *A complete length space $\mathcal{L}$ is* CBB(κ) *if and only if for any three-point set V_3 and any four-point set $V_4 \supset V_3$ in $\mathcal{L}$, any short map $f\colon V_3 \to \mathbb{M}^2(\kappa)$ can be extended to a short map $F\colon V_4 \to \mathbb{M}^2(\kappa)$ (so $f = F|_{V_3}$).*

The only-if part of Theorem 10.1 can be obtained as a corollary of Kirszbraun's Theorem 10.14. We present another, more elementary proof; using the following analogue of Alexandrov's Lemma 6.3.

We say that two triangles with a common vertex *do not overlap* if their convex hulls intersect only at the common vertex.

10.2. Overlap lemma. *Let $[\tilde{x}^1\tilde{x}^2\tilde{x}^3]$ be a triangle in $\mathbb{M}^2(\kappa)$. Let $\tilde{p}^1$, $\tilde{p}^2$, $\tilde{p}^3$ be points in $\mathbb{M}^2(\kappa)$ such that, for any permutation $\{i, j, k\}$ of $\{1, 2, 3\}$, we have*

(i) $|\tilde{p}^i - \tilde{x}^k| = |\tilde{p}^j - \tilde{x}^k|$,

(ii) *$\tilde{p}^i$ and $\tilde{x}^i$ lie in the same closed half-space determined by $[\tilde{x}^j\tilde{x}^k]$,*

If no pair of triangles $[\tilde{p}^i\tilde{x}^j\tilde{x}^k]$ overlap, then

$$\measuredangle\tilde{p}^1 + \measuredangle\tilde{p}^2 + \measuredangle\tilde{p}^3 > 2\cdot\pi,$$

where $\measuredangle\tilde{p}^i := \measuredangle\left[\tilde{p}^i\,{}^{\tilde{x}^k}_{\tilde{x}^j}\right]$ for a permutation $\{i, j, k\}$ of $\{1, 2, 3\}$.

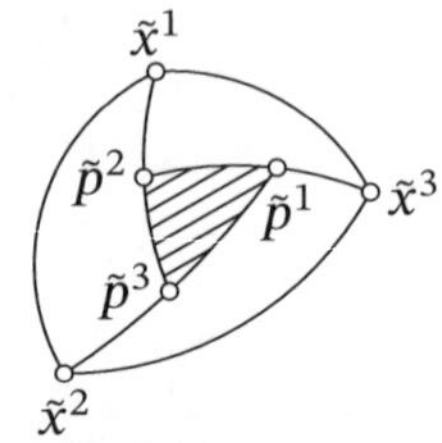

Remarks. If $\kappa\leqslant 0$, then the overlap lemma can be proved without using condition (i). This follows since the sum of external angles for the hexagon $[\tilde p^1\tilde x^2\tilde p^3\tilde x^1\tilde p^2\tilde x^3]$ and its area is $2\cdot\pi-\kappa\cdot a$, where a denotes the area of the hexagon.

The diagram shows that condition (i) is essential in case $\kappa>0$.

Proof. Rotate the triangle $[\tilde p^3\tilde x^1\tilde x^2]$ around $\tilde x^1$ to make $[\tilde x^1\tilde p^3]$ coincide with $[\tilde x^1\tilde p^2]$. Let $\dot x^2$ denote the image of $\tilde x^2$ after rotation. Note that

$$\measuredangle\left[\tilde x^1\,{}^{\tilde x^3}_{\dot x^2}\right]=\min\{\measuredangle\left[\tilde x^1\,{}^{\tilde x^2}_{\tilde p^3}\right]+\measuredangle\left[\tilde x^1\,{}^{\tilde p^2}_{\tilde x^3}\right],\,2\cdot\pi-(\measuredangle\left[\tilde x^1\,{}^{\tilde x^2}_{\tilde p^3}\right]+\measuredangle\left[\tilde x^1\,{}^{\tilde p^2}_{\tilde x^3}\right])\}.$$

By (ii), the triangles $[\tilde p^3\tilde x^1\tilde x^2]$ and $[\tilde p^2\tilde x^3\tilde x^1]$ do not overlap if and only if

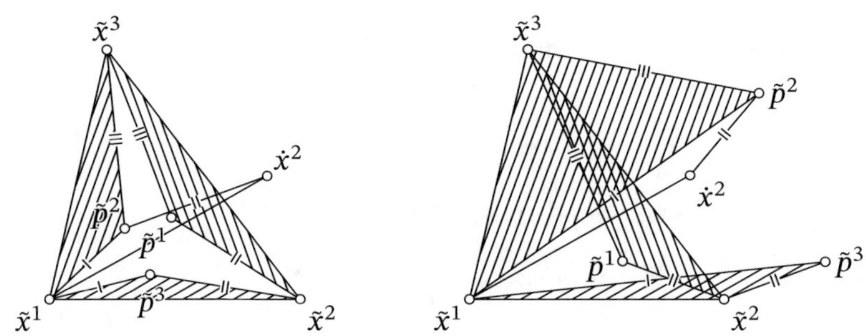

(1)
$$\measuredangle\left[\tilde x^1\,{}^{\tilde x^3}_{\tilde x^2}\right]>\measuredangle\left[\tilde x^1\,{}^{\tilde x^3}_{\dot x^2}\right]$$

and

(2)
$$2\cdot\pi>\measuredangle\left[\tilde x^1\,{}^{\tilde x^2}_{\tilde p^3}\right]+\measuredangle\left[\tilde x^1\,{}^{\tilde p^2}_{\tilde x^3}\right]+\measuredangle\left[\tilde x^1\,{}^{\tilde x^2}_{\tilde x^3}\right].$$

Condition (1) holds if and only if $|\tilde x^2-\tilde x^3|>|\dot x^2-\tilde x^3|$, which in turn holds if and only if

(3)
$$\begin{aligned}\measuredangle\tilde p^1&>\measuredangle\left[\tilde p^2\,{}^{\tilde x^3}_{\dot x^2}\right]\\&=\min\{\measuredangle\tilde p^3+\measuredangle\tilde p^2,2\cdot\pi-(\measuredangle\tilde p^3+\measuredangle\tilde p^2)\}.\end{aligned}$$

The inequality follows since the corresponding hinges have the same pairs of sidelengths. (The two pictures show that both possibilities for the minimum can occur.)

Now assume $\measuredangle\tilde p^1+\measuredangle\tilde p^2+\measuredangle\tilde p^3\leqslant 2\cdot\pi$. Then (3) implies

$$\measuredangle\tilde p^i>\measuredangle\tilde p^j+\measuredangle\tilde p^k.$$

Since no pair of triangles overlap, the same holds for any permutation (i,j,k) of $(1,2,3)$. Therefore

$$\measuredangle\tilde p^1+\measuredangle\tilde p^2+\measuredangle\tilde p^3>2\cdot(\measuredangle\tilde p^1+\measuredangle\tilde p^2+\measuredangle\tilde p^3),$$

a contradiction. □

Proof of Theorem 10.1.

If part. Assume $\mathcal{L}$ is geodesic. Consider $x^1, x^2, x^3 \in \mathcal{L}$ such that the model triangle $[\tilde{x}^1\tilde{x}^2\tilde{x}^3] = \tilde{\triangle}^\kappa(x^1x^2x^3)$ is defined. Choose $p \in\,]x^1x^2[$. Applying the short map extension property with $V_3 = \{x^1, x^2, x^3\}$, $V_4 = \{x^1, x^2, x^3, p\}$ and the map $f\colon x^i \mapsto \tilde{x}^i$, we obtain the point-on-side comparison (see 8.14(b)).

In case $\mathcal{L}$ is not geodesic, pass to its ultrapower $\mathcal{L}^\omega$. Note that the short map extension property survives for $\mathcal{L}^\omega$ and recall that $\mathcal{L}^\omega$ is geodesic (see Observation 4.8). Thus, from above, $\mathcal{L}^\omega$ is a complete length CBB(κ) space. By Proposition 8.4, $\mathcal{L}$ is a complete length CBB(κ) space.

Only-if part. Assume the contrary: $\mathcal{L}$ is complete and CBB(κ), and x^1, x^2, x^3, $p \in \mathcal{L}$ and $\tilde{x}^1, \tilde{x}^2, \tilde{x}^3 \in \mathbb{M}^2(\kappa)$ are such that $|\tilde{x}^i - \tilde{x}^j| \leqslant |x^i - x^j|$ for all i, j but there is no point $\tilde{p} \in \mathbb{M}^2(\kappa)$ such that $|\tilde{p} - \tilde{x}^i| \leqslant |p - x^i|$ for all i.

Note that in this case all comparison triangles $\tilde{\triangle}^\kappa(px^ix^j)$ are defined. This is always true if $\kappa \leqslant 0$. If $\kappa > 0$, and say $\tilde{\triangle}^\kappa(px^1x^2)$ is undefined, then

$$\begin{aligned}|p - x^1| + |p - x^2| &\geqslant 2\cdot\varpi^\kappa - |x^1 - x^2| \geqslant \\ &\geqslant 2\cdot\varpi^\kappa - |\tilde{x}^1 - \tilde{x}^2| \geqslant \\ &\geqslant |\tilde{x}^1 - \tilde{x}^3| + |\tilde{x}^2 - \tilde{x}^3|.\end{aligned}$$

Then the last inequality must be an equality. Thus we may extend by taking $\tilde{p}$ on $[\tilde{x}^1\tilde{x}^3]$ or $[\tilde{x}^2\tilde{x}^3]$. For each $i \in \{1, 2, 3\}$, consider a point $\tilde{p}^i \in \mathbb{M}^2(\kappa)$ such that $|\tilde{p}^i - \tilde{x}^i|$ is minimal among points satisfying $|\tilde{p}^i - \tilde{x}^j| \leqslant |p - x^j|$ for all $j \neq i$. Clearly, every $\tilde{p}^i$ is inside the triangle $[\tilde{x}^1\tilde{x}^2\tilde{x}^3]$ (that is, in $\mathrm{Conv}(\tilde{x}^1, \tilde{x}^2, \tilde{x}^3)$), and $|\tilde{p}^i - \tilde{x}^i| > |p - x^i|$ for each i. Since the function $x \mapsto \tilde{\measuredangle}^\kappa\{x; a, b\}$ is increasing, it follows that

(i) $|\tilde{p}^i - \tilde{x}^j| = |p - x^j|$ for $i \neq j$;

(ii) no pair of triangles from $[\tilde{p}^1\tilde{x}^2\tilde{x}^3]$, $[\tilde{p}^2\tilde{x}^3\tilde{x}^1]$, $[\tilde{p}^3\tilde{x}^1\tilde{x}^2]$ overlap in $[\tilde{x}^1\tilde{x}^2\tilde{x}^3]$.

As follows from Overlap Lemma 10.2, in this case

$$\measuredangle\left[\tilde{p}^1\,{}^{\tilde{x}^2}_{\tilde{x}^3}\right] + \measuredangle\left[\tilde{p}^2\,{}^{\tilde{x}^3}_{\tilde{x}^1}\right] + \measuredangle\left[\tilde{p}^3\,{}^{\tilde{x}^1}_{\tilde{x}^2}\right] > 2\cdot\pi.$$

Since $|\tilde{x}^i - \tilde{x}^j| \leqslant |x^i - x^j|$, we have

$$\measuredangle\left[\tilde{p}^k\,{}^{\tilde{x}^i}_{\tilde{x}^j}\right] \leqslant \tilde{\measuredangle}^\kappa\left(p\,{}^{x^i}_{x^j}\right)$$

if (i, j, k) is a permutation of $(1, 2, 3)$. Therefore

$$\tilde{\measuredangle}^\kappa\left(p\,{}^{x^1}_{x^2}\right) + \tilde{\measuredangle}^\kappa\left(p\,{}^{x^2}_{x^3}\right) + \tilde{\measuredangle}^\kappa\left(p\,{}^{x^3}_{x^1}\right) > 2\cdot\pi,$$

contradicting the CBB(κ) comparison (see Definition 8.2). $\square$

10.3. Theorem. *Assume any pair of points at distance $< \varpi^\kappa$ in the metric space $\mathcal{U}$ are joined by a unique geodesic. Then $\mathcal{U}$ is* $\mathrm{CAT}(\kappa)$ *if and only if for any three-point set V_3 with perimeter $< 2\cdot\varpi^\kappa$ and any four-point set $V_4 \supset V_3$ in $\mathbb{M}^2(\kappa)$, any short map $f\colon V_3 \to \mathcal{U}$ can be extended to a short map $F\colon V_4 \to \mathcal{U}$.*

Note that the only-if part of Theorem 10.3 does not follow directly from Kirszbraun's theorem, since the desired extension is in $\mathcal{U}$—not its completion.

10.4. Lemma. *Let $x^1, x^2, x^3, y^1, y^2, y^3 \in \mathbb{M}(\kappa)$ be points such that $|x^i - x^j| \geqslant |y^i - y^j|$ for all i, j. Then there is a short map $\Phi\colon \mathbb{M}(\kappa) \to \mathbb{M}(\kappa)$ such that $\Phi(x^i) = y^i$ for all i; moreover, one can choose Φ so that*

$$\operatorname{Im}\Phi \subset \operatorname{Conv}(y^1, y^2, y^3).$$

We only give an idea of the proof of this lemma; alternatively, it can be obtained as a corollary of Kirszbraun's Theorem 10.14.

Idea of the proof. The map Φ can be constructed as a composition of an isometry of $\mathbb{M}(\kappa)$ and the following folding map: given a half-space H in $\mathbb{M}(\kappa)$, consider the map $\mathbb{M}(\kappa) \to H$ that is the identity on H and reflects all points outside of H into H. This map is a path isometry; in particular, it is short.

The last part of the lemma can be proved by composing this map with folding maps along the sides of triangle $[y^1y^2y^3]$, and passing to a partial limit. □

Proof of Theorem 10.3.

If part. The point-on-side comparison (see 9.14(b)) follows by taking $V_3 = \{\tilde{x}, \tilde{y}, \tilde{p}\}$ and $V_4 = \{\tilde{x}, \tilde{y}, \tilde{p}, \tilde{z}\}$ where $z \in \,]xy[$. It is only necessary to observe that $F(\tilde{z}) = z$ by uniqueness of $[xy]$.

Only-if part. Let $V_3 = \{\tilde{x}^1, \tilde{x}^2, \tilde{x}^3\}$ and $V_4 = \{\tilde{x}^1, \tilde{x}^2, \tilde{x}^3, \tilde{p}\}$.

Set $y^i = f(\tilde{x}^i)$ for all i. We need to find a point $q \in \mathcal{U}$ such that $|y^i - q| \leqslant |\tilde{x}^i - \tilde{p}|$ for all i.

Let D be the convex set in $\mathbb{M}^2(\kappa)$ bounded by the model triangle $[\tilde{y}^1\tilde{y}^2\tilde{y}^3] = \tilde{\triangle}^\kappa y^1y^2y^3$; that is, $D = \operatorname{Conv}(\tilde{y}^1, \tilde{y}^2, \tilde{y}^3)$.

Note that $|\tilde{y}^i - \tilde{y}^j| = |y^i - y^j| \leqslant |\tilde{x}^i - \tilde{x}^j|$ for all i, j. Applying Lemma 10.4, we get a short map $\Phi\colon \mathbb{M}(\kappa) \to D$ such that $\Phi\colon \tilde{x}^i \mapsto \tilde{y}^i$.

Further, by Majorization Theorem 9.56, there is a short map $F\colon D \to \mathcal{U}$ such that $\tilde{y}^i \mapsto y^i$ for all i.

Thus one can take $q = F \circ \Phi(\tilde{p})$. □

10.5. Exercise. Assume $\mathcal{X}$ is a complete length space that satisfies the following condition: any four-point subset admits a distance-preserving map to the Euclidean three-space.

Prove that $\mathcal{X}$ is isometric to a closed convex subset of a Hilbert space.

10.6. Exercise. Let $\mathcal{F}_s$ be the metric on the five-point set $\{p, q, x, y, z\}$ for which $|p - q| = s$ and all the remaining distances are equal to 1. For which values of s does the space $\mathcal{F}_s$ admit a distance-preserving map into:

(a) a complete length CAT(0) space?

(b) a complete length CBB(0) space?

The following exercise describes the first known definition of spaces with curvature bounded below; it was given by Abraham Wald [**159**].

10.7. Exercise. Let $\mathcal{L}$ be a metric space, and let $\kappa \leqslant 0$. Prove that $\mathcal{L}$ is CBB(κ) if and only if any quadruple of points $p, q, r, s \in \mathcal{L}$ admits a distance-preserving embedding into $\mathbb{M}^2(\mathrm{K})$ for some $\mathrm{K} \geqslant \kappa$.

Is the same true for $\kappa > 0$? What is the difference?

B. $(1 + n)$-point comparison

The following theorem gives a more sensitive analogue of the CBB(κ) comparison (see Definition 8.2).

10.8. $(1 + n)$-**point comparison.** *Let $\mathcal{L}$ be a complete length* CBB(κ) *space. Then for any array $(p, x^1, \dots, x^n)$ of points in $\mathcal{L}$ there is a model array $(\tilde{p}, \tilde{x}^1, \dots, \tilde{x}^n)$ in $\mathbb{M}^n(\kappa)$ such that*

(a) $|\tilde{p} - \tilde{x}^i| = |p - x^i|$ *for all i.*

(b) $|\tilde{x}^i - \tilde{x}^j| \geqslant |x^i - x^j|$ *for all i, j.*

Proof. It is sufficient to show that given $\varepsilon > 0$ there is an array $(\tilde{p}, \tilde{x}^1, \dots, \tilde{x}^n)$ in $\mathbb{M}^n(\kappa)$ such that

$$|\tilde{x}^i - \tilde{x}^j| \geqslant |x^i - x^j| \quad \text{and} \quad |\tilde{p} - \tilde{x}^i| \lessgtr |p - x^i|| \pm \varepsilon.$$

Then one can pass to a limit array for $\varepsilon \to 0+$.

According to Theorem 8.11, the set $\mathrm{Str}(x^1, \dots, x^n)$ is dense in $\mathcal{L}$. Thus there is a point $p' \in \mathrm{Str}(\tilde{x}^1, \dots, \tilde{x}^n)$ such that $|p' - p| \leqslant \varepsilon$. According to Corollary 13.40, $\mathrm{T}_{p'}$ contains a subcone E isometric to a Euclidean space and containing all vectors $\log[p'x^i]$. Passing to a subspace if necessary, we may assume that $\dim E \leqslant n$.

Mark a point $\tilde{p} \in \mathbb{M}^n(\kappa)$ and choose a distance-preserving map $\iota \colon E \to \mathrm{T}_{\tilde{p}} \mathbb{M}^n(\kappa)$. Let

$$\tilde{x}^i = \exp_{\tilde{p}} \circ \iota(\log[p'x^i]).$$

Thus $|\tilde{p} - \tilde{x}^i| = |p' - x^i|$. Since $|p - p'| \leqslant \varepsilon$, we get

$$|\tilde{p} - \tilde{x}^i| \lessgtr |p - x^i| \pm \varepsilon.$$

From the hinge comparison (see 8.14(c)) we have

$$\tilde{\measuredangle}^\kappa\left(\tilde{p}\,{}^{\tilde{x}^i}_{\tilde{x}^j}\right)=\measuredangle\left[\tilde{p}\,{}^{\tilde{x}^i}_{\tilde{x}^j}\right]=\measuredangle\left[p'\,{}^{x^i}_{x^j}\right]\geqslant\tilde{\measuredangle}^\kappa\left(p'\,{}^{x^i}_{x^j}\right),$$

and thus

$$|\tilde{x}^i-\tilde{x}^j|\geqslant|x^i-x^j|.$$

□

10.9. Exercise. Let $(p,x_1,\dots,x_n)$ be a point array in a CBB(0) space. Consider the $n{\times}n$-matrix M with components

$$m_{i,j}=\tfrac{1}{2}\cdot(|x_i-p|^2+|x_j-p|^2-|x_i-x_j|^2).$$

Show that

(1) $$\boldsymbol{s}\cdot M\cdot\boldsymbol{s}^\top\geqslant0$$

for any vector $\boldsymbol{s}=(s_1,\dots,s_n)$ with nonnegative components.

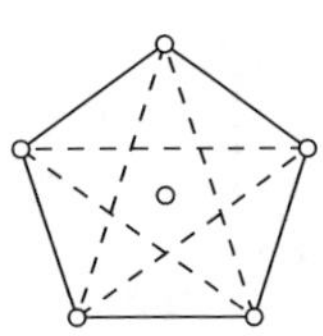

The above exercise describes the so-called *Lang–Schroeder–Sturm inequality*; it was discovered by Urs Lang and Viktor Schroeder [99] and rediscovered by Karl-Theodor Sturm [153]. It turns out to be weaker than $(1+n)$-point comparison. An example can be constructed by perturbing the six-point metric isometric to a regular pentagon with its center, making its sides slightly longer and diagonals slightly shorter [103]. In particular, this inequality in general metric spaces (not necessarily length spaces) does not imply the inequality in the following exercise.

10.10. Exercise. Let $\mathcal{L}$ be a complete length CBB(κ) space. Show that for any points p,x_1,x_2,x_3,x_4,x_5 in $\mathcal{L}$ we have

$$\tilde{\measuredangle}^\kappa\left(p\,{}^{x_1}_{x_5}\right)+\tilde{\measuredangle}^\kappa\left(p\,{}^{x_2}_{x_1}\right)+\tilde{\measuredangle}^\kappa\left(p\,{}^{x_3}_{x_2}\right)+\tilde{\measuredangle}^\kappa\left(p\,{}^{x_4}_{x_3}\right)+\tilde{\measuredangle}^\kappa\left(p\,{}^{x_5}_{x_4}\right)\leqslant4\cdot\pi,$$

assuming that the left-hand side is defined.

10.11. Exercise. Give an example of a metric on a finite set that satisfies the comparison inequality

$$\tilde{\measuredangle}^0\left(p\,{}^{x_1}_{x_2}\right)+\tilde{\measuredangle}^0\left(p\,{}^{x_2}_{x_3}\right)+\tilde{\measuredangle}^0\left(p\,{}^{x_3}_{x_1}\right)\leqslant2\cdot\pi$$

for any quadruple of points (p,x_1,x_2,x_3), but is not isometric to a subset of an Alexandrov space with curvature $\geqslant0$.

10.12. Exercise. Let $\mathcal{L}$ be a complete length CBB(κ) space. Assume that a point array $(a^0,a^1,\dots,a^k)$ in $\mathcal{L}$ is *κ-strutting* (Definition 15.1) for a point $p\in\mathcal{L}$. Show that there are points $\tilde{p}$, $\tilde{a}^0,\dots,\tilde{a}^m$ in $\mathbb{M}^{m+1}(\kappa)$ such that

$$|\tilde{p}-\tilde{a}^i|=|p-a^i|\quad\text{and}\quad|\tilde{a}^i-\tilde{a}^j|=|a^i-a^j|$$

for all i and j.

C. Helly's theorem

10.13. Helly's theorem. *Let $\mathcal{U}$ be a complete length* CAT(0) *space, and let $\{K_\alpha\}_{\alpha\in\mathcal{A}}$ be an arbitrary collection of closed bounded convex subsets in $\mathcal{U}$.*

If

$$\bigcap_{\alpha\in\mathcal{A}} K_\alpha = \varnothing,$$

then there is a finite index array $(\alpha_1, \alpha_2, \dots, \alpha_n)$ in $\mathcal{A}$ such that

$$\bigcap_i K_{\alpha_i} = \varnothing.$$

Remarks.

(i) In general, none of the K_α may be compact; otherwise the statement is trivial.

(ii) If $\mathcal{U}$ is a Hilbert space (not necessarily separable), then Helly's theorem is equivalent to the following statement: if a convex bounded set is closed in the ordinary topology, then it is compact in the weak topology. One can define *weak topology* in an arbitrary metric space by taking exteriors of closed ball as prebase. Then Helly's theorem implies the analogous statement for complete length CAT(0) spaces (compare to [**119**]).

(iii) For a finite-dimensional version of Helly's theorem, see 14.18.

We present the proof of Urs Lang and Viktor Schroeder [**99**].

Proof of Theorem 10.13. Assume the contrary. Then for any finite set $F \subset \mathcal{A}$,

$$K_F := \bigcap_{\alpha\in F} K_\alpha \neq \varnothing.$$

We will construct a point z such that $z \in K_\alpha$ for each α. Thus we will arrive at a contradiction since

$$\bigcap_{\alpha\in\mathcal{A}} K_\alpha = \varnothing.$$

Choose a point $p \in \mathcal{U}$, and let $r = \sup\{\operatorname{dist}_{K_F} p\}$ where F runs over all finite subsets of $\mathcal{A}$. Let p_F^* be the closest point on K_F to p; according to Closest-Point Projection Lemma 9.74, p_F^* exists and is unique.

Take a nested sequence of finite subsets $F_1 \subset F_2 \subset \cdots$ of $\mathcal{A}$, such that $\operatorname{dist}_{K_{F_n}} p \to r$.

Let us show that the sequence $p_{F_n}^*$ is Cauchy. If not, then for fixed $\varepsilon > 0$, we can choose two subsequences y_n' and y_n'' of $p_{F_n}^*$ such that $|y_n' - y_n''| \geqslant \varepsilon$. Let

z_n be the midpoint of $[y'_n y''_n]$. From the point-on-side comparison (see 8.14(b)), there is $\delta > 0$ such that

$$|p - z_n| \leqslant \max\{|p - y'_n|, |p - y''_n|\} - \delta.$$

Thus

$$\overline{\lim_{n\to\infty}} |p - z_n| < r.$$

On the other hand, from convexity, each K_{F_n} contains all z_k with sufficiently large k, a contradiction.

Thus, $p^*_{F_n}$ converges and we can take $z = \lim_n p^*_{F_n}$. Clearly

$$|p - z| = r.$$

Repeat the above arguments for the sequence $F'_n = F_n \cup \{\alpha\}$. As a result, we get another point z' such that $|p - z| = |p - z'| = r$ and $z, z' \in K_{F_n}$ for all n. Thus, if $z \neq z'$ the midpoint $\hat{z}$ of $[zz']$ would belong to all K_{F_n}, and from comparison, we would have $|p - \hat{z}| < r$, a contradiction.

Thus, $z' = z$; in particular $z \in K_\alpha$ for each $\alpha \in \mathcal{A}$. □

D. Kirszbraun's theorem

A slightly weaker version of the following theorem was proved by Urs Lang and Viktor Schroeder [**99**].

10.14. Kirszbraun's theorem. *Let $\mathcal{L}$ be a complete length* CBB(κ) *space, let $\mathcal{U}$ be a complete length* CAT(κ) *space, let $Q \subset \mathcal{L}$ be an arbitrary subset, and let $f : Q \to \mathcal{U}$ be a short map. Assume that there is $z \in \mathcal{U}$ such that $f(Q) \subset \subset \mathrm{B}[z, \frac{\varpi^\kappa}{2}]_{\mathcal{U}}$. Then $f : Q \to \mathcal{U}$ can be extended to a short map $F : \mathcal{L} \to \mathcal{U}$ (that is, there is a short map $F : \mathcal{L} \to \mathcal{U}$ such that $F|_Q = f$).*

The condition $f(Q) \subset \mathrm{B}[z, \frac{\varpi^\kappa}{2}]$ trivially holds for any $\kappa \leqslant 0$ since in this case $\varpi^\kappa = \infty$. The following example shows that this condition is needed for $\kappa > 0$.

Conjecture 10.22 (if true) gives an equivalent condition for the existence of a short extension; it states that the following example is the only obstacle.

10.15. Example. Let $\mathbb{S}^m_+$ be a closed m-dimensional unit hemisphere. Denote its boundary, which is isometric to $\mathbb{S}^{m-1}$, by $\partial\mathbb{S}^m_+$. Clearly, $\mathbb{S}^m_+$ is CBB(1) and $\partial\mathbb{S}^m_+$ is CAT(1), but the identity map $\partial\mathbb{S}^m_+ \to \partial\mathbb{S}^m_+$ cannot be extended to a short map $\mathbb{S}^m_+ \to \partial\mathbb{S}^m_+$ (there is no place for the pole).

There is also a direct generalization of this example to a hemisphere in a Hilbert space of arbitrary cardinal dimension.

First we prove this theorem in the case $\kappa \leqslant 0$ (see 10.17). In the proof of the more complicated case $\kappa > 0$, we use the case $\kappa = 0$. The following lemma is the main ingredient in the proof.

10.16. Finite+one lemma. *Let $\kappa \leqslant 0$, $\mathcal{L}$ be a complete length* CBB(κ) *space, and let $\mathcal{U}$ be a complete length* CAT(κ) *space. Suppose $x^1, x^2, \dots, x^n$ in $\mathcal{L}$ and $y^1, y^2, \dots, y^n$ in $\mathcal{U}$ are such that $|x^i - x^j| \geqslant |y^i - y^j|$ for all i, j.*

Then for any $p \in \mathcal{L}$, there is $q \in \mathcal{U}$ such that $|y^i - q| \leqslant |x^i - p|$ for each i.

Proof. It is sufficient to prove the lemma only for $\kappa = 0$ and -1. The proofs of these two cases are identical, only the formulas differ. In the proof, we assume $\kappa = 0$ and provide the formulas for $\kappa = -1$ in the footnotes.

From the $(1+n)$-point comparison (Section 10.8), there is a model configuration $\tilde{p}, \tilde{x}^1, \tilde{x}^2, \dots, \tilde{x}^n$ in $\mathbb{M}^n(\kappa)$ such that $|\tilde{p} - \tilde{x}^i| = |p - x^i|$ and $|\tilde{x}^i - \tilde{x}^j| \geqslant |x^i - x^j|$ for all i, j. It follows that we can assume that $\mathcal{L} = \mathbb{M}^n(\kappa)$.

For each i, consider functions $f^i : \mathcal{U} \to \mathbb{R}$ and $\tilde{f}^i : \mathbb{M}^n(\kappa) \to \mathbb{R}$ defined as follows:[1]

$$(A)^0 \qquad f^i = \tfrac{1}{2}\cdot \mathrm{dist}^2_{y^i}, \quad \tilde{f}^i = \tfrac{1}{2}\cdot \mathrm{dist}^2_{\tilde{x}^i}.$$

Consider the function arrays

$$\boldsymbol{f} = (f^1, f^2, \dots, f^n) \colon \mathcal{U} \to \mathbb{R}^n \quad \text{and} \quad \tilde{\boldsymbol{f}} = (\tilde{f}^1, \tilde{f}^2, \dots, \tilde{f}^n) \colon \mathbb{M}^n(\kappa) \to \mathbb{R}^n.$$

Define

$$\begin{aligned} \operatorname{Up} \boldsymbol{f}(\mathcal{U}) &= \left\{ \boldsymbol{v} \in \mathbb{R}^{k+1} : \exists\, \boldsymbol{w} \in \boldsymbol{f}(\mathcal{U}) \text{ such that } \boldsymbol{v} \succcurlyeq \boldsymbol{w} \right\}, \\ \operatorname{Min} \boldsymbol{f}(\mathcal{U}) &= \left\{ \boldsymbol{v} \in \boldsymbol{f}(\mathcal{U}) : \text{if } \boldsymbol{v} \succcurlyeq \boldsymbol{w} \in \boldsymbol{f}(\mathcal{U}), \text{ then } \boldsymbol{w} = \boldsymbol{v} \right\}. \end{aligned}$$

(See Definition 14.1.) Note it is sufficient to prove that $\tilde{\boldsymbol{f}}(\tilde{p}) \in \operatorname{Up} \boldsymbol{f}(\mathcal{U})$. Clearly, $(f^i)'' \geqslant 1$. Thus by Theorem 14.3(a), the set $\operatorname{Up} \boldsymbol{f}(\mathcal{U}) \subset \mathbb{R}^n$ is convex.

Arguing by contradiction, let us assume that $\tilde{\boldsymbol{f}}(\tilde{p}) \notin \operatorname{Up} \boldsymbol{f}(\mathcal{U})$.

Then there exists a supporting hyperplane $\alpha_1 \cdot x_1 + \dots + \alpha_n \cdot x_n = c$ to $\operatorname{Up} \boldsymbol{f}(\mathcal{U})$, separating it from $\tilde{\boldsymbol{f}}(\tilde{p})$. According to Lemma 14.6(b), $\alpha_i \geqslant 0$ for each i. So we may assume that $(\alpha_1, \alpha_2, \dots, \alpha_n) \in \Delta^{n-1}$ (that is, $\alpha_i \geqslant 0$ for each i) and $\sum \alpha_i = 1$ and

$$\sum_i \alpha_i \cdot \tilde{f}^i(\tilde{p}) < \inf \left\{ \sum_i \alpha_i \cdot f^i(q) : q \in \mathcal{U} \right\}.$$

The latter contradicts the following claim.

---

[1] In case $\kappa = -1$,

$$(A)^- \qquad f^i = \cosh \circ \mathrm{dist}_{y^i}, \qquad \tilde{f}^i = \cosh \circ \mathrm{dist}_{\tilde{x}^i}.$$

(1) *Given* $\boldsymbol{\alpha} = (\alpha_1, \alpha_2, \dots, \alpha_n) \in \Delta^{n-1}$, *let*

$$h = \sum_i \alpha_i \cdot f^i \quad h\colon \mathcal{U} \to \mathbb{R} \qquad z = \operatorname{MinPoint} h \in \mathcal{U},$$

$$\tilde h = \sum_i \alpha_i \cdot \tilde f^i \quad \tilde h\colon \mathbb{M}^n(\kappa) \to \mathbb{R} \quad \tilde z = \operatorname{MinPoint} \tilde h \in \mathbb{M}^n(\kappa).$$

Then $h(z) \leqslant \tilde h(\tilde z)$.

Proof of the claim. Note that $\boldsymbol{d}_z h \geqslant 0$. Thus, for each i, we have[2]

$$\begin{aligned} 0 &\leqslant (\boldsymbol{d}_z h)(\uparrow_{[zy^i]}) = \\ &= -\sum_j \alpha_j \cdot |z - y^j| \cdot \cos\measuredangle\left[z\,{}^{y^i}_{y^j}\right] \leqslant \\ &\leqslant -\sum_j \alpha_j \cdot |z - y^j| \cdot \cos\tilde\measuredangle^0\left(z\,{}^{y^i}_{y^j}\right) = \qquad (B)^0 \\ &= -\tfrac{1}{2\cdot|z-y^i|} \cdot \sum_j \alpha_j \cdot \left[|z-y^i|^2 + |z-y^j|^2 - |y^i - y^j|^2\right]. \end{aligned}$$

In particular,[3]

$$\sum_i \alpha_i \cdot \left[\sum_j \alpha_j \cdot \left[|z-y^i|^2 + |z-y^j|^2 - |y^i - y^j|^2\right]\right] \leqslant 0, \qquad (C)^0$$

or[4]

$$2\cdot h(z) \leqslant \sum_{i,j} \alpha_i \cdot \alpha_j \cdot |y^i - y^j|^2. \qquad (D)^0$$

[2]In case $\kappa = -1$, the same calculations give

$$0 \leqslant \dots \leqslant -\tfrac{1}{\sinh|z-y^i|} \cdot \sum_j \alpha_j \cdot \left[\cosh|z-y^i| \cdot \cosh|z-y^j| - \cosh|y^i - y^j|\right]. \qquad (B)^-$$

[3]In case $\kappa = -1$, the same calculations give

$$\sum_i \alpha_i \cdot \left[\sum_j \alpha_j \cdot \left[\cosh|z-y^i| \cdot \cosh|z-y^j| - \cosh|y^i - y^j|\right]\right] \leqslant 0. \qquad (C)^-$$

[4]In case $\kappa = -1$,

$$(h(z))^2 \leqslant \sum_{i,j} \alpha_i \cdot \alpha_j \cdot \cosh|y^i - y^j|. \qquad (D)^-$$

Note that if $\mathcal{U} \overset{\text{iso}}{=\!=} \mathbb{M}^n(\kappa)$, then all inequalities in (B, C, D) are sharp. Thus the same argument as above, repeated for $\tilde{x}^1, \tilde{x}^2, \dots, \tilde{x}^n$ in $\mathbb{M}^n(\kappa)$, gives[5]

$$(E)^0 \qquad 2\cdot\tilde{h}(\tilde{z}) = \sum_{i,j} \alpha_i\cdot\alpha_j\cdot|\tilde{x}^i - \tilde{x}^j|^2.$$

Note that

$$|\tilde{x}^i - \tilde{x}^j| \geqslant |x^i - x^j| \geqslant |y^i - y^j|$$

for all i, j. Thus, (D) and (E) imply the claim. □

10.17. Kirszbraun's theorem for nonpositive bound. *Let $\kappa \leqslant 0$, let $\mathcal{L}$ be a complete length* CBB(κ) *space, let $\mathcal{U}$ be a complete length* CAT(κ) *space, let $Q \subset \mathcal{L}$ be arbitrary subset, and let $f : Q \to \mathcal{U}$ be a short map. Then there is a short extension $F : \mathcal{L} \to \mathcal{U}$ of f; that is, there is a short map $F : \mathcal{L} \to \mathcal{U}$ such that $F|_Q = f$.*

Remark. If $\mathcal{U}$ is proper, then we do not need Helly's Theorem 10.13; compactness of closed balls in $\mathcal{U}$ is sufficient in this case.

Proof of Theorem 10.17. By Zorn's lemma, we can assume that $Q \subset \mathcal{L}$ is a maximal set; that is, $f : Q \to \mathcal{U}$ does not admit a short extension to any larger set $Q' \supset Q$.

Let us argue by contradiction. Assume that $Q \neq \mathcal{L}$; choose $p \in \mathcal{L} \setminus Q$. Then

$$\bigcap_{x\in Q} \overline{\mathrm{B}}[f(x), |p - x|] = \varnothing.$$

Since $\kappa \leqslant 0$, the balls are convex; thus, by Helly's Theorem 10.13, one can choose points $x^1, x^2, \dots, x^n$ in Q such that

$$(2) \qquad \bigcap_{i} \overline{\mathrm{B}}[y^i, |x^i - p|] = \varnothing,$$

where $y^i = f(x^i)$. Finally note that (2) contradicts the finite+one lemma (Lemma 10.16). □

Proof of Kirszbraun's Theorem 10.14. The case $\kappa \leqslant 0$ is already proved in Theorem 10.17. Thus it remains to prove the theorem only in case $\kappa > 0$. After rescaling we can assume that $\kappa = 1$ and therefore $\varpi^\kappa = \pi$.

Since $\overline{\mathrm{B}}[z, \pi/2]_{\mathcal{U}}$ is a complete length CAT(κ) space, we can assume $\mathcal{U} = \overline{\mathrm{B}}[z, \pi/2]_{\mathcal{U}}$. In particular, $\operatorname{diam} \mathcal{U} \leqslant \pi$.

[5] In case $\kappa = -1$,

$$(E)^- \qquad (\tilde{h}(\tilde{z}))^2 = \sum_{i,j} \alpha_i\cdot\alpha_j\cdot \cosh|\tilde{x}^i - \tilde{x}^j|.$$

Further, any two points $x, y \in \mathcal{U}$ such that $|x - y| < \pi$ are joined by a unique geodesic; if $|x - y| = \pi$, then the concatenation of $[xz]$ and $[zy]$ as a geodesic from x to y. Hence $\mathcal{U}$ is geodesic.

We may also assume that $\operatorname{diam} \mathcal{L} \leqslant \pi$. Otherwise $\mathcal{L}$ is one-dimensional (see 8.44); in this case the result follows since $\mathcal{U}$ is geodesic.

Assume the theorem is false. Then there is a set $Q \subset \mathcal{L}$, a short map $f\colon Q \to \mathcal{U}$, and $p \in \mathcal{L} \setminus Q$ such that

$$\bigcap_{x \in Q} \overline{\mathrm{B}}[f(x), |x - p|] = \varnothing. \tag{3}$$

We will apply Theorem 10.17 for $\kappa = 0$ to the Euclidean cones $\mathring{\mathcal{L}} = \operatorname{Cone} \mathcal{L}$ and $\mathring{\mathcal{U}} = \operatorname{Cone} \mathcal{U}$. Note that

- $\mathring{\mathcal{U}}$ is a complete length CAT(0) space (see Theorem 11.7(a)),
- since $\operatorname{diam} \mathcal{L} \leqslant \pi$, we have $\mathring{\mathcal{L}}$ is CBB(0) (see Theorem 11.6(a)).

Further, we will view the spaces $\mathcal{L}$ and $\mathcal{U}$ as unit spheres in $\mathring{\mathcal{L}}$ and $\mathring{\mathcal{U}}$, respectively. In the cones $\mathring{\mathcal{L}}$ and $\mathring{\mathcal{U}}$ we will use "$|{*}|$" for distance to the tip, denoted by 0, "$\cdot$" for cone multiplication, "$\measuredangle(x,y)$" for $\measuredangle\left[0\,{}^{x}_{y}\right]$, and "$\langle x, y\rangle$" for $|x|\cdot|y|\cdot\cos\measuredangle\left[0\,{}^{x}_{y}\right]$. In particular,

- $|x - y|_{\mathcal{L}} = \measuredangle(x, y)$ for any $x, y \in \mathcal{L}$,
- $|x - y|_{\mathcal{U}} = \measuredangle(x, y)$ for any $x, y \in \mathcal{U}$,
- for any $y \in \mathcal{U}$, we have

$$\measuredangle(z, y) \leqslant \tfrac{\pi}{2}. \tag{4}$$

Let $\mathring{Q} = \operatorname{Cone} Q \subset \mathring{\mathcal{L}}$ and let $\mathring{f}\colon \mathring{Q} \to \mathring{\mathcal{U}}$ be the natural cone extension of f; that is, $y = f(x) \Rightarrow t\cdot y = \mathring{f}(t\cdot x)$ for $t \geqslant 0$. Clearly $\mathring{f}$ is short.

Applying Theorem 10.17 for $\mathring{f}$, we get a short extension map $\mathring{F}\colon \mathring{\mathcal{L}} \to \mathring{\mathcal{U}}$. Let $s = \mathring{F}(p)$. Then

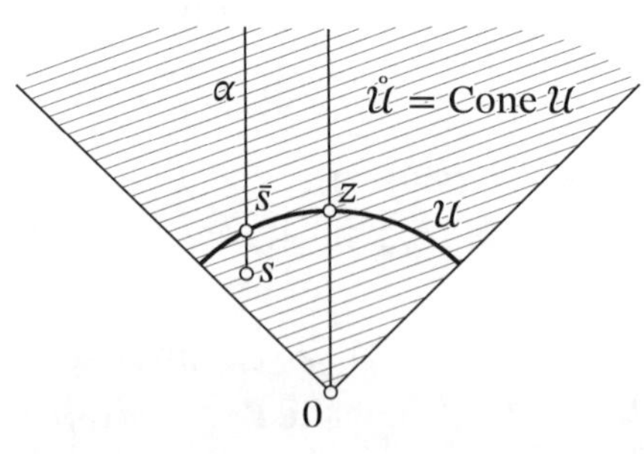

$$|s - \mathring{f}(w)| \leqslant |p - w| \tag{5}$$

for any $w \in \mathring{Q}$. In particular, $|s| \leqslant 1$. Applying (5) for $w = t\cdot x$ and $t \to \infty$, we have

$$\langle f(x), s\rangle \geqslant \cos\measuredangle(p, x) \tag{6}$$

for any $x \in Q$.

By comparison, the geodesics $\operatorname{geod}_{[s\,t\cdot z]}$ converge as $t \to \infty$ and the limit is a half-line; denote it by $\alpha\colon [0, \infty) \to \mathring{\mathcal{U}}$. From (4), the function $t \mapsto \langle f(x), \alpha(t)\rangle$

is nondecreasing. From (6), for the necessarily unique point $\bar{s}$ on the half-line α such that $|\bar{s}| = 1$, we also have

$$\langle f(x), \bar{s}\rangle \geqslant \cos \measuredangle(p, x) \quad \text{or, equivalently,} \quad \measuredangle(\bar{s}, f(x)) \leqslant \measuredangle(p, f(x))$$

for any $x \in Q$, in contradiction to (3). □

10.18. Exercise. Let $\mathcal{U}$ be CAT(0). Assume there are two point arrays, $(x^0, x^1, \dots, x^k)$ in $\mathcal{U}$ and $(\tilde{x}^0, \tilde{x}^1, \dots, \tilde{x}^k)$ in $\mathbb{E}^m$, such that $|x^i - x^j|_{\mathcal{U}} = |\tilde{x}^i - \tilde{x}^j|_{\mathbb{E}^m}$ for each i, j, and for any point $z_0 \in \mathcal{U}$ there is $i > 0$ such that $|z_0 - x_i| \geqslant |x_0 - x_i|$.

Prove that there is a subset $Q \subset \mathcal{L}$ isometric to a convex set in $\mathbb{E}^m$ and containing all the points x^i.

10.19. Exercise. Let $(x^0, x^1, \dots, x^k)$ in $\mathcal{L}$ and $(\tilde{x}^0, \tilde{x}^1, \dots, \tilde{x}^k)$ in $\mathbb{E}^m$ be two point arrays in complete length CBB(0) space $\mathcal{L}$. Assume that $|x^i - x^j|_{\mathcal{L}} = |\tilde{x}^i - \tilde{x}^j|_{\mathbb{E}^m}$ for each i, j and $\tilde{x}^0$ lies in the interior of $\operatorname{Conv}(\tilde{x}^1, \dots, \tilde{x}^k)$.

Prove that there is a subset $Q \subset \mathcal{L}$ isometric to a convex set in $\mathbb{E}^m$ and containing all the points x^i.

The following statement we call $(2n+2)$-*point comparison.*

10.20. Exercise. Let $\mathcal{U}$ be a complete length CAT(κ) space. Consider $x, y \in \in \mathcal{U}$ and an array $((p^1, q^1), (p^2, q^2), \dots, (p^n, q^n))$ of pairs of points in $\mathcal{U}$, such that there is a model configuration $\tilde{x}$, $\tilde{y}$ and array of pairs $((\tilde{p}^1, \tilde{q}^1), (\tilde{p}^2, \tilde{q}^2), \dots \dots, (\tilde{p}^n, \tilde{q}^n))$ in $\mathbb{M}^3(\kappa)$ with the following properties:

(a) $[\tilde{x}\tilde{p}^1\tilde{q}^1] = \tilde{\triangle}^\kappa x p^1 q^1$ and $[\tilde{y}\tilde{p}^n\tilde{q}^n] = \tilde{\triangle}^\kappa y p^n q^n$;

(b) the simplex $\tilde{p}^i\tilde{p}^{i+1}\tilde{q}^i\tilde{q}^{i+1}$ is a model simplex of $p^i p^{i+1} q^i q^{i+1}$ for all i, that is,

$$|\tilde{p}^i - \tilde{q}^i| = |p^i - q^i|,$$
$$|\tilde{p}^i - \tilde{p}^{i+1}| = |p^i - p^{i+1}|,$$
$$|\tilde{q}^i - \tilde{q}^{i+1}| = |q^i - q^{i+1}|,$$
$$|\tilde{p}^i - \tilde{q}^{i+1}| = |p^i - q^{i+1}|,$$

and

$$|\tilde{p}^{i+1} - \tilde{q}^i| = |p^{i+1} - q^i|.$$

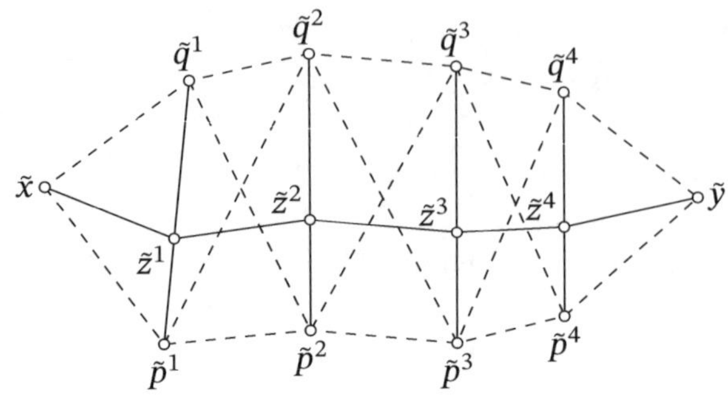

Then for any choice of n points $\tilde z^i \in [\tilde p^i \tilde q^i]$, we have

$$|\tilde x - \tilde z^1| + |\tilde z^1 - \tilde z^2| + \dots + |\tilde z^{n-1} - \tilde z^n| + |\tilde z^n - \tilde y| \geqslant |x - y|.$$

E. Remarks

The following problem and its relatives were mentioned by Michael Gromov [**71**, 1.19].

10.21. Open problem. *Find a necessary and sufficient condition for a finite metric space to admit distance-preserving embeddings into*

(a) *some length* CBB(κ) *space,*

(b) *some length* CAT(κ) *space.*

A metric on a finite set $\{a^1, a^2, \dots, a^n\}$, can be described by the matrix with components

$$s^{ij} = |a^i - a^j|^2,$$

which we will call the *associated matrix*. The set of associated matrices of all metrics that admit a distance-preserving map into a CBB(0) or a CAT(0) space form a convex cone. The latter follows since the rescalings and products of CBB(0) (or CAT(0)) spaces are CBB(0) (or CAT(0), respectively). This convexity gives a bit of hope that the cone admits an explicit description.

For the five-point CAT(0) case, the (2+2)-comparison is a necessary and sufficient condition. This was proved by Tetsu Toyoda [**157**]; another proof was found by Nina Lebedeva and the third author [**105**]. For the five-point CBB(0) case, the (1+4)-comparison is a necessary and sufficient condition; it was proved by Nina Lebedeva and the third author [**104**]. Starting from the six-point case, only some necessary and some sufficient conditions are known; for more on the subject see [**8, 103, 105**].

The following conjecture (if true) would give the right generality for Kirszbraun's Theorem 10.14. It states that Example 10.15 is the only obstacle to extending short maps.

10.22. Conjecture. *Assume $\mathcal{L}$ is a complete length* CBB(1) *space, $\mathcal{U}$ is a complete length* CAT(1) *space, $Q \subset \mathcal{L}$ is a proper subset, and $f : Q \to \mathcal{U}$ is a short map that does not admit a short extension to any bigger set $Q' \supset Q$. Then:*

(a) *Q is isometric to a sphere in a Hilbert space (of finite or cardinal dimension). Moreover, there is a point $p \in \mathcal{L}$ such that $|p - q| = \frac{\pi}{2}$ for any $q \in Q$.*

(b) *The map $f : Q \to \mathcal{U}$ is a distance-preserving map and there is no point $p' \in \mathcal{U}$ such that $|p' - q'| = \frac{\pi}{2}$ for any $q' \in f(Q)$.*

Curvature-free analogues. Let us present a collection of exercises on curvature-free analogues of Kirszbraun's theorem. It is worthwhile to know these results despite they are far from Alexandrov geometry.

10.23. Exercise. Let $\mathcal{X}$ and $\mathcal{Y}$ be metric spaces, $A \subset \mathcal{X}$, and $f : A \to \mathcal{Y}$ be a short map. Assume $\mathcal{Y}$ is compact and for any finite set $F \subset \mathcal{X}$ there is a short map $F \to \mathcal{Y}$ that agrees with f on $F \cap A$. Then there is a short map $\mathcal{X} \to \mathcal{Y}$ that agrees with f on A.

10.24. Exercise. We say that a metric space $\mathcal{X}$ is *injective* if for an arbitrary metric space $\mathcal{Z}$ and a subset $Q \subset \mathcal{Z}$, any short map $Q \to \mathcal{X}$ can be extended as a short map $\mathcal{Z} \to \mathcal{X}$.

(a) Prove that any metric space $\mathcal{X}$ admits a distance-preserving embedding into an injective metric space.

(b) Use this to construct an analogue of convex hull in the category of metric spaces; this is called the *injective hull*.

Chapter 11

Warped products

The warped product is a construction that produces a new metric space, denoted by $\mathcal{B} \times_f \mathcal{F}$, from two metric spaces base $\mathcal{B}$ and fiber $\mathcal{F}$, and a function $f\colon \mathcal{B} \to \mathbb{R}_{\geqslant 0}$.

Many important constructions such as direct products, cones, spherical suspensions, and joins can be defined using warped products.

A. Definition

First we define the warped product for length spaces, and then we expand the definition to allow for arbitrary fiber $\mathcal{F}$.

Let $\mathcal{B}$ and $\mathcal{F}$ be length spaces, and let $f\colon \mathcal{B} \to [0,\infty)$ be a continuous function.

For any path $\gamma\colon [0,1] \to \mathcal{B} \times \mathcal{F}$, we write $\gamma = (\gamma_{\mathcal{B}}, \gamma_{\mathcal{F}})$, where $\gamma_{\mathcal{B}}$ is the projection of γ to $\mathcal{B}$, and $\gamma_{\mathcal{F}}$ is the projection to $\mathcal{F}$. If $\gamma_{\mathcal{B}}$ and $\gamma_{\mathcal{F}}$ are Lipschitz, set

$$\text{length}_f \gamma := \int_0^1 \sqrt{v_{\mathcal{B}}^2 + (f \circ \gamma_{\mathcal{B}})^2 \cdot v_{\mathcal{F}}^2} \cdot dt, \tag{1}$$

where $\int$ is Lebesgue integral, and $v_{\mathcal{B}}$ and $v_{\mathcal{F}}$ are the speeds of $\gamma_{\mathcal{B}}$ and $\gamma_{\mathcal{F}}$, respectively. (Note that $\text{length}_f \gamma \geqslant \text{length}\, \gamma_{\mathcal{B}}$.)

Consider the pseudometric on $\mathcal{B} \times_f \mathcal{F}$ defined by

$$|x - y| := \inf\left\{ \text{length}_f \gamma : \gamma(0) = x, \gamma(1) = y \right\},$$

where the exact lower bound is taken for all Lipschitz paths $\gamma\colon [0,1]\to \mathcal{B}\times\mathcal{F}$. The corresponding metric space is called the *warped product with base $\mathcal{B}$, fiber $\mathcal{F}$, and warping function f*; it will be denoted by $\mathcal{B}\times_f\mathcal{F}$.

The points in $\mathcal{B}\times_f\mathcal{F}$ can be described by corresponding pairs $(p,\phi)\in \in\mathcal{B}\times\mathcal{F}$. Note that if $f(p)=0$ for $p\in\mathcal{B}$, then $(p,\phi)=(p,\psi)$ for any $\phi,\psi\in\mathcal{F}$.

We do not claim that every Lipschitz curve in $\mathcal{B}\times_f\mathcal{F}$ may be reparametrized as the image of a Lipschitz curve in $\mathcal{B}\times\mathcal{F}$; in fact this is not true.

11.1. Proposition. *The warped product $\mathcal{B}\times_f\mathcal{F}$ satisfies the following.*

(a) *The projection $(p,\phi_0)\mapsto p$ is a submetry. Moreover, its restriction to any horizontal leaf $\mathcal{B}\times\{\phi_0\}$ is an isometry to $\mathcal{B}$.*

(b) *If $f(p_0)\neq 0$, the projection $(p_0,\phi)\mapsto\phi$ of the vertical leaf $\{p_0\}\times\mathcal{F}$, with its length metric, is a homothety onto $\mathcal{F}$ with multiplier $\frac{1}{f(p_0)}$.*

(c) *If f achieves its (local) minimum at p_0, then the inclusion of the vertical leaf $\{p_0\}\times\mathcal{F}$ in $\mathcal{B}\times_f\mathcal{F}$ is (locally) distance-preserving.*

Proof. Claim (b) follows from the f-length formula (1).

Also, by (1), the projection of $\mathcal{B}\times_f\mathcal{F}$ onto $\mathcal{B}\times\{\phi_0\}$ given by $(p,\phi)\mapsto(p,\phi_0)$ is length-nonincreasing; hence (a).

Suppose p_0 is a local minimum point of f. Then the projection $(p,\phi)\mapsto \mapsto(p_0,\phi)$ of a neighborhood of the vertical leaf $\{p_0\}\times\mathcal{F}$ to $\{p_0\}\times\mathcal{F}$ is length-nonincreasing.

If p_0 is a global minimum point of f, then the same holds for the projection of whole space. Hence (c). □

Note that any horizontal leaf is weakly convex, but does not have to be convex even if $\mathcal{B}\times_f\mathcal{F}$ is a geodesic space. The latter follows since vanishing of the warping function f allows geodesics to bifurcate into distinct horizontal leaves. For instance, if there is a geodesic with the ends in the zero set

$$Z=\left\{(p,\phi)\in\mathcal{B}\times_f\mathcal{F}\ :\ f(p)=0\right\},$$

then there is a geodesic with the same endpoints in each horizontal leaf.

11.2. Proposition. *Suppose $\mathcal{B}$ and $\mathcal{F}$ are length spaces and $f\colon\mathcal{B}\to[0,\infty)$ is a continuous function. Then the warped product $\mathcal{B}\times_f\mathcal{F}$ is a length space.*

Proof. It is sufficient to show that for any $\alpha\colon[0,1]\to\mathcal{B}\times_f\mathcal{F}$ there is a path $\beta\colon[0,1]\to\mathcal{B}\times\mathcal{F}$ with the same endpoints such that

$$\operatorname{length}\alpha\geqslant\operatorname{length}_f\beta.$$

If $f \circ \alpha_{\mathcal{B}}(t) > 0$ for any t, then the vertical projection $\alpha_{\mathcal{F}}$ is defined. In this case, let $\beta(t) = (\alpha_{\mathcal{B}}(t), \alpha_{\mathcal{F}}(t)) \in \mathcal{B} \times \mathcal{F}$. Clearly

$$\operatorname{length} \alpha = \operatorname{length}_f \beta.$$

If $f \circ \alpha_{\mathcal{B}}(t_0) = 0$ for some t_0, let β be the concatenation of three curves in $\mathcal{B} \times \mathcal{F}$, namely,

(1) the horizontal curve $(\alpha_{\mathcal{B}}(t), \phi)$ for $t \leqslant t_0$,

(2) a vertical path in form (s, ϕ) to (s, ψ), and

(3) the horizontal curve $(\alpha_{\mathcal{B}}(t), \psi)$ for $t \geqslant t_0$.

By item (1), the f-length of the middle path in the concatenation is vanishing; therefore the f-length of α cannot be smaller than length of $\alpha_{\mathcal{B}}$, that is,

$$\operatorname{length}_f \alpha \geqslant \operatorname{length} \alpha_{\mathcal{B}} = \operatorname{length}_f \beta.$$

The statement follows. □

The following theorem states that distance in a warped product is fiber-independent, in the sense that distances may be calculated by substituting for $\mathcal{F}$ a different length space:

11.3. Fiber-independence theorem. *Consider length spaces $\mathcal{B}$, $\mathcal{F}$, and $\check{\mathcal{F}}$, and a locally Lipschitz function $f\colon \mathcal{B} \to \mathbb{R}_{\geqslant 0}$. Assume $p, q \in \mathcal{B}$, $\phi, \psi \in \mathcal{F}$, and $\check{\phi}, \check{\psi} \in \check{\mathcal{F}}$. Then*

$$\begin{gathered} |\phi - \psi|_{\mathcal{F}} \geqslant |\check{\phi} - \check{\psi}|_{\check{\mathcal{F}}} \\ \Downarrow \\ |(p, \phi) - (q, \psi)|_{\mathcal{B} \times_f \mathcal{F}} \geqslant |(p, \check{\phi}) - (q, \check{\psi})|_{\mathcal{B} \times_f \check{\mathcal{F}}}. \end{gathered}$$

In particular,

$$|(p, \phi) - (q, \psi)|_{\mathcal{B} \times_f \mathcal{F}} = |(p, 0) - (q, \ell)|_{\mathcal{B} \times_f \mathbb{R}},$$

where $\ell = |\phi - \psi|_{\mathcal{F}}$.

Proof. Let γ be a path in $(\mathcal{B} \times \mathcal{F})$.

Since $|\phi - \psi|_{\mathcal{F}} \geqslant |\check{\phi} - \check{\psi}|_{\check{\mathcal{F}}}$, there is a Lipschitz path $\gamma_{\check{\mathcal{F}}}$ from $\check{\phi}$ to $\check{\psi}$ in $\check{\mathcal{F}}$ such that

$$(\operatorname{speed} \gamma_{\mathcal{F}})(t) \geqslant (\operatorname{speed} \gamma_{\check{\mathcal{F}}})(t)$$

for almost all $t \in [0, 1]$. Consider the path $\check{\gamma} = (\gamma_{\mathcal{B}}, \gamma_{\check{\mathcal{F}}})$ from $(p, \check{\phi})$ to $(q, \check{\psi})$ in $\mathcal{B} \times_f \check{\mathcal{F}}$. Clearly

$$\operatorname{length}_f \gamma \geqslant \operatorname{length}_f \check{\gamma}.$$

□

11.4. Exercise. Let $\mathcal{B}$ and $\mathcal{F}$ be length spaces and let $f, g\colon \mathcal{B} \to \mathbb{R}_{\geqslant 0}$ be two locally Lipschitz nonnegative functions. Assume $f(b) \leqslant g(b)$ for any $b \in \mathcal{B}$. Show that $\mathcal{B} \times_f \mathcal{F} \leqslant \mathcal{B} \times_g \mathcal{F}$; that is, there is a distance-noncontracting map $\mathcal{B} \times_f \mathcal{F} \to \mathcal{B} \times_g \mathcal{F}$.

B. Extended definitions

The fiber-independence theorem implies that

$$|(p,\phi)-(q,\psi)|_{\mathcal{B}\times_f\mathcal{F}} = |(p,0)-(q,|\phi-\psi|_{\mathcal{F}})|_{\mathcal{B}\times_f\mathbb{R}}$$

for any $(p,\phi),(q,\psi) \in \mathcal{B} \times \mathcal{F}$. In particular, if $\iota\colon A \to \check{A}$ is an isometry between two subsets $A \subset \mathcal{F}$ and $\check{A} \subset \check{\mathcal{F}}$ in length spaces $\mathcal{F}$ and $\check{\mathcal{F}}$, and $\mathcal{B}$ is a length space, then for any warping function $f\colon \mathcal{B} \to \mathbb{R}_{\geqslant 0}$, the map ι induces an isometry between the sets $\mathcal{B} \times_f A \subset \mathcal{B} \times_f \mathcal{F}$ and $\mathcal{B} \times_f \check{A} \subset \mathcal{B} \times_f \check{\mathcal{F}}$.

This observation allows us to define the warped product $\mathcal{B}\times_f\mathcal{F}$ where the fiber $\mathcal{F}$ does not carry its length metric. Indeed we can use Kuratowsky embedding to realize $\mathcal{F}$ as a subspace in a length space, say $\mathcal{F}'$. Therefore we can take the warped product $\mathcal{B} \times_f \mathcal{F}'$ and identify $\mathcal{B} \times_f \mathcal{F}$ with its subspace consisting of all pairs (b,ϕ) such that $\phi \in \mathcal{F}$. According to the fiber-independence theorem, Theorem 11.3, the resulting space does not depend on the choice of $\mathcal{F}'$.

C. Examples

Direct product. The simplest example is the *direct product* $\mathcal{B}\times\mathcal{F}$, which can also be written as the warped product $\mathcal{B} \times_1 \mathcal{F}$. That is, for p, $q \in \mathcal{B}$ and $\phi, \psi \in \mathcal{F}$, the latter metric simplifies to

$$|(p,\phi)-(q,\psi)| = \sqrt{|p-q|^2+|\phi-\psi|^2}.$$

Cones. The *Euclidean cone* $\operatorname{Cone}\mathcal{F}$ over a metric space $\mathcal{F}$ can be written as the warped product $[0,\infty) \times_{\mathrm{id}} \mathcal{F}$. That is, for s, $t \in [0,\infty)$ and $\phi,\psi \in \mathcal{F}$, the metric is given by the cosine rule

$$|(s,\phi)-(t,\psi)| = \sqrt{s^2+t^2-2\cdot s\cdot t\cdot \cos\alpha},$$

where $\alpha = \min\{\pi, |\phi-\psi|\}$. (See Section 6.E.)

Instead of the Euclidean cosine rule, we may use the cosine rule in $\mathbb{M}^2(\kappa)$:

$$|(s,\phi)-(t,\psi)| = \tilde{\gamma}^{\kappa}\{\alpha; s,t\}.$$

This way we get *κ-cones* over $\mathcal{F}$, denoted by $\operatorname{Cone}^\kappa \mathcal{F} = [0,\infty) \times_{\operatorname{sn}^\kappa} \mathcal{F}$ for $\kappa \leqslant 0$ and $\operatorname{Cone}^\kappa \mathcal{F} = [0,\varpi^\kappa] \times_{\operatorname{sn}^\kappa} \mathcal{F}$ for $\kappa > 0$.

The 1-cone $\operatorname{Cone}^1 \mathcal{F}$ is also called the *spherical suspension* over $\mathcal{F}$; it is also denoted by $\operatorname{Susp}\mathcal{F}$. That is,

$$\operatorname{Susp}\mathcal{F} = [0,\pi] \times_{\sin} \mathcal{F}.$$

11.5. Exercise. Let $\mathcal{F}$ be a length space and let $A \subset \mathcal{F}$. Show that $\operatorname{Cone}^\kappa A$ is convex in $\operatorname{Cone}^\kappa \mathcal{F}$ if and only if A is π-convex in $\mathcal{F}$.

Doubling. The doubling space $\mathcal{W}$ of a metric space $\mathcal{V}$ on a closed subset $A \subset \mathcal{V}$ can be also defined as a special type of warped product. Consider the fiber $\mathbb{S}^0$ consisting of two points with distance 2 from each other. Then

$$\mathcal{W} \stackrel{\text{iso}}{=\!=} \mathcal{V} \times_{\operatorname{dist}_A} \mathbb{S}^0;$$

that is, $\mathcal{W}$ is isometric to the warped product with base $\mathcal{V}$, fiber $\mathbb{S}^0$, and warping function dist_A.

D. One-dimensional base

The following theorems provide conditions for the spaces and functions in a warped product with a one-dimensional base to have curvature bounds. These theorems are originally due to Valerii Berestovskii [**23**]. They are "baby" cases of the characterization of curvature bounds in warped products given in [**6**, **7**].

11.6. Theorem.

(a) *If $\mathcal{L}$ is a complete length* CBB(1) *space and* $\operatorname{diam}\mathcal{L} \leqslant \pi$, *then*

$$\begin{aligned} \operatorname{Susp}\mathcal{L} &= [0,\pi] \times_{\sin} \mathcal{L} && \textit{is}\ \ \mathrm{CBB}(1),\\ \operatorname{Cone}\mathcal{L} &= [0,\infty) \times_{\mathrm{id}} \mathcal{L} && \textit{is}\ \ \mathrm{CBB}(0),\\ \operatorname{Cone}^{-1}\mathcal{L} &= [0,\infty) \times_{\sinh} \mathcal{L} && \textit{is}\ \ \mathrm{CBB}(-1). \end{aligned}$$

Moreover, the converse also holds in each of the three cases.

(b) *If $\mathcal{L}$ is a complete length* CBB(0) *space, then*

$$\begin{aligned} &\mathbb{R} \times \mathcal{L} && \textit{is a complete length}\ \mathrm{CBB}(0)\ \textit{space},\\ &\mathbb{R} \times_{\exp} \mathcal{L} && \textit{is a complete length}\ \mathrm{CBB}(-1)\ \textit{space}. \end{aligned}$$

Moreover, the converse also holds in each of the two cases.

(c) *$\mathcal{L}$ is a complete length* CBB(−1) *space, if and only if the warped product $\mathbb{R} \times_{\cosh} \mathcal{L}$ is a complete length* CBB(−1) *space.*

11.7. Theorem. *Let $\mathcal{L}$ be a metric space.*

(a) *If $\mathcal{L}$ is* CAT(1)*, then*

$$\begin{aligned} \operatorname{Susp}\mathcal{L} &= [0,\pi]\times_{\sin}\mathcal{L} && \textit{is}\quad \mathrm{CAT}(1),\\ \operatorname{Cone}\mathcal{L} &= [0,\infty)\times_{\mathrm{id}}\mathcal{L} && \textit{is}\quad \mathrm{CAT}(0),\\ \operatorname{Cone}^{-1}\mathcal{L} &= [0,\infty)\times_{\sinh}\mathcal{L} && \textit{is}\quad \mathrm{CAT}(-1). \end{aligned}$$

Moreover, the converse also holds in each of the three cases.

(b) *If $\mathcal{L}$ is a complete length* CAT(0) *space, then $\mathbb{R}\times\mathcal{L}$ is* CAT(0) *and $\mathbb{R}\times_{\exp}\mathcal{L}$ is* CAT(−1)*. Moreover, the converse also holds in each of the two cases.*

(c) *$\mathcal{L}$ is* CAT(−1) *if and only if $\mathbb{R}\times_{\cosh}\mathcal{L}$ is* CAT(−1)*.*

In the proof of Theorems 11.6 and 11.7 we will use Proposition 11.8.

11.8. Proposition.

(a)

$$\begin{aligned} \operatorname{Susp}\mathbb{S}^{m-1} &= [0,\pi]\times_{\sin}\mathbb{S}^{m-1} \overset{\text{iso}}{=\!=} \mathbb{S}^m,\\ \operatorname{Cone}\mathbb{S}^{m-1} &= [0,\infty)\times_{\mathrm{id}}\mathbb{S}^{m-1} \overset{\text{iso}}{=\!=} \mathbb{E}^m,\\ \operatorname{Cone}^{-1}\mathbb{S}^{m-1} &= [0,\infty)\times_{\sinh}\mathbb{S}^{m-1} \overset{\text{iso}}{=\!=} \mathbb{M}^m(-1). \end{aligned}$$

(b)

$$\begin{aligned} \mathbb{R}\times\mathbb{E}^{m-1} &\overset{\text{iso}}{=\!=} \mathbb{E}^m,\\ \mathbb{R}\times_{\exp}\mathbb{E}^{m-1} &\overset{\text{iso}}{=\!=} \mathbb{M}^m(-1). \end{aligned}$$

(c)

$$\mathbb{R}\times_{\cosh}\mathbb{M}^{m-1}(-1) \overset{\text{iso}}{=\!=} \mathbb{M}^m(-1).$$

The proof is left to the reader.

Proof of Theorems 11.6 and 11.7. Each proof is based on the fiber-independence theorem (see 11.3) and the corresponding statement in Proposition 11.8.

Let us prove the last statement in 11.7(a); the remaining statements of this theorem are similar. Choose an arbitrary quadruple of points

$$(s,\phi),(t^1,\phi^1),(t^2,\phi^2),(t^3,\phi^3)\in[0,\infty)\times_{\sinh}\mathcal{L}.$$

Since $\operatorname{diam}\mathcal{L}\leqslant\pi$, the (1+3)-point comparison (see 10.8) provides a quadruple of points $\psi,\psi^1,\psi^2,\psi^3\in\mathbb{S}^3$ such that

$$|\psi-\psi^i|_{\mathbb{S}^3}=|\phi-\phi^i|_{\mathcal{L}}$$

and

$$|\psi^i-\psi^j|_{\mathbb{S}^3}\geqslant|\phi^i-\phi^j|_{\mathcal{L}}$$

for all i and j.

According to Proposition 11.8(a),

$$\operatorname{Cone}^{-1}\mathbb{S}^3 = [0,\infty)\times_{\sinh}\mathbb{S}^3 \overset{\text{iso}}{=\!=} \mathbb{M}^4(-1).$$

Consider the quadruple of points

$$(s,\psi),(t^1,\psi^1),(t^2,\psi^2),(t^3,\psi^3)\in\operatorname{Cone}^{-1}\mathbb{S}^3 = \mathbb{M}^4(-1).$$

By the fiber-independence theorem (see 11.3),

$$|(s,\psi)-(t^i,\psi^i)|_{[0,\infty)\times_{\sinh}\mathbb{S}^3} = |(s,\phi)-(t^i,\phi^i)|_{[0,\infty)\times_{\sinh}\mathcal{L}}$$

and

$$|(t^i,\psi^i)-(t^j,\psi^j)|_{[0,\infty)\times_{\sinh}\mathbb{S}^3} \geqslant |(t^i,\phi^i)-(t^j,\phi^j)|_{[0,\infty)\times_{\sinh}\mathcal{L}}$$

for all i and j. Since four points of $\mathbb{M}^4(-1)$ lie in an isometric copy of $\mathbb{M}^3(-1)$, it remains to apply Exercise 8.3.

Now let us prove the converse to Proposition 11.8(a). Choose a quadruple $\phi,\phi^1,\phi^2,\phi^3\in\mathcal{L}$ with all distances smaller than $\frac{\pi}{2}$. Choose small $s>0$ and let t_i be the hypotenuse in a hyperbolic right triangle with angle $|\phi-\phi^i|_{\mathcal{L}}$ and the adjacent side s. Observe that the $\mathbb{M}(-1)$ model angles of the quadruple (s,ϕ), (t^1,ϕ^1), (t^2,ϕ^2), (t^3,ϕ^3) in $[0,\infty)\times_{\sinh}\mathcal{L}$ at (s,ϕ) are the same as the $\mathbb{M}(1)$ model angles of the quadruple ϕ, ϕ^1, ϕ^2, $\phi^3\in\mathcal{L}$ at ϕ. Whence the CBB(1) comparison holds for $\phi,\phi^1,\phi^2,\phi^3$; in particular, $\mathcal{L}$ is locally CBB(1). It remains to apply the globalization theorem (see 8.31).

The proof of Theorem 11.7 is nearly identical, but one has to apply the (2+2)-comparison (Exercise 9.5). □

11.9. Exercise. The *spherical join* $\mathcal{U}\star\mathcal{V}$ of two metric spaces $\mathcal{U}$ and $\mathcal{V}$ is defined as the unit sphere equipped with the angle metric in the product of Euclidean cones $\operatorname{Cone}\mathcal{U}\times\operatorname{Cone}\mathcal{V}$.

Assume $\mathcal{U}$ and $\mathcal{V}$ are nonempty spaces.

(a) Show that $\mathcal{U}\star\mathcal{V}$ is CAT(1) if and only if $\mathcal{U}$ and $\mathcal{V}$ are CAT(1).

(b) Suppose $\mathcal{U}$ and $\mathcal{V}$ have $\operatorname{diam}\leqslant\pi$. Show that $\mathcal{U}\star\mathcal{V}$ is CBB(1) if and only if $\mathcal{U}$ and $\mathcal{V}$ are CBB(1).

E. Remarks

Let us formulate general results on curvature bounds of warped products proved by the first author and Richard Bishop [**7**].

11.10. Theorem. *Let $\mathcal{B}$ be a complete finite-dimensional* CBB(κ) *length space, and let $f\colon \mathcal{B} \to \mathbb{R}_{\geqslant 0}$ be a locally Lipschitz function. Denote by $Z \subset \mathcal{B}$ the zero set of the restriction of f to the boundary $\partial\mathcal{B}$ of $\mathcal{B}$.*

Suppose that $\mathcal{W}$ is doubling of $\mathcal{B}$ along the closure of $\partial\mathcal{B} \setminus Z$, and $\bar f\colon \mathcal{W} \to \mathbb{R}_{\geqslant 0}$ is the natural extension of f. Assume that $\mathcal{W}$ is CBB(κ) *and $\bar f'' + \kappa\cdot \bar f \leqslant 0$.*

Suppose $\mathcal{F}$ is a complete finite-dimensional CBB(κ') *space. Then the warped product $\mathcal{B} \times_f \mathcal{F}$ is* CBB(κ) *in the following two cases:*

(a) *If $Z = \varnothing$ and $\kappa' \geqslant \kappa\cdot f^2(b)$ for any $b \in \mathcal{B}$.*
(b) *If $Z \neq \varnothing$ and $|d_z f|^2 \leqslant \kappa'$ for any $z \in Z$.*

We mention that in the setting of this theorem, f necessarily vanishes only at boundary points if f is not identically 0.

11.11. Theorem. *Let $\mathcal{B}$ be a complete* CAT(κ) *length space, and let the function $f\colon \mathcal{B} \to \mathbb{R}_{\geqslant 0}$ satisfy $f'' + \kappa\cdot f \geqslant 0$, where f is Lipschitz on bounded sets or B is locally compact. Denote by $Z \subset \mathcal{B}$ the zero set of f. Suppose $\mathcal{F}$ is a complete* CAT(κ') *space. Then the warped product $\mathcal{B} \times_f \mathcal{F}$ is* CAT(κ) *in the following two cases:*

(a) *If $Z = \varnothing$ and $\kappa' \leqslant \kappa\cdot f^2(b)$ for any $b \in \mathcal{B}$.*
(b) *If $Z \neq \varnothing$ and $[(d_z f)\uparrow_{[zb]}]^2 \geqslant \kappa'$ for any minimizing geodesic $[zb]$ from Z to a point $b \in \mathcal{B}$ and $\kappa' \leqslant \kappa\cdot f^2(b)$ for any $b \in \mathcal{B}$ such that $\operatorname{dist}_Z b \geqslant \frac{\varpi^\kappa}{2}$.*

Chapter 12

Polyhedral spaces

A. Definitions

12.1. Definition. A length space $\mathcal{P}$ is called a *piecewise* $\mathbb{M}(\kappa)$ if it admits a finite triangulation τ such that an arbitrary simplex σ in τ is isometric to a simplex in the model space $\mathbb{M}^{\dim\sigma}(\kappa)$.

By *triangulation* of a piecewise $\mathbb{M}(\kappa)$ space we will understand a triangulation as in the definition. If we do not wish to specify κ, we will say that $\mathcal{P}$ is a *polyhedral space.*

By rescaling we can assume that $\kappa = 1, 0$, or -1.

(a) Piecewise $\mathbb{M}(1)$ spaces will also be called *spherical polyhedral spaces*;

(b) Piecewise $\mathbb{M}(0)$ spaces will also be called *Euclidean polyhedral spaces*;

(c) Piecewise $\mathbb{M}(-1)$ spaces will also be called *hyperbolic polyhedral spaces.*

Note that according to the above definition, all polyhedral spaces are compact. However, most of the statements below admit straightforward generalizations to *locally polyhedral space*; that is, complete length spaces, any point of which admits a closed neighborhood isometric to a polyhedral space. The latter class of spaces includes in particular infinite covers of polyhedral spaces.

The dimension of a polyhedral space $\mathcal{P}$ is defined as the maximal dimension of a simplex in one (and therefore any) triangulation of $\mathcal{P}$.

Links. Let $\mathcal{P}$ be a polyhedral space and let σ be a simplex in its triangulation τ.

The simplexes that contain σ form an abstract simplicial complex called the *link* of σ, denoted by Link_σ. If $m = \dim\sigma$, then the set of vertexes of Link_σ is formed by the $(m+1)$-simplexes that contain σ; the set of its edges are formed by the $(m+2)$-simplexes that contain σ, and so on.

The link Link_σ can be identified with the subcomplex of τ formed by all the simplexes σ' such that $\sigma \cap \sigma' = \varnothing$ but both σ and σ' are faces of a simplex of τ.

The points in Link_σ can be identified with the normal directions to σ at a point in the interior of σ. The angle metric between directions makes Link_σ into a spherical polyhedral space. We will always consider the link with this metric.

Tangent space and space of directions. Let τ be a triangulation of a polyhedral space $\mathcal{P}$. If a point $p \in \mathcal{P}$ lies in the interior of a k-simplex σ of τ, then the tangent space $\mathrm{T}_p\,\mathcal{P}$ is naturally isometric to

$$\mathbb{E}^k \times (\mathrm{Cone}\,\mathrm{Link}_\sigma).$$

Equivalently, the space of directions Σ_p can be isometrically identified with the kth spherical suspension over Link_σ; that is,

$$\Sigma_p \overset{\mathrm{iso}}{=\!=} \mathrm{Susp}^k(\mathrm{Link}_\sigma).$$

If $\mathcal{P}$ is an m-dimensional polyhedral space, then for any $p \in \mathcal{P}$ the space of directions Σ_p is a spherical polyhedral space of dimension at most $m-1$.

In particular, for any point p in the interior of a simplex σ, the isometry class of Link_σ and $k = \dim \sigma$ determine the isometry class of Σ_p and the other way around.

A small neighborhood of p is isometric to a neighborhood of the tip of the κ-cone over Σ_p. In fact, if this property holds at any point of a compact length space $\mathcal{P}$, then $\mathcal{P}$ is a piecewise $\mathbb{M}(\kappa)$ space [**102**].

B. Curvature bounds

Recall that ℓ-simply connected spaces are defined in Definition 9.66.

The following theorem provides a combinatorial description of polyhedral spaces with curvature bounded above.

12.2. Theorem. *Let $\mathcal{P}$ be a piecewise $\mathbb{M}(\kappa)$ space and let τ be a triangulation of $\mathcal{P}$. Then*

(a) *$\mathcal{P}$ is locally* CAT(κ) *if and only if any connected component of the link of any simplex σ in τ is $(2\cdot\pi)$-simply connected. Equivalently, if and only if any closed local geodesic in* Link_σ *has length at least $2\cdot\pi$.*

(b) *$\mathcal{P}$ is a complete length* CAT(κ) *space if and only if $\mathcal{P}$ is $(2\cdot\varpi^\kappa)$-simply connected and any connected component of the link of any simplex σ in τ is $(2\cdot\pi)$-simply connected.*

Proof. The only-if part is evident by the generalized Hadamard–Cartan theorem (see 9.67), Exercise 9.72, and Theorem 11.6.

To prove the if part, let us apply induction on $\dim \mathcal{P}$. The *base* case $\dim \mathcal{P} = 0$ is evident.

Induction Step. Assume that the theorem is proved in the case $\dim \mathcal{P} < m$. Suppose $\dim \mathcal{P} = m$.

Fix a point $p \in \mathcal{P}$. A neighborhood of p is isometric to a neighborhood of the tip in the κ-cone over Σ_p. By Theorem 11.6(a), it is sufficient to show that

(1) Σ_p *is* CAT(1).

Note that Σ_p is a spherical polyhedral space and its links are isometric to links of $\mathcal{P}$. By the induction hypothesis, Σ_p is locally CAT(1). Applying the generalized Hadamard–Cartan theorem 9.67, we get (1).

Part (b) follows from the generalized Hadamard–Cartan theorem. □

12.3. Exercise. Show that any metric tree is CAT(κ) for any κ.

12.4. Exercise. Show that if in a Euclidean polyhedral space $\mathcal{P}$ any two points can be connected by a unique geodesic, then $\mathcal{P}$ is CAT(0).

The following theorem provides a combinatorial description of polyhedral spaces with curvature bounded below.

12.5. Theorem. *Let $\mathcal{P}$ be a piecewise $\mathbb{M}(\kappa)$ space and let τ be a triangulation of $\mathcal{P}$. Then $\mathcal{P}$ is* CBB(κ) *if and only if the following conditions hold.*

(a) *τ is* pure; *that is, any simplex in τ is a face of some simplex of dimension exactly m.*
(b) *The link of any simplex of dimension $m-1$ is formed by a single point or two points.*
(c) *Any link of any simplex of dimension $m-2$ has diameter at most π.*
(d) *The link of any simplex of dimension $\leqslant m-2$ is connected.*

Remarks. Condition (d) can be reformulated in the following way:

(d)′ Any path $\gamma\colon [0,1] \to \mathcal{P}$ can be approximated by paths γ_n that cross only simplexes of dimension m and $m-1$.

Further, modulo the other conditions, condition (c) is equivalent to the following:

(c)′ The link of any simplex of dimension $m-2$ is isometric to a circle of length $\leqslant 2\cdot\pi$ or a closed real interval of length $\leqslant \pi$.

Proof. We will prove the if part. The only-if part is similar and is left to the reader.

We apply induction on m. The *base* case $m = 1$ follows from assumption (b).

Step. Assume that the theorem is proved for polyhedral spaces of dimension less than m. Suppose $\dim \mathcal{P} = m$.

According to Globalization Theorem 8.31, it is sufficient to show that $\mathcal{P}$ is locally CBB(κ).

Fix $p \in \mathcal{P}$. Note that a spherical neighborhood of p is isometric to a spherical neighborhood of the tip of the tangent κ-cone

$$\mathrm{T}_p^\kappa = \mathrm{Cone}^\kappa(\Sigma_p).$$

Hence it is sufficient to show that

(2) T_p^κ *is* CBB(κ) *for any* $p \in \mathcal{P}$.

By Theorem 11.6(a), the latter is equivalent to

(3) $\operatorname{diam} \Sigma_p \leqslant \pi$ *and* Σ_p *is* CBB(1).

If $m = 2$, then (3) follows from (b).

To prove the case $m \geqslant 3$, note that Σ_p is an $(m-1)$-dimensional spherical polyhedral space and all the conditions of the theorem hold for Σ_p. It remains to apply the induction hypothesis. □

12.6. Exercise. Assume $\mathcal{P}$ is a piecewise $\mathbb{M}(\kappa)$ space and $\dim \mathcal{P} \geqslant 2$. Show that

(a) if $\mathcal{P}$ is CBB(κ'), then $\kappa' \leqslant \kappa$ and $\mathcal{P}$ is CBB(κ),

(b) if $\mathcal{P}$ is CAT(κ'), then $\kappa' \geqslant \kappa$ and $\mathcal{P}$ is CAT(κ).

C. Flag complexes

12.7. Definition. A simplicial complex $\mathcal{S}$ is *flag* if whenever $\{v_0, \dots, v_k\}$ is a set of distinct vertexes of $\mathcal{S}$ that are pairwise joined by edges, then the vertexes $v_0, \dots, v_k$ span a k-simplex in $\mathcal{S}$.

If the above condition is satisfied for $k = 2$, then we say $\mathcal{S}$ satisfies the *no-triangle condition.*

Note that every flag complex is determined by its 1-skeleton.

12.8. Proposition. *A simplicial complex $\mathcal{S}$ is flag if and only if $\mathcal{S}$, as well as all the links of all its simplexes, satisfy the no-triangle condition.*

From the definition of flag complex we get the following:

12.9. Observation. Any link of a flag complex is flag.

Proof of Proposition 12.8. By Observation 12.9, the no-triangle condition holds for any flag complex and all its links.

Now assume a complex $\mathcal{S}$ and all its links satisfy the no-triangle condition. It follows that $\mathcal{S}$ includes a 2-simplex for each triangle. Applying the same observation for each edge we get that $\mathcal{S}$ includes a 3-simplex for any complete graph with four vertexes. Repeating this observation for triangles, 4-simplexes, 5-simplexes, and so on we get that $\mathcal{S}$ is flag. □

Right-angled triangulation. A triangulation of a spherical polyhedral space is called *right-angled* if each simplex of the triangulation is isometric to a spherical simplex all of whose angles are right. Similarly, we say that a simplicial complex is equipped with a *right-angled spherical metric* if it is a length metric and each simplex is isometric to a spherical simplex all of whose angles are right.

Spherical polyhedral CAT(1) spaces glued from right-angled simplexes admit the following characterization discovered by Michael Gromov [**76**, p. 122].

12.10. Flag condition. *Assume that a spherical polyhedral space $\mathcal{P}$ admits a right-angled triangulation τ. Then $\mathcal{P}$ is* CAT(1) *if and only if τ is flag.*

Proof.

Only-if part. Assume there are three vertexes v_1, v_2, and v_3 of τ that are pairwise joined by edges but do not span a simplex. Note that in this case

$$\measuredangle\left[v_1\,{}^{v_2}_{v_3}\right] = \measuredangle\left[v_2\,{}^{v_3}_{v_1}\right] = \measuredangle\left[v_3\,{}^{v_1}_{v_1}\right] = \pi.$$

Equivalently,

(1) *The concatenation of the geodesics $[v_1v_2]$, $[v_2v_3]$, and $[v_3v_1]$ forms a closed local geodesic in $\mathcal{P}$.*

Now assume that $\mathcal{P}$ is CAT(1). Then by Theorem 11.6(a), $\mathrm{Link}_\sigma\mathcal{P}$ is a compact length CAT(1) space for every simplex σ in τ.

Each of these links is a right-angled spherical complex and by Theorem 12.2, none of these links can contain a geodesic circle of length less than $2\cdot\pi$.

Therefore Proposition 12.8 and (1) imply the only-if part.

If part. By Observation 12.9 and Theorem 12.2, it is sufficient to show that any closed local geodesic γ in a flag complex $\mathcal{S}$ with right-angled metric has length at least $2\cdot\pi$.

Fix a flag complex $\mathcal{S}$. Recall that the *star* of a vertex v (briefly $\overline{\mathrm{Star}}_v$) is formed by all the simplexes containing v. Similarly, Star_v, the open star of v, is the union of all simplexes containing v with faces opposite v removed.

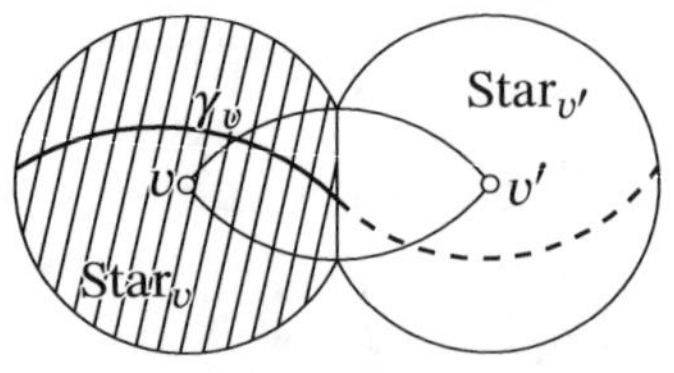

Choose a simplex σ that contains a point of γ. Let v be a vertex of σ. Set $f(t) = \cos|v - \gamma(t)|$. Note that

$$f''(t) + f(t) = 0$$

if $f(t) > 0$. Since the zeroes of sine are π apart, γ spends time π on every visit to Star_v.

After leaving Star_v, the local geodesic γ must enter another simplex, say σ', which has a vertex v' not joined to v by an edge.

Since τ is flag, the open stars Star_v and $\mathrm{Star}_{v'}$ do not overlap. The same argument as above shows that γ spends time π on every visit to $\mathrm{Star}_{v'}$. Therefore the total length of γ is at least $2\cdot\pi$. ☐

12.11. Exercise. Show that the barycentric subdivision of any simplicial complex is flag. Conclude that any finite simplicial complex is homeomorphic to a compact length CAT(1) space.

12.12. Exercise. Let p be a point in a product of metric trees. Show that a closed geodesic in the space of directions Σ_p has length either $2\cdot\pi$ or at least $3\cdot\pi$.

12.13. Exercise. Assume that a spherical polyhedral space $\mathcal{P}$ admits a triangulation τ such that all edge lengths of all simplexes in τ are at least $\frac{\pi}{2}$. Show that $\mathcal{P}$ is CAT(1) if τ is flag.

12.14. Exercise. Let $\phi_1, \phi_2, \dots, \phi_k : \mathcal{U} \to \mathcal{U}$ be commuting short retractions of a complete length CAT(0) space; that is,

- $\phi_i \circ \phi_i = \phi_i$ for each i;
- $\phi_i \circ \phi_j = \phi_j \circ \phi_i$ for any i and j;
- $|\phi_i(x) - \phi_i(y)|_{\mathcal{U}} \leqslant |x - y|_{\mathcal{U}}$ for each i and any $x, y \in \mathcal{U}$.

Set $A_i = \operatorname{Im}\phi_i$ for all i. Note that each A_i is a weakly convex set.

Assume Γ is a finite graph (without loops and multiple edges) with edges labeled by $1, 2, \dots, n$. Denote by $\mathcal{U}^\Gamma$ the space obtained by taking a copy of $\mathcal{U}$ for each vertex of Γ and gluing two such copies along A_i if the corresponding vertexes are joined by an edge labeled by i.

Show that $\mathcal{U}^\Gamma$ is CAT(0).

The space of trees. The following construction is given by Louis Billera, Susan Holmes, and Karen Vogtmann [**27**].

Let $\mathcal{T}_n$ be the set of all metric trees with n end-vertexes labeled by $a_1, \dots, a_n$. To describe one tree in $\mathcal{T}_n$, we may fix a topological tree τ with end vertexes a_1, $\dots, a_n$ and all the other vertexes of degree 3, and prescribe the lengths of $2\cdot n-3$ edges. If the length of an edge is 0, we assume that edge degenerates; such a tree can be also described using a different topological tree τ'. The subset of $\mathcal{T}_n$ corresponding to the given topological tree τ can be identified with a convex closed cone in $\mathbb{R}^{2\cdot n-3}$. Equip each such subset with the metric induced from $\mathbb{R}^{2\cdot n-3}$ and consider the length metric on $\mathcal{T}_n$ induced by these metrics.

12.15. Exercise. Show that $\mathcal{T}_n$ with the described metric is CAT(0).

Cubical complexes. The definition of a cubical complex mostly repeats the definition of a simplicial complex, with simplexes replaced by cubes.

Formally, a *cubical complex* is defined as a subcomplex of the unit cube in Euclidean space of large dimension; that is, a collection of faces of the cube, that with each face contains all its subfaces. Each cube face in this collection will be called a *cube* of the cubical complex.

Note that according to this definition, any cubical complex is finite, that is, contains a finite number of cubes.

The union of all the cubes in a cubical complex $\mathcal{Q}$ will be called its *underlying space*; it will be denoted by $\mathcal{Q}$ or by $\underline{\mathcal{Q}}$ if we need to emphasize that we are talking about a topological space, not a complex. A homeomorphism from $\underline{\mathcal{Q}}$ to a topological space $\mathcal{X}$ is called a *cubulation of* $\mathcal{X}$.

The underlying space of a cubical complex $\mathcal{Q}$ will be always considered with the length metric induced from $\mathbb{R}^N$. In particular, with this metric, each cube of $\mathcal{Q}$ is isometric to the unit cube of the same dimension.

It is straightforward to construct a triangulation of $\underline{\mathcal{Q}}$ such that each simplex is isometric to a Euclidean simplex. In particular, $\underline{\mathcal{Q}}$ is a Euclidean polyhedral space.

The link of each cube in a cubical complex admits a natural right-angled triangulation; each simplex corresponds to an adjusted cube.

12.16. Exercise. Show that a cubical complex $\mathcal{Q}$ is locally CAT(0) if and only if the link of each vertex in $\mathcal{Q}$ is flag.

D. Remarks

The condition on polyhedral CAT(κ) spaces given in Theorem 12.2 might look easy to use, but in fact, it is hard to check even in simple cases. For example, the description of those coverings of $\mathbb{S}^3$ that branch at three great circles and are CAT(1) requires quite a bit of work; an answer is given by Ruth Charney and Michael Davis [**49**].

Analogues of the flag condition for spherical Coxeter simplexes could resolve the following problem.

12.17. Braid space problem. *Consider $\mathbb{C}^n$ with coordinates $z_1, \dots, z_n$. Let us remove from $\mathbb{C}^n$ the complex hyperplanes $z_i = z_j$ for all $i \neq j$, pass to the universal cover, and consider the completion $\mathcal{B}_n$ of the obtained space.*

Is it true that $\mathcal{B}_n$ is CAT(0) *for any n?*

The above question has an affirmative answer for $n \leqslant 3$ and is open for all $n \geqslant 4$ [**49**, **123**].

Recall that by the Hadamard–Cartan theorem (see 9.65), any complete length CAT(0) space is contractible. Therefore any complete length locally CAT(0) space is *aspherical*; that is, has contractible universal cover.

This observation can be used together with Exercise 12.16 to construct examples of exotic aspherical spaces; for example, compact topological manifolds with universal cover not homeomorphic to a Euclidean space. A survey on the subject is given by Michael Davis [**58**]; a more elementary introduction to the subject is given by the authors [**10**, Chapter 3].

The flag condition also leads to the so-called *hyperbolization* procedure, a flexible tool for constructing aspherical spaces; a good survey on the subject is given by Ruth Charney and Michael Davis [**50**].

The CAT(0) property of a cube complex admits interesting (and useful) geometric descriptions if one replaces the ℓ^2-metric with a natural ℓ^1- or ℓ^∞-metric on each cube. The following statement was proved by Brian Bowditch [**32**].

12.18. Theorem. *The following three conditions are equivalent.*

(a) *A cube complex Q equipped with ℓ^2-metric is* CAT(0).

(b) *A cube complex Q equipped with ℓ^∞-metric is* injective; *that is, for any metric space $\mathcal{X}$ with a subset A, any short map $A \to (Q, \ell^\infty)$ can be extended to a short map $\mathcal{X} \to (Q, \ell^\infty)$.*

(c) *A cube complex Q equipped with ℓ^1-metric is* median; *that is, for any three points x, y, z there is a unique point m (called the* median *of x, y, and z) that lies on some geodesics $[xy]$, $[xz]$ and $[yz]$.*

Part 3

Structure and tools

Chapter 13

First order differentiation

A. Ultratangent space

The following theorem is often used together with the observation that the ultralimit of any sequence of length spaces is geodesic (see Observation 4.9).

13.1. Theorem.

(a) *If $\mathcal{L}$ is a complete length* CBB(κ) *space and $p \in \mathcal{L}$, then* T_p^ω *is* CBB(0).

(b) *If $\mathcal{U}$ is a complete length* CAT(κ) *space and $p \in \mathcal{U}$, then* T_p^ω *is* CAT(0).

The proofs of both parts are nearly identical.

Proof.

(a). Since $\mathcal{L}$ is a complete length CBB(κ) space, then its blowup $n\cdot\mathcal{L}$ (see Section 6.H) is a complete length CBB(κ/n^2) space. By Proposition 8.4, the ω-blowup $\omega\cdot\mathcal{L}$ is CBB(0) and so is T_p^ω as a metric component of $\omega\cdot\mathcal{L}$.

(b). Since $\mathcal{U}$ is a complete length CAT(κ) space, then its blowup $n\cdot\mathcal{U}$ is CAT(κ/n^2). By Proposition 9.7, $\omega\cdot\mathcal{U}$ is CAT(0) and so is T_p^ω as a metric component of $\omega\cdot\mathcal{U}$. □

Recall that the tangent space T_p can be considered as a subset of T_p^ω (see Theorem 6.17). Therefore we have the following:

13.2. Corollary.

(a) *If $\mathcal{L}$ is a complete length* CBB(κ) *space and $p \in \mathcal{L}$, then* T_p *is* CBB(0). *Moreover,* T_p *satisfies the* $(1+n)$*-point comparison* (*see* 10.8).

(b) *If $\mathcal{U}$ is a complete length* CAT(κ) *space and $p \in \mathcal{U}$, then* T_p *is* CAT(0). *Moreover,* T_p *satisfies the* $(2n+2)$*-comparison (Exercise* 10.20).

13.3. Proposition. *Assume $\mathcal{Z}$ is a complete length* CBB *or* CAT *space and that $f\colon \mathcal{Z} \multimap \mathbb{R}$ is a semiconcave locally Lipscitz subfunction. Then for any $p \in$ $\in \operatorname{Dom} f$, the ultradifferential $\mathbf{d}_p^\omega\colon \mathrm{T}_p^\omega \to \mathbb{R}$ is a concave function.*

Proof. Fix a geodesic $[x^\omega y^\omega]$ in T_p^ω.

It is sufficient to show that for any subarc $[\bar{x}^\omega \bar{y}^\omega]$ of $[x^\omega y^\omega]$ that does not contain the ends there is a sequence of geodesics $[\bar{x}^n \bar{y}^n]$ in $n\cdot\mathcal{Z}$ converging to $[\bar{x}^\omega \bar{y}^\omega]$.

Choose any sequences $\bar{x}^n, \bar{y}^n \in n\cdot\mathcal{Z}$ such that $\bar{x}^n \to \bar{x}^\omega, \bar{y}^n \to \bar{y}^\omega$ as $n \to \omega$. We can assume that there is a geodesic $[\bar{x}^n \bar{y}^n]_{n\cdot\mathcal{Z}}$ for any n; see Theorem 8.11 and 9.8. Note that $[\bar{x}^n \bar{y}^n]$ converges to $[\bar{x}^\omega \bar{y}^\omega]$ as $n \to \omega$. The latter holds trivially in the CAT case, and the CBB case follows from Corollary 8.38. □

B. Length property of tangent space

13.4. Theorem. *Let $\mathcal{U}$ be a complete length* CAT(κ) *space and let $p \in \mathcal{U}$. Then* $\mathrm{T}_p\,\mathcal{U}$ *is a length space.*

This theorem together with Corollary 13.2 imply the following.

13.5. Corollary. *For any point p in a complete length* CAT(κ) *space, the tangent space* T_p *is a complete length* CAT(0) *space.*

Proof of Theorem 13.4. Since $\mathrm{T}_p = \operatorname{Cone}\Sigma_p$, it is sufficient to show that for any hinge $\left[p\,{}^x_y\right]$ such that $\measuredangle\left[p\,{}^x_y\right] < \pi$ and any $\varepsilon > 0$, there is $z \in \mathcal{U}$ such that

$$\measuredangle\left[p\,{}^x_z\right] < \tfrac{1}{2}\cdot\measuredangle\left[p\,{}^x_y\right] + \varepsilon \quad\text{and}\quad \measuredangle\left[p\,{}^y_z\right] < \tfrac{1}{2}\cdot\measuredangle\left[p\,{}^x_y\right] + \varepsilon. \tag{1}$$

Fix a small $\delta > 0$. Let $\bar{x} \in \,]px]$ and $\bar{y} \in \,]py]$ denote the points such that $|p - \bar{x}| = |p - \bar{y}| = \delta$. Let z denote the midpoint between $\bar{x}$ and $\bar{y}$.

Since δ is small, we can assume that

$$\tilde{\measuredangle}^\kappa\left(p\,{}^{\bar{x}}_{\bar{y}}\right) < \measuredangle\left[p\,{}^x_y\right] + \varepsilon.$$

By Alexandrov's Lemma 6.3, we have

$$\tilde{\measuredangle}^\kappa\left(p\,{}^{\bar{x}}_{z}\right) + \tilde{\measuredangle}^\kappa\left(p\,{}^{\bar{y}}_{z}\right) < \tilde{\measuredangle}^\kappa\left(p\,{}^{\bar{x}}_{\bar{y}}\right).$$

By construction,

$$\tilde{\measuredangle}^\kappa\left(p\,{}^{\bar{x}}_{z}\right) = \tilde{\measuredangle}^\kappa\left(p\,{}^{\bar{y}}_{z}\right).$$

Applying the angle comparison (see 9.14(c)), we get (1). □

The following example was constructed by Stephanie Halbeisen [**80**]. It shows that an analogous statement does not hold for CBB spaces. If the dimension is finite, such examples do not exist; for proper spaces the question is open; see Open Question 13.41.

13.6. Example. There is a complete length CBB space $\check{\mathcal{L}}$ with a point $p \in \check{\mathcal{L}}$ such that the space of directions $\Sigma_p\check{\mathcal{L}}$ is not a π-length space, and therefore the tangent space $\mathrm{T}_p\,\check{\mathcal{L}}$ is not a length space.

Construction. Let $\mathbb{H}$ be a Hilbert space formed by infinite sequences of real numbers $\boldsymbol{x} = (x_0, x_1, \dots)$ with the ℓ^2-norm $|\boldsymbol{x}|^2 = \sum_i (x_i)^2$. Fix $\varepsilon = 0.001$ and consider two functions $f, \check{f} : \mathbb{H} \to \mathbb{R}$:

$$f(\boldsymbol{x}) = |\boldsymbol{x}|,$$

$$\check{f}(\boldsymbol{x}) = \max\left\{|\boldsymbol{x}|, \max_{n\geqslant 1}\{(1+\varepsilon)\cdot x_n - \tfrac{1}{n}\}\right\}.$$

Both of these functions are convex and Lipschitz, therefore their graphs in $\mathbb{H}\times\mathbb{R}$ equipped with its length metric form infinite-dimensional Alexandrov spaces, say $\mathcal{L}$ and $\check{\mathcal{L}}$ (this is proved formally in Lemma 13.7).

Let p be the origin of $\mathbb{H} \times \mathbb{R}$. Note that $\check{\mathcal{L}} \cap \mathcal{L}$ is a star-shaped subset of $\mathbb{H}$ with center at p. Further, $\check{\mathcal{L}} \setminus \mathcal{L}$ consists of a countable number of disjoint sets

$$\Omega_n = \left\{ (\boldsymbol{x}, \check{f}(\boldsymbol{x})) \in \check{\mathcal{L}} : (1+\varepsilon)\cdot x_n - \tfrac{1}{n} > |\boldsymbol{x}| \right\}.$$

Note that $|\Omega_n - p| > \frac{1}{n}$ for each n. It follows that for any geodesic $[pq]$ in $\check{\mathcal{L}}$, a small subinterval $[p\bar{q}] \subset [pq]$ is a straight line segment in $\mathbb{H} \times \mathbb{R}$, and also a geodesic in $\mathcal{L}$. Thus we can treat $\Sigma_p\mathcal{L}$ and $\Sigma_p\check{\mathcal{L}}$ as one set, with two angle metrics $\measuredangle$ and $\check{\measuredangle}$. Let us denote by $\measuredangle_{\mathbb{H}\times\mathbb{R}}$ the angle in $\mathbb{H} \times \mathbb{R}$.

The space $\mathcal{L}$ is isometric to the Euclidean cone over $\Sigma_p\mathcal{L}$ with vertex at p; $\Sigma_p\mathcal{L}$ is isometric to a sphere in Hilbert space with radius $\frac{1}{\sqrt{2}}$. In particular, $\measuredangle$ is the length metric of $\measuredangle_{\mathbb{H}\times\mathbb{R}}$ on $\Sigma_p\mathcal{L}$.

Therefore in order to show that $\check{\measuredangle}$ does not define a length metric on $\Sigma_p\mathcal{L}$, it is sufficient to construct a pair of directions (ξ_+, ξ_-) such that

$$\check{\measuredangle}(\xi_+, \xi_-) < \measuredangle(\xi_+, \xi_-).$$

Set $\boldsymbol{e}_0 = (1, 0, 0, \dots)$, $\boldsymbol{e}_1 = (0, 1, 0, \dots), \dots \in \mathbb{H}$. Consider the following two half-lines in $\mathbb{H} \times \mathbb{R}$:

$$\gamma_+(t) = \tfrac{t}{\sqrt{2}}\cdot(\boldsymbol{e}_0, 1) \quad\text{and}\quad \gamma_-(t) = \tfrac{t}{\sqrt{2}}\cdot(-\boldsymbol{e}_0, 1), \quad t \in [0, +\infty).$$

They form unit-speed geodesics in both $\mathcal{L}$ and $\check{\mathcal{L}}$. Let $\xi_\pm$ be the directions of $\gamma_\pm$ at p. Denote by σ_n the half-planes in $\mathbb{H}$ spanned by $\boldsymbol{e}_0$ and $\boldsymbol{e}_n$; that is, $\sigma_n = \{\, x\cdot\boldsymbol{e}_0 + y\cdot\boldsymbol{e}_n : y \geqslant 0\,\}$. Consider a sequence of two-dimensional sectors $Q_n = \check{\mathcal{L}}\cap(\sigma_n\times\mathbb{R})$. For each n, the sector Q_n intersects Ω_n and is bounded by two geodesic half-lines $\gamma_\pm$. Note that $Q_n \xrightarrow[\mathrm{GH}]{} Q$, where Q is a solid Euclidean angle in $\mathbb{E}^2$ with angle measure $\beta < \measuredangle(\xi_+,\xi_-) = \frac{\pi}{\sqrt{2}}$. Indeed, Q_n is path-isometric to the subset of $\mathbb{E}^3$ described by

$$y \geqslant 0 \quad \text{and} \quad z = \max\left\{\sqrt{x^2+y^2}, (1+\varepsilon)\cdot y - \tfrac{1}{n}\right\}$$

with length metric. Thus its limit Q is path-isometric to the subset of $\mathbb{E}^3$ described by

$$y \geqslant 0 \quad \text{and} \quad z = \max\left\{\sqrt{x^2+y^2}, (1+\varepsilon)\cdot y\right\}$$

with length metric. In particular, for any $t, \tau \geqslant 0$,

$$\begin{aligned}|\gamma_+(t) - \gamma_-(\tau)|_{\mathcal{L}} &\leqslant \lim_{n\to\infty} |\gamma_+(t) - \gamma_-(\tau)|_{Q_n} \\ &= \tilde{\gamma}^0\{\beta; t, \tau\}.\end{aligned}$$

That is, $\check{\measuredangle}(\xi_+,\xi_-) \leqslant \beta < \measuredangle(\xi_+,\xi_-)$. $\square$

13.7. Lemma. *Let $\mathbb{H}$ be a Hilbert space, let $f\colon \mathbb{H} \to \mathbb{R}$ be a convex Lipschitz function, and let $S \subset \mathbb{H}\times\mathbb{R}$ be the graph of f equipped with the length metric. Then S is* CBB(0).

Proof. Recall that for a subset $X \subset \mathbb{H}\times\mathbb{R}$, we will denote by $|{*}-{*}|_X$ the length metric on X.

By the theorem of Sergei Buyalo [**46**], sharpened by the authors in [**9**], any convex hypersurface in a Euclidean space, equipped with the length metric, is nonnegatively curved. Thus it is sufficient to show that for any four-point set $\{x_0, x_1, x_2, x_3\} \subset S$, there is a finite-dimensional subspace $E \subset \mathbb{H}\times\mathbb{R}$ such that $\{x_i\} \in E$ and $|x_i - x_j|_{S\cap E}$ is arbitrarily close to $|x_i - x_j|_S$.

Clearly $|x_i - x_j|_{S\cap E} \geqslant |x_i - x_j|_S$; thus it is sufficient to show that for given $\varepsilon > 0$ one can choose E so that

$$|x_i - x_j|_{S\cap E} < |x_i - x_j|_S + \varepsilon. \tag{2}$$

For each pair (x_i, x_j), choose a polygonal line β_{ij} connecting x_i, x_j that lies under S (that is, outside of $\operatorname{Conv} S$) in $\mathbb{H}\times\mathbb{R}$ and has length at most $|x_i - x_j|_S + \varepsilon$. Let E be the affine hull of all the vertexes in all β_{ij}. Thus

$$|x_i - x_j|_{S\cap E} \leqslant \operatorname{length}\beta_{ij}$$

and (2) follows. $\square$

13.8. Exercise. Construct a noncompact complete geodesic CBB(0) space that contains no half-lines.

C. Rademacher theorem

At the end of this section we give an extension of the Rademacher theorem (see Section 3.D) to CBB and CAT spaces (see 13.12); it was proved by Alexander Lytchak [**110**]. The following proposition is the one-dimensional case of the extended Rademacher theorem.

Recall that differentiable curves are defined in Definition 6.12.

13.9. Proposition. *Let $\alpha\colon \mathbb{I} \to \mathcal{Z}$ be a locally Lipschitz curve in a complete length space. Suppose that $\mathcal{Z}$ is either* CBB *or* CAT. *Then α is differentiable almost everywhere.*

The following two lemmas provide sufficient conditions for existence of the one-sided derivative of a curve in CBB and CAT spaces. The proofs of both lemmas are similar.

13.10. Lemma. *Let $\alpha\colon \mathbb{I} \to \mathcal{L}$ be a 1-Lipschitz curve in a* CBB *space. Suppose that for some $t_0 \in \mathbb{I}$ and any $\varepsilon > 0$, there is a point p such that $|\alpha(t_0) - p| < \varepsilon$ and*

$$\varliminf_{t\to t_0+} \frac{\operatorname{dist}_p\circ\alpha(t) - \operatorname{dist}_p\circ\alpha(t_0)}{t - t_0} > 1 - \varepsilon.$$

Then the right derivative $\alpha^+(t_0)$ is defined and $|\alpha^+(t_0)| = 1$.

Proof. Without loss of generality, we may assume that $t_0 = 0$. Set $x = \alpha(0)$. Fix a sequence of points $p_n \to x$ such that

$$\varliminf_{t\to 0+} \frac{|p_n - \alpha(t)| - |p_n - x|}{t} \to 1$$

as $n \to \infty$.

Observe that there are sequences $\delta_n \to 0+$ and $t_n \to 0+$ such that

$$\tilde\measuredangle^\kappa\left(x\,{}^{\alpha(s)}_{p_n}\right) > \pi - \delta_n \quad\text{and}\quad (1-\delta_n)\cdot s < |\alpha(s) - x| \leqslant s \tag{1}$$

for any $s \in (0, t_n]$.

For each n, choose $q_n \in \operatorname{Str}(x)$ sufficiently close to $\alpha(t_n)$ such that the inequality

$$\tilde\measuredangle^\kappa\left(x\,{}^{q_n}_{p_n}\right) > \pi - \delta_n$$

still holds (see Definition 8.10).

Set $\gamma_n = \operatorname{geod}_{[xq_n]}$. By comparison,

$$\begin{aligned}
\tilde\measuredangle^\kappa\left(x\,{}^{\alpha(s)}_{\gamma_n(s)}\right) &\leqslant 2\cdot\pi - \tilde\measuredangle^\kappa\left(x\,{}^{p_n}_{\gamma_n(s)}\right) - \tilde\measuredangle^\kappa\left(x\,{}^{\alpha(s)}_{p_n}\right) \leqslant \\
&\leqslant 2\cdot\pi - \tilde\measuredangle^\kappa\left(x\,{}^{q_n}_{p_n}\right) - \tilde\measuredangle^\kappa\left(x\,{}^{\alpha(s)}_{p_n}\right) < \\
&< 2\cdot\delta_n.
\end{aligned}$$

Therefore (1) implies that

$$|\gamma_n(s) - \alpha(s)| < 10\cdot\delta_n\cdot(s)$$

if s is a sufficiently small and positive. That is, $\alpha^+(0)$ is defined (see Definition 6.9). □

13.11. Lemma. *Let $\alpha\colon \mathbb{I} \to \mathcal{U}$ be a 1-Lipschitz curve in a* CAT *space. Suppose that for some $t_0 \in \mathbb{I}$ and any $\varepsilon > 0$ there is a point q such that $|\alpha(t_0) - q| < \varepsilon$ and*

$$\varlimsup_{t\to t_0+} \frac{\operatorname{dist}_q\circ\alpha(t) - \operatorname{dist}_q\circ\alpha(t_0)}{t - t_0} < -1 + \varepsilon.$$

Then the right derivative $\alpha^+(t_0)$ is defined and $|\alpha^+(t_0)| = 1$.

Proof. Without loss of generatiy we may assume that $t_0 = 0$. Set $x = \alpha(0)$. Fix a sequence of points $q_n \to x$ such that

$$\varlimsup_{t\to 0+} \frac{|q_n - \alpha(t)| - |q_n - x|}{t} \to -1$$

as $n \to \infty$.

Observe that there are sequences $\delta_n \to 0+$ and $t_n \to 0+$ such that

(2) $$\tilde{\measuredangle}^\kappa\left(x\,{}^{\alpha(s)}_{q_n}\right) < \delta_n \quad\text{and}\quad (1-\delta_n)\cdot s < |\alpha(s) - x| \leqslant s$$

for any $s \in (0, t_n]$.

Without loss of generality, we may assume that $|x - q_n| < \varpi^\kappa$ for any n; in particular, the geodesic $\gamma_n = \operatorname{geod}_{[xq_n]}$ is uniquely defined.

By comparison,

$$\tilde{\measuredangle}^\kappa\left(x\,{}^{\alpha(s)}_{\gamma_n(s)}\right) \leqslant \tilde{\measuredangle}^\kappa\left(x\,{}^{\alpha(s)}_{q_n}\right) < \\ < \delta_n.$$

Therefore (2) implies that

$$|\gamma_n(s) - \alpha(s)| < 10\cdot\delta_n\cdot s$$

if s is a sufficiently small and positive. That is, $\alpha^+(0)$ is defined (see Definition 6.9). □

Proof of 13.9. By the standard Rademacher theorem, we may assume that α has an arc-length parametrization. In particular, α is 1-Lipschitz.

Recall that by Theorem 3.10,

(3) $$\operatorname{speed}_s \alpha \overset{\text{a.e.}}{=\!=} 1.$$

Fix a countable dense set $T \subset \mathbb{I}$; given $t \in T$, let

$$h_t(s) = |\alpha(t) - \alpha(s)|.$$

Note that h_t is 1-Lipschitz for each $t \in T$. Therefore, by the standard Rademacher theorem and countability of T for almost all $s \in \mathbb{I}$, $h_t'(s)$ is defined for all $t \in T$.

Let

$$w^+(s) := \varlimsup_{\substack{t \in T \\ t \to s-}} \{h_t'(s)\}.$$

Let us show that

$$w^+(s) \overset{\text{a.e.}}{=\!=} 1. \tag{4}$$

Note that once this is proved, Lemma 13.10 implies the proposition in the CBB case.

For a small $\varepsilon > 0$, denote by N_ε^+ the set of all points $s \in \mathbb{I}$ such that $w^+(s) < 1 - \varepsilon$. Note that the sets N_ε^+ are measurable.

Suppose N_ε^+ has positive measure. Let $s_0 \in N_\varepsilon^+$ be a Lebesgue point of α. We may assume that $\operatorname{speed}_{s_0} \alpha = 1$ and $h_t'(s_0)$ is defined for any $t \in T$. Suppose $t \in T$ is sufficiently close to s_0 and $t < s_0$. Since $\operatorname{speed}_{s_0} \alpha = 1$, we have

$$h_t(s_0) \geqslant (s_0 - t)\cdot(1 - \varepsilon^2). \tag{5}$$

Further, there is a set $A \subset [t, s_0]$ with measure at least $(1-\varepsilon)\cdot|s_0 - t|$ such that

$$h_t'(s) < 1 - \varepsilon$$

for any $s \in A_n$. Since h_t is 1-Lipschitz, we have

$$\begin{aligned} h_t(s_0) &= \int_{[t,s_0]\backslash A} h_t'(s)\cdot \mathbf{d}s + \int_A h_t'(s)\cdot \mathbf{d}s \leqslant \\ &\leqslant (s_0 - t)\cdot[\varepsilon + (1-\varepsilon)^2]. \end{aligned}$$

The latter contradicts (5). Thus $w^+(s) \geqslant 1 - \varepsilon$ almost everywhere. Since $\varepsilon > 0$ is arbitrary, (4) follows.

In the same way we can show that

$$w^-(s) \overset{\text{a.e.}}{=\!=} -1, \tag{6}$$

where

$$w^-(s) := \varliminf_{\substack{t \in T \\ t \to s+}} \{h_t'(s)\}.$$

Then Lemma 13.11 implies the proposition in the CAT case. □

13.12. Extended Rademacher theorem. *Let $f : \mathbb{E}^m \multimap \mathcal{Z}$ be a locally Lipschitz submap from a Euclidean space to a complete length space $\mathcal{Z}$. Suppose that $\mathcal{Z}$ is either* CBB *or* CAT. *Then the differential $\mathbf{d}_x f$ is defined at almost all points $x \in \operatorname{Dom} f$.*

Moreover the differential $\boldsymbol{d}_x f$ *is* linear *at almost all* x *in the following sense: the image* $\operatorname{Im} f$ *is a convex subcone of* $\mathrm{T}_{f(x)}\mathcal{Z}$*, and there is an isometry* ι *from* $\operatorname{Im} f$ *to a Euclidean space such that the composition* $\iota\circ\boldsymbol{d}_x f$ *is linear.*

The proof is a reduction to the one-dimensional case (Proposition 13.9) by standard arguments [**95**, **114**].

Proof. Without loss of generality, we may assume that $\operatorname{Dom} f$ is bounded and f is Lipschitz.

Fix a countable dense set of vectors $\{v_i\}$ in $\mathbb{E}^m$. Fix v_i and a point $p\in$ $\in\operatorname{Dom} f$. By Proposition 13.9, the value $\boldsymbol{d}_x f(v_i)$ is defined at $x = p + t\cdot v_i$ for almost all t such that $x\in\operatorname{Dom} f$. It follows that $\boldsymbol{d}_x f(v_i)$ is defined for every i on a set A of full measure in $\operatorname{Dom} f$. Since the metric differential of f is defined almost everywhere (see 3.11), we have that $\boldsymbol{d}_x f(v)$ is defined for any v on a set B of full measure in $\operatorname{Dom} f$.

Applying the definitions of metric differential and differential (see 3.11) and Definition 6.15, we obtain that the image of $\boldsymbol{d}_x f$ is a weakly convex set in $\mathrm{T}_{f(x)}$. It follows that $\operatorname{Im}\boldsymbol{d}_x f$ is CBB(0) or CAT(0) if the space $\mathcal{Z}$ is CBB or CAT, respectively. It remains to apply Exercises 8.15 or 9.16 if the space $\mathcal{Z}$ is CBB or CAT, respectively. □

D. Differential

13.13. Exercise. Let $\mathcal{U}$ be a complete length CAT(κ) space and let $p, q\in\mathcal{U}$. Assume $|p-q| < \varpi^\kappa$. Show that

$$(\boldsymbol{d}_q \operatorname{dist}_p)(v) = -\langle\uparrow_{[qp]}, v\rangle.$$

13.14. Exercise. Let $\mathcal{L}$ be a length CBB(κ) space and let $p, q\in\mathcal{L}$ be distinct points. Assume $q\in\operatorname{Str}(p)$ or $p\in\operatorname{Str}(q)$. Show that

$$(\boldsymbol{d}_q \operatorname{dist}_p)(v) = -\langle\uparrow_{[qp]}, v\rangle.$$

13.15. Lemma. *Let* $\mathcal{U}$ *be a complete length* CAT *space, let* $f\colon \mathcal{U}\multimap\mathbb{R}$ *be a locally Lipschitz semiconcave subfunction, and let* $p\in\operatorname{Dom} f$*. Then* $\boldsymbol{d}_p f$ *is a Lipschitz concave function on* $\mathrm{T}_p\mathcal{U}$*.*

Proof. Recall that the tangent space $\mathrm{T}_p = \mathrm{T}_p\mathcal{U}$ can be considered as a subspace of the ultratangent space T_p^ω (see Theorem 6.17). Since T_p^ω is CAT(0), Theorem 13.4 implies that T_p is a convex set in T_p^ω.

By Proposition 13.3, $\boldsymbol{d}_p^\omega f$ is a concave function on T_p^ω. It remains to apply that $\boldsymbol{d}_p f = (\boldsymbol{d}_p^\omega f)|_{\mathrm{T}_p}$ (Proposition 6.16(c)). □

As was shown in Halbeisen's example (Section 13.B), a CBB space might have tangent spaces that are not length spaces; thus concavity of the differential

$\boldsymbol{d}_p f$ of a semiconcave function f is meaningless. Nevertheless, as the following lemma says, the differential $\boldsymbol{d}_p f$ of a semiconcave function always satisfies a weaker property similar to concavity (compare to [**139**, Prop. 136] and [**122**, 4.2]). In the finite-dimensional case, $\boldsymbol{d}_p f$ is concave.

13.16. Lemma. *Let $\mathcal{Z}$ be a complete length space, let $f\colon \mathcal{Z} \hookrightarrow \mathbb{R}$ be a locally Lipschitz semiconcave subfunction, and let $p \in \operatorname{Dom} f$. Suppose that $\mathcal{Z}$ is either* CBB *or* CAT. *Then for any $u, v \in \mathrm{T}_p$, we have*

$$s\cdot\sqrt{|u|^2 + 2\cdot\langle u, v\rangle + |v|^2} \geqslant (\boldsymbol{d}_p f)(u) + (\boldsymbol{d}_p f)(v),$$

where

$$s = \sup\left\{ (\boldsymbol{d}_p f)(\xi) : \xi \in \Sigma_p \right\}.$$

Proof. If $\mathcal{Z}$ is CAT, then the statement follows from Lemma 13.15. Indeed, let z be the midpoint of a geodesic $[uv]_{\mathrm{T}_p}$. Observe that

$$2\cdot|z| = \sqrt{|u|^2 + 2\cdot\langle u, v\rangle + |v|^2}.$$

Since $\boldsymbol{d}_p f$ is concave, we have that

$$2\cdot\boldsymbol{d}_p f(z) \geqslant \boldsymbol{d}_p f(u) + \boldsymbol{d}_p f(v).$$

It remains to choose $\xi \in \Sigma_p$ so that $\xi\cdot|z| = z$ and observe that $s \geqslant \boldsymbol{d}_p(\xi)$.

Now assume $\mathcal{Z}$ is CBB. We can assume that $\alpha = \measuredangle(u, v) > 0$, otherwise the statement is trivial. Moreover, we can assume that $\exp_p(t\cdot u)$ and $\exp_p(t\cdot v)$ are defined for all small $t > 0$; the latter follows since geodesic space of directions Σ'_p is dense in Σ_p.

Prepare a model configuration of five points: $\tilde p$, $\tilde u$, $\tilde v$, $\tilde q$, $\tilde w \in \mathbb{E}^2$ such that

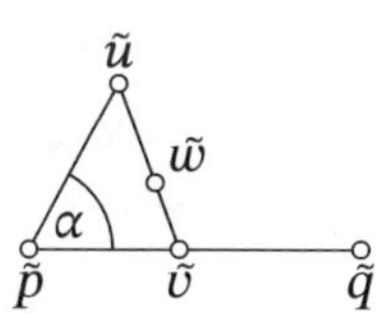

- $\measuredangle\left[\tilde p\,{}^{\tilde u}_{\tilde v}\right] = \alpha$,
- $|\tilde p - \tilde u| = |u|$,
- $|\tilde p - \tilde v| = |v|$,
- $\tilde q$ lies on an extension of $[\tilde p\tilde v]$ so that $\tilde v$ is the midpoint of $[\tilde p\tilde q]$,
- $\tilde w$ is the midpoint between $\tilde u$ and $\tilde v$.

Note that

$$|\tilde p - \tilde w| = \tfrac12\cdot\sqrt{|u|^2 + 2\cdot\langle u, v\rangle + |v|^2}.$$

Assume that $\mathcal{Z}$ is geodesic.

For all small $t > 0$, construct points $u_t, v_t, q_t, w_t \in \mathcal{Z}$ as follows:

(a) $v_t = \exp_p(t\cdot v)$, $\quad q_t = \exp_p(2\cdot t\cdot v)$;

(b) $u_t = \exp_p(t\cdot u)$;

(c) w_t is the midpoint of $[u_t v_t]$.

Clearly $|p-u_t| = t\cdot|u|$, $|p-v_t| = t\cdot|v|$, $|p-q_t| = 2\cdot t\cdot|v|$. Since $\measuredangle(u,v)$ is defined, we have $|u_t-v_t| = t\cdot|\tilde u-\tilde v|+o(t)$ and $|u_t-q_t| = t\cdot|\tilde u-\tilde q|+o(t)$ (see Theorem 8.14(c) and Section 8.A).

From the point-on-side and hinge comparisons (see 8.14, items (b) and (c)), we have

$$\tilde\measuredangle^\kappa\left(v_t\,{}^{p}_{w_t}\right) \geqslant \tilde\measuredangle^\kappa\left(v_t\,{}^{p}_{u_t}\right) \geqslant \measuredangle\left[\tilde v\,{}^{\tilde p}_{\tilde u}\right] + \tfrac{o(t)}{t}$$

and

$$\tilde\measuredangle^\kappa\left(v_t\,{}^{q_t}_{w_t}\right) \geqslant \tilde\measuredangle^\kappa\left(v_t\,{}^{q_t}_{u_t}\right) \geqslant \measuredangle\left[\tilde v\,{}^{\tilde q}_{\tilde u}\right] + \tfrac{o(t)}{t}.$$

Clearly, $\measuredangle\left[\tilde v\,{}^{\tilde p}_{\tilde u}\right]+\measuredangle\left[\tilde v\,{}^{\tilde q}_{\tilde u}\right] = \pi$. From the adjacent angle comparison (see 8.14(a)), $\tilde\measuredangle^\kappa\left(v_t\,{}^{p}_{u_t}\right) + \tilde\measuredangle^\kappa\left(v_t\,{}^{u_t}_{q_t}\right) \leqslant \pi$. Hence $\tilde\measuredangle^\kappa\left(v_t\,{}^{p}_{w_t}\right) \to \measuredangle\left[\tilde v\,{}^{\tilde p}_{\tilde w}\right]$ as $t\to 0+$ and thus

$$|p-w_t| = t\cdot|\tilde p-\tilde w| + o(t).$$

Since f is λ-concave, we have

$$\begin{aligned}2\cdot f(w_t) &\geqslant f(u_t)+f(v_t)+\tfrac{\lambda}{4}\cdot|u_t-v_t|^2 = \\ &= 2\cdot f(p) + t\cdot[(\mathbf{d}_pf)(u)+(\mathbf{d}_pf)(v)] + o(t).\end{aligned}$$

Applying λ-concavity of f, we have

$$(\mathbf{d}_pf)(\uparrow_{[pw_t]}) \geqslant \frac{t\cdot[(\mathbf{d}_pf)(u)+(\mathbf{d}_pf)(v)]+o(t)}{2\cdot t\cdot|\tilde p-\tilde w|+o(t)}. \tag{1}$$

The lemma follows.

Finally, if $\mathcal{Z}$ is not geodesic, one needs to make two adjustments in the above construction. Namely:

(i) For the geodesic $[u_tv_t]$ to be defined, in (b) one has to take $u_t \in \operatorname{Str}(v_t)$, $u_t \approx \exp_p(t\cdot u)$; more precisely,

$$|u_t - \exp_p(t\cdot u)| = o(t).$$

Thus instead of $|p-u_t| = t\cdot|u|$, we have

$$|p-u_t| = t\cdot|u| + o(t),$$

and this is sufficient for the rest of proof.

(ii) The direction $\uparrow_{[pw_t]}$ might be undefined. Thus in estimate (1), instead of $\uparrow_{[pw_t]}$ one should take $\uparrow_{[pw_t']}$ for some point $w_t' \in \operatorname{Str}(p)$ near w_t (that is, $|w_t-w_t'| = o(t)$). □

E. Definition of gradient

13.17. Definition of gradient. *Let $\mathcal{X}$ be a length space with defined angles and let $f\colon \mathcal{X} \circ\!\!\to \mathbb{R}$ be a subfunction. Suppose for a point $p \in \operatorname{Dom} f$ the differential $\boldsymbol{d}_p f\colon \mathrm{T}_p \to \mathbb{R}$ is defined.*

A tangent vector $g \in \mathrm{T}_p$ is called a gradient of f at p *(briefly, $g = \nabla_p f$) if*

(a) $(\boldsymbol{d}_p f)(w) \leqslant \langle g, w\rangle$ *for any $w \in \mathrm{T}_p$, and*

(b) $(\boldsymbol{d}_p f)(g) = \langle g, g\rangle$.

13.18. Example. Consider the Euclidean plane with standard (x, y)-coordinates. Then the function $f\colon (x, y) \mapsto -|x| - |y|$ is concave; its gradient field is sketched in the figure.

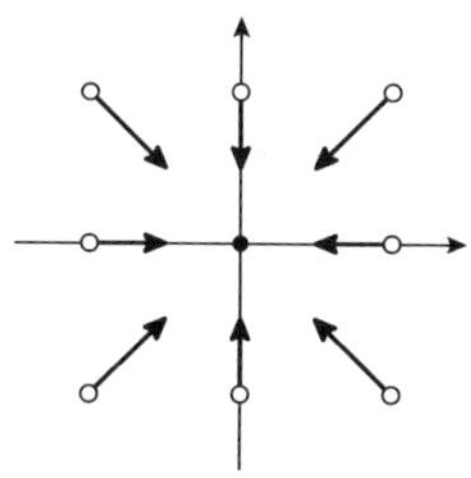

If a point does not lie on an axis, then its gradient has length $\sqrt{2}$ and takes one of four values $(\pm 1, \pm 1)$ depending on the quadrant of the point. At the origin the gradient vanishes, and on the on the remaining parts of the x-axis and y-axis it is $(\pm 1, 0)$ and $(0, \pm 1)$, respectively.

13.19. Exercise. Let $\mathcal{U}$ be a complete length CAT(0) space. Show that

$$\nabla_p(-\operatorname{dist}_q) = \uparrow_{[pq]}$$

for any pair of distinct points $p, q \in \mathcal{U}$.

13.20. Existence and uniqueness of the gradient. *Let $\mathcal{Z}$ be a complete length space and $f\colon \mathcal{Z} \circ\!\!\to \mathbb{R}$ be a locally Lipschitz and semiconcave subfunction. Suppose that $\mathcal{Z}$ is either* CBB *or* CAT. *Then for any point $p \in \operatorname{Dom} f$, there is a unique gradient $\nabla_p f \in \mathrm{T}_p$.*

Proof.

Uniqueness. If $g, g' \in \mathrm{T}_p$ are two gradients of f, then

$$\langle g, g\rangle = (\boldsymbol{d}_p f)(g) \leqslant \langle g, g'\rangle, \qquad \langle g', g'\rangle = (\boldsymbol{d}_p f)(g') \leqslant \langle g, g'\rangle.$$

Therefore,

$$|g - g'|^2 = \langle g, g\rangle - 2\cdot\langle g, g'\rangle + \langle g', g'\rangle \leqslant 0.$$

It follows that $g = g'$.

Existence. Note first that if $\boldsymbol{d}_p f \leqslant 0$, then one can take $\nabla_p f = 0$.

Otherwise, if $s = \sup\{\, (\boldsymbol{d}_p f)(\xi) : \xi \in \Sigma_p \,\} > 0$, it is sufficient to show that there is $\overline{\xi} \in \Sigma_p$ such that

$$(\boldsymbol{d}_p f)\left(\overline{\xi}\right) = s. \tag{1}$$

Indeed, suppose $\bar{\xi}$ exists. Then applying Lemma 13.16 for $u = \bar{\xi}$, $v = \varepsilon \cdot w$ with $\varepsilon \to 0+$, we get

$$(\boldsymbol{d}_p f)(w) \leqslant \langle w, s \cdot \bar{\xi} \rangle$$

for any $w \in \mathrm{T}_p$; that is, $s \cdot \bar{\xi}$ is the gradient at p.

Take a sequence of directions $\xi_n \in \Sigma_p$, such that $(\boldsymbol{d}_p f)(\xi_n) \to s$. Applying Lemma 13.16 for $u = \xi_n$, $v = \xi_m$, we get

$$s \geqslant \frac{(\boldsymbol{d}_p f)(\xi_n) + (\boldsymbol{d}_p f)(\xi_m)}{\sqrt{2 + 2 \cdot \cos \measuredangle(\xi_n, \xi_m)}}.$$

Therefore $\measuredangle(\xi_n, \xi_m) \to 0$ as $n, m \to \infty$; that is, the sequence ξ_n is Cauchy. Clearly $\bar{\xi} = \lim_n \xi_n$ satisfies (1). □

13.21. Exercise. Let p be a point in a complete length CBB space. Show that

$$|\nabla_x \operatorname{dist}_p| = 1$$

for x in a dense G-delta subset.

F. Calculus of gradient

The next lemma states that the gradient points in the direction of maximal slope; moreover, if the slope in the given direction is almost maximal, then it is almost the direction of the gradient.

13.22. Lemma. *Let $\mathcal{Z}$ be a complete length space, let $f : \mathcal{Z} \multimap \mathbb{R}$ be locally Lipschitz and semiconcave, and let $p \in \operatorname{Dom} f$. Suppose that $\mathcal{Z}$ is either* CBB *or* CAT.

Assume $|\nabla_p f| > 0$; let $\bar{\xi} = \frac{1}{|\nabla_p f|} \cdot \nabla_p f$. Then:

(a) *If for some $v \in \mathrm{T}_p$, we have*

$$|v| \leqslant 1 + \varepsilon \quad \textit{and} \quad (\boldsymbol{d}_p f)(v) > |\nabla_p f| \cdot (1 - \varepsilon),$$

then

$$|\bar{\xi} - v| < 100 \cdot \sqrt{\varepsilon}.$$

(b) *If $v_n \in \mathrm{T}_p$ is a sequence of vectors such that*

$$\varlimsup_{n\to\infty} |v_n| \leqslant 1 \quad \textit{and} \quad \varliminf_{n\to\infty} (\boldsymbol{d}_p f)(v_n) \geqslant |\nabla_p f|,$$

then

$$\lim_{n\to\infty} v_n = \bar{\xi}.$$

(c) *$\bar{\xi}$ is the unique maximum direction for the restriction $\boldsymbol{d}_p f|_{\Sigma_p}$. In particular,*

$$|\nabla_p f| = \sup\{\, \boldsymbol{d}_p f : \xi \in \Sigma_p f \,\}.$$

Proof. According to the definition of gradient,

$$\begin{aligned}|\nabla_p f|\cdot(1-\varepsilon) < (\mathbf{d}_p f)(v) &\leqslant \\ &\leqslant \langle v, \nabla_p f\rangle = \\ &= |v|\cdot|\nabla_p f|\cdot \cos\measuredangle(\nabla_p f, v).\end{aligned}$$

Thus $|v| > 1-\varepsilon$ and $\cos\measuredangle(\nabla_p f, v) > \frac{1-\varepsilon}{1+\varepsilon}$. Hence (a).

Statements (b) and (c) follow directly from (a). □

As a corollary of the above lemma and Proposition 3.18 we obtain the following:

13.23. Chain rule. *Let $\mathcal{Z}$ be a complete length space, let $f\colon \mathcal{Z} \hookrightarrow \mathbb{R}$ be a semiconcave function, and let $\phi\colon \mathbb{R} \to \mathbb{R}$ be a nondecreasing semiconcave function. Suppose that $\mathcal{Z}$ is either* CBB *or* CAT. *Then $\phi\circ f$ is semiconcave and $\nabla_x(\phi\circ f) = \phi^+(f(x))\cdot\nabla_x f$ for any $x\in \operatorname{Dom} f$.*

The following inequalities describe an important property of the *gradient vector field.*

13.24. Lemma. *Let $\mathcal{Z}$ be a complete length space, $f\colon \mathcal{Z}\hookrightarrow \mathbb{R}$ satisfy $f''+\kappa\cdot f\leqslant\lambda$ for some $\kappa,\lambda\in\mathbb{R}$. Let $[pq]\subset\operatorname{Dom} f$, and let $\ell = |p-q|$. Suppose that $\mathcal{Z}$ is either* CBB *or* CAT. *Then*

$$\langle\uparrow_{[pq]}, \nabla_p f\rangle \geqslant \frac{f(q) - f(p)\cdot \operatorname{cs}^\kappa\ell - \lambda\cdot\operatorname{md}^\kappa\ell}{\operatorname{sn}^\kappa\ell}.$$

In particular,

(a) *if $\kappa = 0$,*

$$\langle\uparrow_{[pq]}, \nabla_p f\rangle \geqslant \left(f(q) - f(p) - \tfrac{\lambda}{2}\cdot\ell^2\right)/\ell;$$

(b) *if $\kappa = 1$, $\lambda = 0$, we have*

$$\langle\uparrow_{[pq]}, \nabla_p f\rangle \geqslant (f(q) - f(p)\cdot\cos\ell)/\sin\ell;$$

(c) *if $\kappa = -1$, $\lambda = 0$, we have*

$$\langle\uparrow_{[pq]}, \nabla_p f\rangle \geqslant (f(q) - f(p)\cdot\cosh\ell)/\sinh\ell.$$

Proof of Lemma 13.24. Note that

$$\operatorname{geod}_{[pq]}(0) = p,\quad \operatorname{geod}_{[pq]}(\ell) = q,\quad (\operatorname{geod}_{[pq]})^+(0) = \uparrow_{[pq]}.$$

Thus,

$$\begin{aligned}\langle\uparrow_{[pq]},\nabla_p f\rangle \geqslant d_p f(\uparrow_{[pq]}) &=\\ &= (f\circ \mathrm{geod}_{[pq]})^+(0)\geqslant\\ &\geqslant \frac{f(q)-f(p)\cdot \mathrm{cs}^\kappa \ell - \lambda\cdot \mathrm{md}^\kappa \ell}{\mathrm{sn}^\kappa \ell}.\end{aligned}$$ □

The following corollary states that the gradient vector field is monotonic in a sense similar to the definition of *monotone operators* [**135**].

13.25. λ-Monotonicity of gradient. *Let $\mathcal{Z}$ be a complete length space, let $f\colon \mathcal{Z}\multimap\mathbb{R}$ be locally Lipschitz and λ-concave, and let $[pq]\subset \operatorname{Dom} f$. Suppose that $\mathcal{Z}$ is either* CBB *or* CAT. *Then*

$$\langle\uparrow_{[pq]},\nabla_p f\rangle + \langle\uparrow_{[qp]},\nabla_q f\rangle \geqslant -\lambda\cdot|p-q|.$$

Proof. Add two inequalities from Lemma 13.24(a). □

13.26. Lemma. *Let $\mathcal{Z}$ be a complete length space, let $f,\ g\colon \mathcal{Z}\multimap\mathbb{R}$, and let $p\in \operatorname{Dom} f\cap \operatorname{Dom} g$. Suppose that $\mathcal{Z}$ is either* CBB *or* CAT.

Then

$$|\nabla_p f - \nabla_p g|^2_{\mathrm{T}_p} \leqslant s\cdot(|\nabla_p f| + |\nabla_p g|),$$

where

$$s = \sup\{\,|(\boldsymbol{d}_p f)(\xi) - (\boldsymbol{d}_p g)(\xi)| \;:\; \xi\in\Sigma_p\,\}.$$

In particular, if $f_n\colon \mathcal{Z}\multimap\mathbb{R}$ is a sequence of locally Lipschitz and semiconcave subfunctions, $p\in \operatorname{Dom} f_n$ for each n, and $\boldsymbol{d}_p f_n$ converges uniformly on Σ_p, then the sequence $\nabla_p f_n\in \mathrm{T}_p$ converges.

Proof. Clearly for any $v\in\mathrm{T}_p$, we have

$$|(\boldsymbol{d}_p f)(v) - (\boldsymbol{d}_p g)(v)| \leqslant s\cdot|v|.$$

From the definition of gradient Definition 13.17 we have

$$\begin{aligned}(\boldsymbol{d}_p f)(\nabla_p g) &\leqslant \langle\nabla_p f,\nabla_p g\rangle, & (\boldsymbol{d}_p g)(\nabla_p f) &\leqslant \langle\nabla_p f,\nabla_p g\rangle,\\ (\boldsymbol{d}_p f)(\nabla_p f) &= \langle\nabla_p f,\nabla_p f\rangle, & (\boldsymbol{d}_p g)(\nabla_p g) &= \langle\nabla_p g,\nabla_p g\rangle.\end{aligned}$$

Therefore,

$$\begin{aligned}|\nabla_p f - \nabla_p g|^2 &= \langle\nabla_p f,\nabla_p f\rangle + \langle\nabla_p g,\nabla_p g\rangle - 2\cdot\langle\nabla_p f,\nabla_p g\rangle \leqslant\\ &\leqslant (\boldsymbol{d}_p f)(\nabla_p f) + (\boldsymbol{d}_p g)(\nabla_p g) - (\boldsymbol{d}_p f)(\nabla_p g) - (\boldsymbol{d}_p g)(\nabla_p f) \leqslant\\ &\leqslant s\cdot(|\nabla_p f| + |\nabla_p g|).\end{aligned}$$ □

13.27. Exercise. Let $\mathcal{L}$ be a complete length CBB(κ) space, let the function $f\colon\mathcal{L}\to\mathbb{R}$ be semiconcave and locally Lipschitz, and let $\alpha\colon\mathbb{I}\to\mathcal{L}$ be a Lipschitz curve. Show that

$$\langle\nabla_{\alpha(t)} f,\alpha^+(t)\rangle = (\boldsymbol{d}_{\alpha(t)} f)(\alpha^+(t))$$

for almost all $t\in\mathbb{I}$.

G. Semicontinuity of |gradient|

In this section we collect a few consequences of the following lemma.

13.28. Ultralimit of |gradient|. *Assume that*

- *$(\mathcal{Z}_n)$ is a sequence of complete length spaces and $(\mathcal{Z}_n, p_n) \to (\mathcal{Z}_\omega, p_\omega)$ as $n \to \omega$. Suppose that all $\mathcal{Z}_n$ are either* CBB *or* CAT.
- *$f_n : \mathcal{Z}_n \multimap \mathbb{R}$ and $f_\omega : \mathcal{Z}_\omega \multimap \mathbb{R}$ are locally Lipschitz and λ-concave, and $f_n \to f_\omega$ as $n \to \omega$.*
- *$x_n \in \operatorname{Dom} f_n$ and $x_n \to x_\omega \in \operatorname{Dom} f_\omega$ as $n \to \omega$.*

Then

$$|\nabla_{x_\omega} f_\omega| \leqslant \lim_{n\to\omega} |\nabla_{x_n} f_n|.$$

Remarks. The inequality might be strict. For example, consider $\mathcal{Z}_n = \mathbb{R}$, $f_n(x) = -|x|$, and $x_n \to 0+$.

From the convergence of gradient curves (proved in 16.17), one can deduce the following slightly stronger statement.

13.29. Proposition. *Assume that*

- *$\mathcal{Z}_n$ is a sequence of complete length spaces and $(\mathcal{Z}_n, p_n) \to (\mathcal{Z}_\omega, p_\omega)$ as $n \to \omega$. Suppose that all $\mathcal{Z}_n$ are either* CBB *or* CAT.
- *$f_n : \mathcal{Z}_n \multimap \mathbb{R}$ and $f_\omega : \mathcal{Z}_\omega \multimap \mathbb{R}$ are locally Lipschitz and λ-concave and $f_n \to f_\omega$ as $n \to \omega$.*

Then

$$|\nabla_{x_\omega} f_\omega| = \inf\{\lim_{n\to\omega} |\nabla_{x_n} f_n|\},$$

where infimum is taken for all sequences $x_n \in \operatorname{Dom} f_n$ such that $x_n \to x_\omega \in$ $\in \operatorname{Dom} f_\omega$ as $n \to \omega$.

Proof of 13.28. Fix an $\varepsilon > 0$ and choose $y_\omega \in \operatorname{Dom} f_\omega$ sufficiently close to x_ω that

$$|\nabla_{x_\omega} f_\omega| - \varepsilon < \frac{f_\omega(y_\omega) - f_\omega(x_\omega)}{|x_\omega - y_\omega|}.$$

Choose $y_n \in \mathcal{Z}_n$ such that $y_n \to y_\omega$ as $n \to \omega$. Since $|x_\omega - y_\omega|$ is sufficiently small, the λ-concavity of f_n implies that

$$|\nabla_{x_\omega} f_\omega| - 2\cdot\varepsilon < (\boldsymbol{d}_{x_n} f_n)(\uparrow_{[x_n y_n]})$$

for ω-almost all n. Hence

$$|\nabla_{x_\omega} f_\omega| - 2\cdot\varepsilon \leqslant \lim_{n\to\omega} |\nabla_{x_n} f_n|.$$

Since $\varepsilon > 0$ is arbitrary, the proposition follows. □

Note that the distance-preserving map $\iota\colon \mathcal{Z} \hookrightarrow \mathcal{Z}^\omega$ induces an embedding

$$\boldsymbol{d}_p\iota\colon \mathrm{T}_p\mathcal{Z} \hookrightarrow \mathrm{T}_p\mathcal{Z}^\omega.$$

Thus, we can (and will) consider $\mathrm{T}_p\mathcal{Z}$ as a subcone of $\mathrm{T}_p\mathcal{Z}^\omega$.

13.30. Corollary. *Let $\mathcal{Z}$ be a complete length space and let $f\colon \mathcal{Z} \multimap \mathbb{R}$ be a locally Lipschitz semiconcave subfunction. Suppose that $\mathcal{Z}$ is either* CBB *or* CAT. *Then*

$$\nabla_x f = \nabla_x f^\omega$$

for any point $x \in \operatorname{Dom} f$.

Proof. Note that

$$\begin{array}{ccc} \mathcal{Z} & \subset & \mathcal{Z}^\omega \\ \cup & & \cup \\ \operatorname{Dom} f & \subset & \operatorname{Dom} f^\omega \end{array}.$$

Applying 13.28 for $\mathcal{Z}_n = \mathcal{Z}$ and $x_n = x$, we get that $|\nabla_x f| \geqslant |\nabla_x f^\omega|$.

On the other hand, $f = f^\omega|_{\mathcal{Z}}$, hence $\boldsymbol{d}_p f = \boldsymbol{d}_p f^\omega|_{\mathrm{T}_p\mathcal{Z}}$. Thus from Lemma 13.22(c), $|\nabla_x f| \leqslant |\nabla_x f^\omega|$. Therefore

$$|\nabla_x f| = |\nabla_x f^\omega| \tag{1}$$

for any $x \in \mathcal{Z}$.

Further,

$$\begin{aligned} |\nabla_x f|^2 &= (\boldsymbol{d}_x f)(\nabla_x f) = \\ &= \boldsymbol{d}_x f^\omega(\nabla_x f) \leqslant \\ &\leqslant \langle \nabla_x f^\omega, \nabla_x f\rangle = \\ &= |\nabla_x f^\omega|\cdot|\nabla_x f|\cdot \cos\measuredangle(\nabla_x f^\omega, \nabla_x f). \end{aligned}$$

Together with (1), this implies $\measuredangle(\nabla_x f^\omega, \nabla_x f) = 0$, and the statement follows. □

13.31. Semicontinuity of |gradient|. *Let $\mathcal{Z}$ be a complete length space and let $f\colon \mathcal{Z} \multimap \mathbb{R}$ be a locally Lipschitz semiconcave subfunction. Suppose that $\mathcal{Z}$ is either* CBB *or* CAT. *Then the function $x \mapsto |\nabla_x f|$ is lower-continuous; that is, for any sequence $x_n \to x \in \operatorname{Dom} f$, we have*

$$|\nabla_x f| \leqslant \varliminf_{n\to\infty} |\nabla_{x_n} f|.$$

Proof. According to Corollary 13.30, $|\nabla_x f| = |\nabla_x f^\omega|$. Applying 13.28 for $x_n \to x$, we obtain

$$\lim_{n\to\omega} |\nabla_{x_n} f| \geqslant |\nabla_x f^\omega| = |\nabla_x f|.$$

The same holds for an arbitrary subsequence of x_n—hence the result. □

H. Polar vectors

Here we give a corollary of Lemma 13.26. It will be used to prove basic properties of the tangent space.

13.32. Antisum lemma. *Let $\mathcal{L}$ be a complete length* CBB *space and let $p \in \mathcal{L}$.*

Given two vectors u, $v \in \mathrm{T}_p$, there is a unique vector $w \in \mathrm{T}_p$ such that

$$\langle u, x\rangle + \langle v, x\rangle + \langle w, x\rangle \geqslant 0$$

for any $x \in \mathrm{T}_p$, and

$$\langle u, w\rangle + \langle v, w\rangle + \langle w, w\rangle = 0.$$

If T_p were a length space, then the lemma would follow from the existence of the gradient (see 13.20), applied to the function $\mathrm{T}_p \to \mathbb{R}$ defined by $x \mapsto$ $\mapsto -(\langle u, x\rangle + \langle v, x\rangle)$. However, the tangent space T_p might be not a length space; see Halbeisen's Example 13.6.

Applying the above lemma for $u = v$, we have the following statement.

13.33. Existence of polar vector. *Let $\mathcal{L}$ be a complete length* CBB *space and let $p \in \mathcal{L}$. Given a vector $u \in \mathrm{T}_p$, there is a unique vector $u^* \in \mathrm{T}_p$ such that $\langle u^*, u^*\rangle + \langle u, u^*\rangle = 0$ and u^* is* polar *to u; that is, $\langle u^*, x\rangle + \langle u, x\rangle \geqslant 0$ for any $x \in \mathrm{T}_p$.*

In particular, for any vector $u \in \mathrm{T}_p$ there is a polar vector $u^* \in \mathrm{T}_p$ such that $|u^*| \leqslant |u|$.

Milka's lemma provides a refinement of this statement; it states that in the finite-dimensional CBB space, any tangent vector u has a polar vector u^* such that $|u^*| = |u|$.

13.34. Example. Let $\mathcal{L}$ be the upper half-space of the Euclidean space $\mathbb{E}^n$; that is, $\mathcal{L} = \{(x_1, \dots, x_n) \in \mathbb{E}^n \mid x_n \geqslant 0\}$. It is a complete length CBB(0) space. For $p = 0$, the tangent space T_p can be canonically identified with $\mathcal{L}$. Then any vector $u = (v_1, \dots, v_n) \in \mathrm{T}_p$ has a unique polar vector such that $|u^*| = |u|$ which is $u^* = (-v_1, \dots, -v_{n-1}, v_n)$. However, if $v_n \neq 0$, then u has other polar vectors, in particular $(-v_1, \dots, -v_{n-1}, 0)$.

It is instructive to solve the following exercise before reading the proof of Lemma 13.32.

13.35. Exercise. Let $\mathcal{L}$ be a complete length CBB(κ) space and let a, b, p be mutually distinct points in $\mathcal{L}$. Prove that

$$(\boldsymbol{d}_p \operatorname{dist}_a)(\nabla_p \operatorname{dist}_b) \leqslant \cos \tilde{\measuredangle}^\kappa \left(p\,{}^a_b\right).$$

Proof of Lemma 13.32. Choose two sequences of points $a_n, b_n \in \mathrm{Str}(p)$ such that $\uparrow_{[pa_n]} \to u/|u|$ and $\uparrow_{[pb_n]} \to v/|v|$ as $n \to \infty$. Consider a sequence of functions

$$f_n = |u| \cdot \mathrm{dist}_{a_n} + |v| \cdot \mathrm{dist}_{b_n}.$$

According to Exercise 13.14,

$$(\boldsymbol{d}_p f_n)(x) = -|u| \cdot \langle \uparrow_{[pa_n]}, x\rangle - |v| \cdot \langle \uparrow_{[pb_n]}, x\rangle.$$

Thus we have the following uniform convergence for all $x \in \Sigma_p$:

$$(\boldsymbol{d}_p f_n)(x) \xrightarrow[n\to\infty]{} -\langle u, x\rangle - \langle v, x\rangle.$$

According to Lemma 13.26, the sequence $\nabla_p f_n$ converges. Let

$$w = \lim_{n\to\infty} \nabla_p f_n.$$

By the definition of gradient,

$$\begin{aligned} \langle w, w\rangle &= \lim_{n\to\infty} \langle \nabla_p f_n, \nabla_p f_n\rangle = \\ &= \lim_{n\to\infty} (\boldsymbol{d}_p f_n)(\nabla_p f_n) = \\ &= -\langle u, w\rangle - \langle v, w\rangle, \end{aligned} \qquad \begin{aligned} \langle w, x\rangle &= \lim_{n\to\infty} \langle \nabla_p f_n, x\rangle \geqslant \\ &\geqslant \lim_{n\to\infty} (\boldsymbol{d}_p f_n)(x) = \\ &= -\langle u, x\rangle - \langle v, x\rangle. \end{aligned}$$

□

I. Linear subspace of tangent space

13.36. Definition. Let $\mathcal{L}$ be a complete length CBB(κ) space, let $p \in \mathcal{L}$, and let $u, v \in \mathrm{T}_p$. We say that vectors u and v are *opposite* to each other (briefly, $u + v = 0$) if $|u| = |v| = 0$ or $\measuredangle(u, v) = \pi$ and $|u| = |v|$.

The subcone

$$\mathrm{Lin}_p = \left\{ v \in \mathrm{T}_p \; : \; \exists\, w \in \mathrm{T}_p \quad \text{such that} \quad w + v = 0 \right\}$$

will be called the *linear subcone* of T_p.

The reason for the term "linear" will become evident in Theorem 13.39.

13.37. Proposition. *Let $\mathcal{L}$ be a complete length* CBB *space and let $p \in \mathcal{L}$. Given two vectors $u, v \in \mathrm{T}_p$, the following statements are equivalent:*

(a) $u + v = 0$;

(b) $\langle u, x\rangle + \langle v, x\rangle = 0$ *for any* $x \in \mathrm{T}_p$;

(c) $\langle u, \xi\rangle + \langle v, \xi\rangle = 0$ *for any* $\xi \in \Sigma_p$.

Proof. The condition $u + v = 0$ is equivalent to

$$\langle u, u\rangle = -\langle u, v\rangle = \langle v, v\rangle;$$

thus (b) ⇒ (a). Since T_p is isometric to a subset of T_p^ω, the splitting theorem (see 16.22) applied to T_p^ω gives (a) ⇒ (b).

The equivalence (b)⇔(c) is trivial. □

13.38. Proposition. *Let $\mathcal{L}$ be a complete length* CBB *space and let $p \in \mathcal{L}$. Then for any three vectors u, v, $w \in \mathrm{T}_p$, if $u + v = 0$ and $u + w = 0$, then $v = w$.*

Proof. By Proposition 13.37, both v and w satisfy the condition in 13.33. Hence the result. □

Let $u \in \mathrm{Lin}_p$; that is $u + v = 0$ for some $v \in \mathrm{T}_p$. Given $s < 0$, let

$$s \cdot u := (-s) \cdot v.$$

This way we define multiplication of any vector in Lin_p by any real number (positive and negative). Proposition 13.38 implies that such multiplication is uniquely defined.

13.39. Theorem. *Let $\mathcal{L}$ be a complete length* CBB(κ) *space and let $p \in \mathcal{L}$. Then* Lin_p *is a subcone of* T_p *isometric to a Hilbert space.*

Before proving the theorem, let us give a corollary.

13.40. Corollary. *Let $\mathcal{L}$ be a complete length* CBB(κ) *space and let $p \in$ $\in \mathrm{Str}(x_1, \dots, x_n)$. Then there is a subcone $E \subset \mathrm{T}_p$ that is isometric to a Euclidean space such that $\log[px_i] \in E$ for every i.*

Proof. By the definition of $\mathrm{Str}(x_1, \dots, x_n)$ (Definition 8.10), $\log[px_i] \in \mathrm{Lin}_p$ for each i. It remains to apply Theorem 13.39. □

The main difficulty in the proof of Theorem 13.39 comes from the fact that in general T_p is not a length space; see Halbeisen's Example 13.6. If the tangent space were a length space, the statement would follow directly from the splitting theorem (see 16.22). In fact the proof of Theorem 13.39 is very circuitous—we use the construction of the gradient, as well as the splitting theorem, namely its corollary (Corollary 16.23). Thus in order to understand our proof, one needs to read most of Chapter 16.

Proof of Theorem 13.39. First we show that Lin_p is a complete geodesic CBB(0) space.

Recall that T_p^ω is a complete geodesic CBB(0) space (see Observation 4.9 and Theorem 13.1(a)) and Lin_p is a closed subset of T_p^ω. Thus, it is sufficient to show that the metric on Lin_p inherited from T_p^ω is a length metric.

Fix two vectors x, $y \in \mathrm{Lin}_p$. Let u and v be such that $u + \frac{1}{2} \cdot x = 0$ and $v + \frac{1}{2} \cdot y = 0$. Apply Lemma 13.32 to the vectors u and v; let $w \in \mathrm{T}_p$ denote the obtained tangent vector.

(1) *w is a midpoint of $[xy]$.*

Indeed, according to Lemma 13.32,

$$\begin{aligned}|w|^2 &= -\langle w,u\rangle - \langle w,v\rangle = \\ &= \tfrac{1}{2}\cdot\langle w,x\rangle + \tfrac{1}{2}\cdot\langle w,y\rangle.\end{aligned}$$

Therefore

$$\begin{aligned}|x-w|^2+|w-y|^2 &= 2\cdot|w|^2+|x|^2+|y|^2-2\cdot\langle w,x\rangle-2\cdot\langle w,y\rangle = \\ &= |x|^2+|y|^2-\langle w,x\rangle-\langle w,y\rangle \leqslant \\ &\leqslant |x|^2+|y|^2+\langle u,x\rangle+\langle v,x\rangle+\langle u,y\rangle+\langle v,y\rangle = \\ &= \tfrac{1}{2}\cdot|x|^2+\tfrac{1}{2}\cdot|y|^2-\langle x,y\rangle = \\ &= \tfrac{1}{2}\cdot|x-y|^2.\end{aligned}$$

Thus $|x-w| = |w-y| = \frac{1}{2}\cdot|x-y|$ and (1) follows. △

Note that for any $v \in \mathrm{Lin}_p$ there is a line ℓ that contains v and 0. Therefore by Corollary 16.23, Lin_p is isometric to a Hilbert space. □

J. Comments

13.41. Open question. Let $\mathcal{L}$ be a proper length CBB(κ) space. Is it true that for any $p \in \mathcal{L}$, the tangent space T_p is a length space?

Chapter 14

Dimension of CAT spaces

In this chapter we discuss constructions introduced by Bruce Kleiner [**96**].

The material of this chapter is used mostly for CAT spaces, but the results in Section 14.A find applications for finite-dimensional CBB spaces as well.

A. The case of complete geodesic spaces

The following construction gives a k-dimensional submanifold for a given *nondegenerate* array of $k+1$ strongly convex functions.

14.1. Definition. For two real arrays $\boldsymbol{v}$, $\boldsymbol{w} \in \mathbb{R}^{k+1}$, $\boldsymbol{v} = (v^0, v^1, \dots, v^k)$, and $\boldsymbol{w} = (w^0, w^1, \dots, w^k)$, we will write $\boldsymbol{v} \succcurlyeq \boldsymbol{w}$ if $v^i \geqslant w^i$ for each i.

Given a subset $Q \subset \mathbb{R}^{k+1}$, denote by $\operatorname{Up} Q$ the smallest upper set containing Q, and by $\operatorname{Min} Q$ the set of minimal elements of Q with respect to $\succcurlyeq$; that is,

$$\operatorname{Up} Q = \left\{ \boldsymbol{v} \in \mathbb{R}^{k+1} : \exists\, \boldsymbol{w} \in Q \text{ such that } \boldsymbol{v} \succcurlyeq \boldsymbol{w} \right\},$$
$$\operatorname{Min} Q = \left\{ \boldsymbol{v} \in Q : \text{if } \boldsymbol{v} \succcurlyeq \boldsymbol{w} \in Q \text{ then } \boldsymbol{w} = \boldsymbol{v} \right\}.$$

14.2. Definition. Let $\boldsymbol{f} = (f^0, f^1, \dots, f^k) \colon \mathcal{X} \to \mathbb{R}^{k+1}$ be a function array on a metric space $\mathcal{X}$. The set

$$\operatorname{Web} \boldsymbol{f} := \boldsymbol{f}^{-1}\left[\operatorname{Min} \boldsymbol{f}(\mathcal{X})\right] \subset \mathcal{X}$$

will be called the *web* of $\boldsymbol{f}$.

Given an array $\boldsymbol{f} = (f^0, f^1, \dots, f^k)$, we denote by $\boldsymbol{f}^{-i}$ the subarray of $\boldsymbol{f}$ with f^i removed; that is,

$$\boldsymbol{f}^{-i} := (f^0, \dots, f^{i-1}, f^{i+1}, \dots, f^k).$$

Clearly $\operatorname{Web}\boldsymbol{f}^{-i} \subset \operatorname{Web}\boldsymbol{f}$. Define the *inner web* of $\boldsymbol{f}$ as

$$\operatorname{InWeb}\boldsymbol{f} = \operatorname{Web}\boldsymbol{f} \Big\backslash \left(\bigcup_i \operatorname{Web}\boldsymbol{f}^{-i}\right).$$

We say that a function array is *nondegenerate* if $\operatorname{InWeb}\boldsymbol{f} \neq \varnothing$.

Example. If $\mathcal{X}$ is a geodesic space, then $\operatorname{Web}(\operatorname{dist}_x, \operatorname{dist}_y)$ is the union of all geodesics from x to y, and

$$\operatorname{InWeb}(\operatorname{dist}_x, \operatorname{dist}_y) = \operatorname{Web}(\operatorname{dist}_x, \operatorname{dist}_y) \setminus \{x, y\}.$$

Barycenters. Let us denote by $\Delta^k \subset \mathbb{R}^{k+1}$ the *standard k-simplex*; that is, $\boldsymbol{x} = (x^0, x^1, \dots, x^k) \in \Delta^k$ if $\sum_{i=0}^k x^i = 1$ and $x^i \geqslant 0$ for all i.

Let $\mathcal{X}$ be a metric space and let $\boldsymbol{f} = (f^0, f^1, \dots, f^k) \colon \mathcal{X} \to \mathbb{R}^{k+1}$ be a function array. Consider the map $\mathfrak{S}_{\boldsymbol{f}} \colon \Delta^k \to \mathcal{X}$ defined by

$$\mathfrak{S}_{\boldsymbol{f}}(\boldsymbol{x}) = \operatorname{MinPoint} \sum_{i=0}^{k} x^i \cdot f^i,$$

where $\operatorname{MinPoint} f$ denotes a point of minimum of f. The map $\mathfrak{S}_{\boldsymbol{f}}$ will be called a *barycentric simplex* of $\boldsymbol{f}$. Note that for a general function array $\boldsymbol{f}$, the value $\mathfrak{S}_{\boldsymbol{f}}(\boldsymbol{x})$ might be undefined or nonuniquely defined.

It is clear from the definition that $\mathfrak{S}_{\boldsymbol{f}^{-i}}$ coincides with the restriction of $\mathfrak{S}_{\boldsymbol{f}}$ to the corresponding facet of Δ^k.

14.3. Theorem. *Let $\mathcal{X}$ be a complete geodesic space and let $\boldsymbol{f} = (f^0, f^1, \dots, f^k) \colon \mathcal{X} \to \mathbb{R}^{k+1}$ be an array of strongly convex and locally Lipschitz functions. Then $\boldsymbol{f}$ defines a $C^{\frac{1}{2}}$-embedding $\operatorname{Web}\boldsymbol{f} \hookrightarrow \mathbb{R}^{k+1}$.*

Moreover,

(a)

$$W = \operatorname{Up}[\boldsymbol{f}(\mathcal{X})] \text{ is a convex closed subset of } \mathbb{R}^{k+1},$$

and

$$S = \partial_{\mathbb{R}^{k+1}} W \text{ is a convex hypersurface in } \mathbb{R}^{k+1}.$$

(b)

$$\boldsymbol{f}(\operatorname{Web}\boldsymbol{f}) = \operatorname{Min} W \subset S$$

and

$$\boldsymbol{f}(\operatorname{InWeb}\boldsymbol{f}) = \operatorname{Interior}_S(\operatorname{Min} W).$$

(c) *The barycentric simplex $\mathfrak{S}_{\boldsymbol{f}} \colon \Delta^k \to \mathcal{X}$ is a uniquely defined Lipshitz map and $\operatorname{Im}\mathfrak{S}_{\boldsymbol{f}} = \operatorname{Web}\boldsymbol{f}$. In particular, $\operatorname{Web}\boldsymbol{f}$ is compact.*

(d) *Let us equip Δ^k with the metric induced by the ℓ^1-norm on $\mathbb{R}^{k+1}$. Then the Lipschitz constant of $\mathfrak{S}_{\boldsymbol{f}}\colon \Delta^k \to \mathcal{U}$ can be estimated in terms of positive lower bounds on $(f^i)''$ and Lipschitz constants of f^i in a neighborhood of* Web $\boldsymbol{f}$ *for all i.*

In particular, by (a) *and* (b), InWeb $\boldsymbol{f}$ *is $C^{\frac12}$-homeomorphic to an open set of $\mathbb{R}^k$.*

The proof is preceded by a few preliminary statements.

14.4. Lemma. *Suppose $\mathcal{X}$ is a complete geodesic space and $f\colon \mathcal{X} \to \mathbb{R}$ is a locally Lipschitz, strongly convex function. Then the minimum point of f is uniquely defined.*

Proof. Without loss of generality, we can assume that f is 1-convex. In particular, the following claim holds:

(1) *if z is a midpoint of the geodesic $[xy]$, then*

$$s \leqslant f(z) \leqslant \tfrac{1}{2}\cdot f(x) + \tfrac{1}{2}\cdot f(y) - \tfrac{1}{8}\cdot|x-y|^2,$$

where s is the infimum of f.

Uniqueness. Assume that x and y are distinct minimum points of f. From (1) we have

$$f(z) < f(x) = f(y),$$

a contradiction.

Existence. Fix a point $p \in \mathcal{X}$, and let $\ell \in \mathbb{R}$ be a Lipschitz constant of f in a neighborhood of p.

Choose a geodesic $[px]$; consider the function $\phi\colon t \mapsto f \circ \mathrm{geod}_{[px]}(t)$. Clearly ϕ is 1-convex and $\phi^+(0) \geqslant -\ell$. Setting $a = |p-x|$, we have

$$\begin{aligned} f(x) = \phi(a) &\geqslant \\ &\geqslant f(p) - \ell\cdot a + \tfrac{1}{2}\cdot a^2 \geqslant \\ &\geqslant f(p) - \tfrac{1}{2}\cdot \ell^2. \end{aligned}$$

In particular,

$$\begin{aligned} s \coloneqq \inf\{\, f(x) : x \in \mathcal{X}\,\} &\geqslant \\ &\geqslant f(p) - \tfrac{1}{2}\cdot\ell^2. \end{aligned}$$

Choose a sequence of points $p_n \in \mathcal{X}$ such that $f(p_n) \to s$. Applying (1) for $x = p_n$, $y = p_m$, we see that p_n is a Cauchy sequence. Thus the sequence p_n converges to a minimum point of f. □

14.5. Definition. Let Q be a closed subset of $\mathbb{R}^{k+1}$. A vector $\boldsymbol{x} = (x^0, x^1, \dots$ $\dots, x^k) \in \mathbb{R}^{k+1}$ is *subnormal* to Q at a point $\boldsymbol{v} \in Q$ if

$$\langle \boldsymbol{x}, \boldsymbol{w} - \boldsymbol{v} \rangle := \sum_i x^i \cdot (w^i - v^i) \geqslant 0$$

for any $\boldsymbol{w} \in Q$.

14.6. Lemma. *Let $\mathcal{X}$ be a complete geodesic space and let $\boldsymbol{f} = (f^0, f^1, \dots, f^k)$: $\mathcal{X} \to \mathbb{R}^{k+1}$ be an array of strongly convex and locally Lipschitz functions. Let $W = \operatorname{Up} \boldsymbol{f}(\mathcal{X})$. Then:*

(a) *W is a closed convex set, bounded below with respect to $\succcurlyeq$.*

(b) *If $\boldsymbol{x}$ is a subnormal vector to W, then $\boldsymbol{x} \succcurlyeq 0$.*

(c) *$S = \partial_{\mathbb{R}^{k+1}} W$ is a complete convex hypersurface in $\mathbb{R}^{k+1}$.*

Proof. Denote by $\bar{W}$ the closure of W.

Convexity of all f^i implies that for any two points $p, q \in \mathcal{X}$ and $t \in [0,1]$ we have

$$(1-t)\cdot \boldsymbol{f}(p) + t \cdot \boldsymbol{f}(q) \succcurlyeq \boldsymbol{f} \circ \operatorname{path}_{[pq]}(t), \tag{2}$$

where $\operatorname{path}_{[pq]}$ denotes a geodesic path from p to q. Therefore W, as well as $\bar{W}$, are convex sets in $\mathbb{R}^{k+1}$.

Let

$$w^i = \min\{\, f^i(x) : x \in \mathcal{X} \,\}.$$

By Lemma 14.4, w^i is finite for each i. Evidently, $\boldsymbol{w} = (w^0, w^1, \dots, w^k)$ is a lower bound of $\bar{W}$ with respect to $\succcurlyeq$.

It is clear that W has nonempty interior, and $W \neq \mathbb{R}^{k+1}$ since W is bounded below. Therefore $S = \partial_{\mathbb{R}^{k+1}} W = \partial_{\mathbb{R}^{k+1}} \bar{W}$ is a complete convex hypersurface in $\mathbb{R}^{k+1}$.

Since $\bar{W}$ is closed and bounded below, we also have

$$\bar{W} = \operatorname{Up}[\operatorname{Min} \bar{W}]. \tag{3}$$

Choose an arbitrary $\boldsymbol{v} \in S$. Let $\boldsymbol{x} \in \mathbb{R}^{k+1}$ be a subnormal vector to $\bar{W}$ at $\boldsymbol{v}$. In particular, $\langle \boldsymbol{x}, \boldsymbol{y} \rangle \geqslant 0$ for any $\boldsymbol{y} \succcurlyeq 0$; that is, $\boldsymbol{x} \succcurlyeq 0$.

Further, according to Lemma 14.4, the function

$$p \mapsto \langle \boldsymbol{x}, \boldsymbol{f}(p) \rangle = \sum_i x^i \cdot f^i(p)$$

has a uniquely defined minimum point, say p. Clearly

$$\boldsymbol{v} \succcurlyeq \boldsymbol{f}(p) \quad \text{and} \quad \boldsymbol{f}(p) \in \operatorname{Min} W. \tag{4}$$

Note that for any $\boldsymbol{u} \in \bar{W}$ there is $\boldsymbol{v} \in S$ such that $\boldsymbol{u} \succcurlyeq \boldsymbol{v}$. Therefore (4) implies

$$\bar{W} \subset \mathrm{Up}[\mathrm{Min}\, W] \subset W.$$

Hence $\bar{W} = W$; that is, W is closed. □

Proof of Theorem 14.3.

(a)+(b). Without loss of generality, we may assume that all f^i are 1-convex.

Given $\boldsymbol{v} = (v^0, v^1, \dots, v^k) \in \mathbb{R}^{k+1}$, consider the function $h_{\boldsymbol{v}} : \mathcal{X} \to \mathbb{R}$ defined by

$$h_{\boldsymbol{v}}(p) = \max_i \{f^i(p) - v^i\}.$$

Note that $h_{\boldsymbol{v}}$ is 1-convex. Let

$$\Phi(\boldsymbol{v}) := \mathrm{MinPoint}\, h_{\boldsymbol{v}}.$$

According to Lemma 14.4, $\Phi(\boldsymbol{v})$ is uniquely defined.

From the definition of web (see Definition 14.2) we have $\Phi \circ \boldsymbol{f}(p) = p$ for any $p \in \mathrm{Web}\, \boldsymbol{f}$; that is, Φ is a left inverse to the restriction $\boldsymbol{f}|_{\mathrm{Web}\, \boldsymbol{f}}$. In particular,

$$\mathrm{Web}\, \boldsymbol{f} = \mathrm{Im}\, \Phi. \tag{5}$$

Given $\boldsymbol{v}, \boldsymbol{w} \in \mathbb{R}^{k+1}$, set $p = \Phi(\boldsymbol{v})$ and $q = \Phi(\boldsymbol{w})$. Since $h_{\boldsymbol{v}}$ and $h_{\boldsymbol{w}}$ are 1-convex, we have

$$h_{\boldsymbol{v}}(q) \geqslant h_{\boldsymbol{v}}(p) + \tfrac{1}{2} \cdot |p - q|^2, \qquad h_{\boldsymbol{w}}(p) \geqslant h_{\boldsymbol{w}}(q) + \tfrac{1}{2} \cdot |p - q|^2.$$

Therefore,

$$\begin{aligned} |p - q|^2 &\leqslant 2 \cdot \sup_{x \in \mathcal{X}} \{|h_{\boldsymbol{v}}(x) - h_{\boldsymbol{w}}(x)|\} \leqslant \\ &\leqslant 2 \cdot \max_i \{|v^i - w^i|\}. \end{aligned}$$

In particular, Φ is $C^{\frac{1}{2}}$-continuous. Hence $\boldsymbol{f}|_{\mathrm{Web}\, \boldsymbol{f}}$ is a $C^{\frac{1}{2}}$-embedding.

As in Lemma 14.6, let $W = \mathrm{Up}\, \boldsymbol{f}(\mathcal{X})$ and $S = \partial_{\mathbb{R}^{k+1}} W$. Then S is a convex hypersurface in $\mathbb{R}^{k+1}$. Clearly $\boldsymbol{f}(\mathrm{Web}\, \boldsymbol{f}) = \mathrm{Min}\, W \subset S$. From the definition of inner web, we have $\boldsymbol{v} \in \boldsymbol{f}(\mathrm{InWeb}\, \boldsymbol{f})$ if and only if $\boldsymbol{v} \in S$ and for any i there is $\boldsymbol{w} = (w^0, w^1, \dots, w^k) \in W$ such that $w^j < v^j$ for all $j \neq i$. Thus $\boldsymbol{f}(\mathrm{InWeb}\, \boldsymbol{f})$ is open in S. That is, $\mathrm{InWeb}\, \boldsymbol{f}$ is $C^{\frac{1}{2}}$-homeomorphic to an open set in a convex hypersurface $S \subset \mathbb{R}^{k+1}$, and hence to an open set of $\mathbb{R}^k$, as claimed.

(c)+(d). Since f^i is 1-convex, for any $\boldsymbol{x} = (x^0, x^1, \dots, x^k) \in \Delta^k$ the convex combination

$$\left(\sum_i x^i \cdot f^i\right) : \mathcal{X} \to \mathbb{R}$$

is also 1-convex. Therefore, according to Lemma 14.4, the barycentric simplex $\mathfrak{S}_{\boldsymbol{f}}$ is uniquely defined on Δ^k.

For $\boldsymbol{x}, \boldsymbol{y} \in \Delta^k$, let

$$\begin{aligned} f_{\boldsymbol{x}} &= \sum_i x^i \cdot f^i, & f_{\boldsymbol{y}} &= \sum_i y^i \cdot f^i, \\ p &= \mathfrak{S}_{\boldsymbol{f}}(\boldsymbol{x}), & q &= \mathfrak{S}_{\boldsymbol{f}}(\boldsymbol{y}), \\ s &= |p - q|. \end{aligned}$$

Note the following:

- The function $\phi(t) = f_{\boldsymbol{x}} \circ \mathrm{geod}_{[pq]}(t)$ has minimum at 0. Therefore $\phi^+(0) \geqslant 0$.
- The function $\psi(t) = f_{\boldsymbol{y}} \circ \mathrm{geod}_{[pq]}(t)$ has minimum at s. Therefore $\psi^-(s) \geqslant 0$.

From 1-convexity of $f_{\boldsymbol{y}}$, we have $\psi^+(0) + \psi^-(s) + s \leqslant 0$.

Let ℓ be a Lipschitz constant for all f^i in a neighborhood $\Omega \ni p$. Then

$$\psi^+(0) \leqslant \phi^+(0) + \ell \cdot \|\boldsymbol{x} - \boldsymbol{y}\|_1,$$

where $\|\boldsymbol{x} - \boldsymbol{y}\|_1 = \sum_{i=0}^k |x^i - y^i|$. That is, given $\boldsymbol{x} \in \Delta^k$, there is a constant ℓ such that

$$\begin{aligned} |\mathfrak{S}_{\boldsymbol{f}}(\boldsymbol{x}) - \mathfrak{S}_{\boldsymbol{f}}(\boldsymbol{y})| &= s \\ &\leqslant \ell \cdot \|\boldsymbol{x} - \boldsymbol{y}\|_1 \end{aligned}$$

for any $\boldsymbol{y} \in \Delta^k$. In particular, there is $\varepsilon > 0$ such that if $\|\boldsymbol{x} - \boldsymbol{y}\|_1 < \varepsilon$, $\|\boldsymbol{x} - \boldsymbol{z}\|_1 < \varepsilon$, then $\mathfrak{S}_{\boldsymbol{f}}(\boldsymbol{y}), \mathfrak{S}_{\boldsymbol{f}}(\boldsymbol{z}) \in \Omega$. Thus the same argument as above implies

$$|\mathfrak{S}_{\boldsymbol{f}}(\boldsymbol{y}) - \mathfrak{S}_{\boldsymbol{f}}(\boldsymbol{z})| \leqslant \ell \cdot \|\boldsymbol{y} - \boldsymbol{z}\|_1$$

for any $\boldsymbol{y}$ and $\boldsymbol{z}$ sufficiently close to $\boldsymbol{x}$; that is, $\mathfrak{S}_{\boldsymbol{f}}$ is locally Lipschitz. Since Δ^k is compact, $\mathfrak{S}_{\boldsymbol{f}}$ is Lipschitz.

Clearly $\mathfrak{S}_{\boldsymbol{f}}(\Delta^k) \subset \mathrm{Web}\,\boldsymbol{f}$. It remains to show that $\mathfrak{S}_{\boldsymbol{f}}(\Delta^k) \supset \mathrm{Web}\,\boldsymbol{f}$. According to Lemma 14.6, $W = \mathrm{Up}\,\boldsymbol{f}(\mathcal{X})$ is a closed convex set in $\mathbb{R}^{k+1}$. Let $p \in \mathrm{Web}\,\boldsymbol{f}$. Clearly $\boldsymbol{f}(p) \in \mathrm{Min}\,W \subset S = \partial_{\mathbb{R}^{k+1}} W$. Let $\boldsymbol{x}$ be a subnormal vector to W at $\boldsymbol{f}(p)x$. According to Lemma 14.6, $\boldsymbol{x} \geqslant 0$. Without loss of generality, we may assume that $\sum_i x^i = 1$; that is, $\boldsymbol{x} \in \Delta^k$. By Lemma 14.4, p is the unique minimum point of $\sum_i x^i \cdot f^i$; that is, $p = \mathfrak{S}_{\boldsymbol{f}}(\boldsymbol{x})$. □

B. The case of CAT spaces

Let $\boldsymbol{a} = (a^0, a^1, \dots, a^k)$ be a point array in a metric space $\mathcal{U}$. Recall that $\operatorname{dist}_{\boldsymbol{a}}$ denotes the distance map

$$(\operatorname{dist}_{a^0}, \operatorname{dist}_{a^1}, \dots, \operatorname{dist}_{a^k})\colon \mathcal{U} \to \mathbb{R}^{k+1},$$

which can be also regarded as a function array. The *radius* of the point array $\boldsymbol{a}$ is defined to be the radius of the set $\{a^0, a^1, \dots, a^k\}$; that is,

$$\operatorname{rad} \boldsymbol{a} = \inf\big\{ r > 0 : \exists z \in \mathcal{U} \text{ such that } a^i \in \mathrm{B}(z, r) \text{ for any } i \big\}.$$

Fix $\kappa \in \mathbb{R}$. Let $\boldsymbol{a} = (a^0, a^1, \dots, a^k)$ be a point array of radius $< \frac{\varpi^\kappa}{2}$ in a metric space $\mathcal{U}$. Consider the function array $\boldsymbol{f} = (f^0, f^1, \dots, f^k)$ where

$$f^i(x) = \operatorname{md}^\kappa |a^i - x|.$$

Assuming the barycentric simplex $\mathfrak{S}_{\boldsymbol{f}}$ is defined, then $\mathfrak{S}_{\boldsymbol{f}}$ is called the *κ-barycentric simplex* for the point array $\boldsymbol{a}$; it will be denoted by $\mathfrak{S}^\kappa_{\boldsymbol{a}}$. The points a^0, a^1, …, a^k are called *vertexes* of the κ-barycentric simplex. Note that once we say the κ-barycentric simplex is defined, we automatically assume that $\operatorname{rad} \boldsymbol{a} < \frac{\varpi^\kappa}{2}$.

14.7. Theorem. *Let $\mathcal{U}$ be a complete length* CAT(κ) *space and let $\boldsymbol{a} = (a^0, a^1, \dots, a^k)$ be a point array with radius $< \frac{\varpi^\kappa}{2}$. Then:*

(a) *The κ-barycentric simplex $\mathfrak{S}^\kappa_{\boldsymbol{a}}\colon \Delta^k \to \mathcal{U}$ is defined. Moreover, $\mathfrak{S}^\kappa_{\boldsymbol{a}}$ is a Lipschitz map, and if Δ^k is equipped with the ℓ^1-metric, then its Lipschitz constant can be estimated in terms of κ and the radius of $\boldsymbol{a}$ (in particular it does not depend on k).*

(b) $\operatorname{Web}(\operatorname{dist}_{\boldsymbol{a}}) = \operatorname{Im} \mathfrak{S}^\kappa_{\boldsymbol{a}}$. *Moreover, if a closed convex set $K \subset \mathcal{U}$ contains all a^i, then* $\operatorname{Web}(\operatorname{dist}_{\boldsymbol{a}}) \subset K$.

(c) *The restriction*[1] $\operatorname{dist}_{\boldsymbol{a}-0}|_{\operatorname{InWeb}(\operatorname{dist}_{\boldsymbol{a}})}$ *is an open $C^{\frac12}$-embedding in $\mathbb{R}^k$. Thus there is an inverse of* $\operatorname{dist}_{\boldsymbol{a}-0}|_{\operatorname{InWeb}(\operatorname{dist}_{\boldsymbol{a}})}$, *say $\Phi\colon \mathbb{R}^k \mathrel{\circ\!\!\to} \mathcal{U}$.*

The subfunction $f = \operatorname{dist}_{a^0} \circ \Phi$ is semiconvex and locally Lipschitz. Moreover, if $\kappa \leqslant 0$, then f is convex.

In particular, $\operatorname{Web}(\operatorname{dist}_{\boldsymbol{a}})$ *is a compact set and* $\operatorname{InWeb}(\operatorname{dist}_{\boldsymbol{a}})$ *is $C^{\frac12}$-homeomorphic to an open subset of $\mathbb{R}^k$.*

14.8. Exercise. Show that any finite isometric group action on a complete length CAT(0) space has a fixed point.

14.9. Definition. The submap $\Phi\colon \mathbb{R}^k \mathrel{\circ\!\!\to} \mathcal{X}$ of Theorem 14.7(c) will be called the $\operatorname{dist}_{\boldsymbol{a}}$*-web embedding* with *brace* $\operatorname{dist}_{a^0}$. The terminology invokes Theorem 14.7(c).

[1] Recall that $\operatorname{dist}_{\boldsymbol{a}-0}$ denotes the array $(\operatorname{dist}_{a^1}, \dots, \operatorname{dist}_{a^k})$.

14.10. Definition. Let $\mathcal{U}$ be a complete length CAT(κ) space and let $\boldsymbol{a} = (a^0, a^1, \dots, a^k)$ be a point array with radius $< \frac{\varpi^\kappa}{2}$. If $\mathrm{InWeb}(\mathrm{dist}_{\boldsymbol{a}})$ is nonempty, then the point array $\boldsymbol{a}$ is called *nondegenerate.*

Lemma 14.12 will provide examples of nondegenerate point arrays, which can be used in Theorem 14.7(c).

14.11. Corollary. *Let $\mathcal{U}$ be a complete length* CAT(κ) *space, let $\boldsymbol{a} = (a^0, a^1, \dots, a^m)$ be a nondegenerate point array of radius $< \frac{\varpi^\kappa}{2}$ in $\mathcal{U}$, and let $\sigma = \mathfrak{S}^\kappa_{\boldsymbol{a}}$ be the corresponding κ-barycentric simplex. Then for some $\boldsymbol{x} \in \Delta^m$, the differential $\mathbf{d}_{\boldsymbol{x}}\sigma$ is linear and the image $\mathrm{Im}\, \mathbf{d}_{\boldsymbol{x}}\sigma$ forms a subcone isometric to an m-dimensional Euclidean space in the tangent cone $\mathrm{T}_{\sigma(\boldsymbol{x})}$.*

Proof. Denote the distance map $\mathrm{dist}_{\boldsymbol{a}\text{-}0}$ by $\tau\colon \mathcal{U} \to \mathbb{R}^m$.

According to Theorem 14.7, σ is Lipschitz and the distance map τ gives an open embedding of $\mathrm{InWeb}(\mathrm{dist}_{\boldsymbol{a}}) = \sigma(\Delta^m) \setminus \sigma(\partial\Delta^m)$. Note that τ is Lipschitz. According to Rademacher's Theorem 13.12, the differential $\mathbf{d}_{\boldsymbol{x}}(\tau \circ \sigma)$ is linear for almost all $\boldsymbol{x} \in \Delta^m$. Further, since $\mathrm{InWeb}(\mathrm{dist}_{\boldsymbol{a}}) \neq \varnothing$, the area formula **[93]** implies that $\mathbf{d}_{\boldsymbol{x}}(\tau \circ \sigma)$ is surjective on a set of positive masure of points $\boldsymbol{x} \in \Delta^m$.

Note that $\mathbf{d}_{\boldsymbol{x}}(\tau \circ \sigma) = (\mathbf{d}_{\sigma(\boldsymbol{x})}\tau) \circ (\mathbf{d}_{\boldsymbol{x}}\sigma)$. Applying Rademacher's theorem again, we have linearity of $\mathbf{d}_{\boldsymbol{x}}\sigma$ for almost all $\boldsymbol{x} \in \Delta^m$; at these points $\mathrm{Im}\, \mathbf{d}_{\boldsymbol{x}}\sigma$ forms a subcone isometric to a Euclidean space in $\mathrm{T}_{\sigma(\boldsymbol{x})}$. Clearly the dimension of $\mathrm{Im}\, \mathbf{d}_{\boldsymbol{x}}(\tau \circ \sigma)$ is at least as big as the dimension of $\mathrm{Im}\, \mathbf{d}_{\boldsymbol{x}}\sigma$. Hence the result. □

Proof of Theorem 14.7. Fix $z \in \mathcal{U}$ and $r < \frac{\varpi^\kappa}{2}$ such that $|z - a^i| < r$ for all i. Note that the set $K \cap \overline{\mathrm{B}}[z, r]$ is convex, closed, and contains all a^i. Applying the theorem on short retract (Exercise 9.76), we get the second part of (b).

The remaining statements are proved first in the case $\kappa \leqslant 0$, and then the remaining case $\kappa > 0$ is reduced to the case $\kappa = 0$.

Case $\kappa \leqslant 0$. Consider the function array $f^i = \mathrm{md}^\kappa \circ \mathrm{dist}_{a^i}$. From the definition of web (see Definition 14.2), it is clear that $\mathrm{Web}(\mathrm{dist}_{\boldsymbol{a}}) = \mathrm{Web}\, \boldsymbol{f}$. Further, by the definition of the κ-barycentric simplex, $\mathfrak{S}^\kappa_{\boldsymbol{a}} = \mathfrak{S}_{\boldsymbol{f}}$.

All the functions f^i are strongly convex (see Theorem 9.25(b)). Therefore (a), (b), and the first statements in (c) follow from Theorem 14.3.

Case $\kappa > 0$. Applying rescaling, we may assume $\kappa = 1$, so $\varpi^\kappa = \varpi^1 = \pi$.

Let $\mathring{\mathcal{U}} = \mathrm{Cone}\, \mathcal{U}$. By Theorem 11.7(a), $\mathring{\mathcal{U}}$ is CAT(0). Let us denote by ι the natural embedding of $\mathcal{U}$ as the unit sphere in $\mathring{\mathcal{U}}$, and by $\mathrm{proj}\colon \mathring{\mathcal{U}} \multimap \mathcal{U}$ the submap defined by $\mathrm{proj}(v) = \iota^{-1}(v/|v|)$ for all $v \neq 0$. Note that there is $z \in \mathcal{U}$ and $\varepsilon > 0$ such that the set

$$K_\varepsilon = \left\{ v \in \mathring{\mathcal{U}} : \langle \iota(z), v \rangle \geqslant \varepsilon \right\}$$

contains all $\iota(a^i)$. Then $0 \notin K_\varepsilon$, and the set K_ε is closed and convex. The latter follows from Exercise 9.28, since $v \mapsto -\langle \iota(z), v\rangle$ is a Busemann function.

Denote by $\iota(\boldsymbol{a})$ the point array $(\iota(a^0), \iota(a^1), \dots, \iota(a^k))$ in $\mathring{\mathcal{U}}$. From the case $\kappa = 0$, we get that $\operatorname{Im} \mathfrak{S}^0_{\iota(\boldsymbol{a})} \subset K_\varepsilon$. In particular, $\operatorname{Im} \mathfrak{S}^0_{\iota(\boldsymbol{a})} \not\ni 0$ and thus $\operatorname{proj} \circ \mathfrak{S}^0_{\iota(\boldsymbol{a})}$ is defined. Direct calculations show

$$\mathfrak{S}^1_{\boldsymbol{a}} = \operatorname{proj} \circ \mathfrak{S}^0_{\iota(\boldsymbol{a})} \quad \text{and} \quad \operatorname{Web}(\operatorname{dist}_{\boldsymbol{a}}) = \operatorname{proj}[\operatorname{Web}(\operatorname{dist}_{\iota(\boldsymbol{a})})].$$

Thus the case $\kappa = 1$ of the theorem is reduced to the case $\kappa = 0$, which is proved already. □

14.12. Lemma. *Let $\boldsymbol{a} = (a^0, a^1, \dots a^k)$ be a point array of radius $< \frac{\varpi^\kappa}{2}$ in a complete length* CAT(κ) *space $\mathcal{U}$, and let $B^i = \overline{\mathrm{B}}[a^i, r^i]$ for an array of positive reals $(r^0, r^1, \dots, r^k)$. Assume that $\bigcap_i B^i = \varnothing$, but $\bigcap_{i\ne j} B^i \ne \varnothing$ for any j. Then $\boldsymbol{a}$ is nondegenerate.*

Proof. Without loss of generality, we may assume that $\mathcal{U}$ is geodesic and $\operatorname{diam} \mathcal{U} < \varpi^\kappa$. If not, choose $z \in \mathcal{U}$ and $r < \frac{\varpi^\kappa}{2}$ so that $|z - a^i| \leqslant r$ for each i, and consider $\overline{\mathrm{B}}[z, r]$ instead of $\mathcal{U}$. The latter can be done since $\overline{\mathrm{B}}[z, r]$ is convex and closed, so $\overline{\mathrm{B}}[z, r]$ is a complete length CAT(κ) space and $\operatorname{Web}(\operatorname{dist}_{\boldsymbol{a}}) \subset \overline{\mathrm{B}}[z, r]$; see Corollary 9.27 and Theorem 14.7(b).

By Theorem 14.7, $\operatorname{Web}(\operatorname{dist}_{\boldsymbol{a}})$ is a compact set; therefore there is a point $p \in \operatorname{Web}(\operatorname{dist}_{\boldsymbol{a}})$ minimizing the function

$$f(x) = \max_i\{\operatorname{dist}_{B^i} x\} = \max\{0, |a^0 - x| - r^0, \dots, |a^k - x| - r^k\}.$$

By the definition of web (see Definition 14.2), p is also the minimum point of f on $\mathcal{U}$. Let us prove the following claim:

(1) $\quad p \notin B^j$ *for any* j.

Indeed, assume the contrary; that is,

(2) $$p \in B^j$$

for some j. Then p is a point of local minimum for the function

$$h^j(x) = \max_{i\ne j}\{\operatorname{dist}_{B^i} x\}.$$

Hence

$$\max_{i\ne j}\{\measuredangle[p\,{}^{x}_{a^i}]\} \geqslant \tfrac{\pi}{2}$$

for any $x \in \mathcal{U}$. From the angle comparison (see Theorem 9.14(c)), it follows that p is a global minimum of h^j and hence

$$p \in \bigcap_{i\ne j} B^i.$$

The latter and (2) contradict $\bigcap_i B^i = \varnothing$. △

From the definition of web, it also follows that

$$\mathrm{Web}(\mathrm{dist}_{\boldsymbol{a}^{-j}}) \subset \bigcup_{i\neq j} B^i.$$

Indeed, if $q \in \bigcap_{i\neq j} B^i$ and $q' \notin \bigcup_{i\neq j} B^i$, then $|a_i - q| < |a_i - q'|$ for any $i \neq j$ therefore $q' \notin \mathrm{Web}(\mathrm{dist}_{\boldsymbol{a}^{-j}})$. Therefore the claim implies that $p \notin$ $\notin \mathrm{Web}(\mathrm{dist}_{\boldsymbol{a}^{-j}})$ for each j; that is, $p \in \mathrm{InWeb}(\mathrm{dist}_{\boldsymbol{a}})$. □

C. Dimension

See Chapter 7 for definitions of various dimension-like invariants of metric spaces.

We start with two examples.

The first example shows that the dimension of complete length CAT spaces is not local; that is, such spaces might have open sets with different linear dimensions.

Such an example can be constructed by gluing at one point two Euclidean spaces of different dimensions. According to Reshetnyak's Gluing Theorem 9.39, this construction gives a CAT(0) space.

The second example provides a complete length CAT space with topological dimension 1 and arbitrarily large Hausdorff dimension. Thus for complete length CAT spaces, one should not expect any relations between topological and Hausdorff dimensions except for the one provided by Szpilrajn's theorem (see 7.5).

To construct the second type of example, note that the completion of any metric tree has topological dimension 1 and is CAT($\varkappa$) for any $\varkappa$. Start with a binary tree Γ and a sequence $\varepsilon_n > 0$ such that $\sum_n \varepsilon_n < \infty$. Define the metric on Γ by prescribing the length of an edge from level n to level $n+1$ to be ε_n. For an appropriately chosen sequence ε_n, the completion of Γ will contain a Cantor set of arbitrarily large Hausdorff dimension.

The following is a version of a theorem proved by Bruce Kleiner [**96**], with an improvement made by Alexander Lytchak [**110**].

14.13. Theorem. *For any complete length* CAT($\varkappa$) *space $\mathcal{U}$, the following statements are equivalent.*

(a) $\mathrm{LinDim}\, \mathcal{U} \geqslant m$.
(b) *For some $z \in \mathcal{U}$ there is an array of $m+1$ balls $B^i = \mathrm{B}(a^i, r^i)$ with a^0, $a^1, \dots, a^m \in \mathrm{B}(z, \frac{\varpi^\varkappa}{2})$ such that*

$$\bigcap_i B^i = \varnothing \quad \textit{and} \quad \bigcap_{i\neq j} B^i \neq \varnothing \quad \textit{for each } j.$$

(c) *There is a* $C^{\frac{1}{2}}$*-embedding* $\Phi\colon \overline{\mathrm{B}}[1]_{\mathbb{E}^m} \hookrightarrow \mathcal{U}$*; that is,* Φ *is bi-Hölder with exponent* $\frac{1}{2}$.

(d) *There is a closed separable set* $K \subset \mathcal{U}$ *such that*

$$\operatorname{TopDim} K \geqslant m.$$

Remarks. Theorem 14.16 gives a stronger version of part (c) in the finite-dimensional case. Namely, a complete length CAT space with linear dimension m admits a bi-Lipschitz embedding Φ of an open set of $\mathbb{R}^m$. Moreover, the Lipschitz constants of Φ can be made arbitrarily close to 1.

14.14. Corollary. *For any separable complete length* CAT *space* $\mathcal{U}$*, we have*

$$\operatorname{TopDim} \mathcal{U} = \operatorname{LinDim} \mathcal{U}.$$

Any simplicial complex can be equipped with a length metric such that each k-simplex is isometric to the standard simplex

$$\Delta^k = \left\{ (x_0, \dots, x_k) \in \mathbb{R}^{k+1} : x_i \geqslant 0, \quad x_0 + \dots + x_k = 1 \right\}$$

with the metric induced by the ℓ^1-norm on $\mathbb{R}^{k+1}$. This metric will be called the *ℓ^1-metric* on the simplicial complex.

14.15. Lemma. *Let* $\mathcal{U}$ *be a complete length* CAT(κ) *space and let* $\rho\colon \mathcal{U} \to \mathbb{R}$ *be a continuous positive function. Then there is a simplicial complex* $\mathcal{N}$ *equipped with* ℓ^1*-metric, a locally Lipschitz map* $\Phi\colon \mathcal{U} \to \mathcal{N}$*, and a Lipschitz map* $\Psi\colon \mathcal{N} \to \mathcal{U}$ *such that:*

(a) *The displacement of the composition* $\Psi \circ \Phi\colon \mathcal{U} \to \mathcal{U}$ *is bounded by* ρ*; that is,*

$$|x - \Psi \circ \Phi(x)| < \rho(x)$$

for any $x \in \mathcal{U}$.

(b) *If* LinDim $\mathcal{U} \leqslant m$*, then the* Ψ*-image of any closed simplex in* $\mathcal{N}$ *coincides with the image of its m-skeleton.*

Proof. Without loss of generality, we may assume that for any x we have $\rho(x) < \, < \rho_0$ for fixed $\rho_0 < \frac{\varpi^\kappa}{2}$.

By Stone's theorem, any metric space is paracompact. Thus, we can choose a locally finite covering $\{ \Omega_\alpha : \alpha \in \mathcal{A} \}$ of $\mathcal{U}$ such that $\Omega_\alpha \subset \mathrm{B}(x, \frac{1}{3}\cdot\rho(x))$ for any $x \in \Omega_\alpha$.

Denote by $\mathcal{N}$ the nerve of the covering $\{\Omega_\alpha\}$; that is, $\mathcal{N}$ is an abstract simplicial complex with vertex set $\mathcal{A}$, such that $\{\alpha^0, \alpha^1, \dots, \alpha^n\} \subset \mathcal{A}$ are vertexes of a simplex if and only if $\Omega_{\alpha^0} \cap \Omega_{\alpha^1} \cap \dots \cap \Omega_{\alpha^n} \neq \varnothing$.

Fix a Lipschitz partition of unity $\phi_\alpha\colon \mathcal{U} \to [0,1]$ subordinate to $\{\Omega_\alpha\}$. Consider the map $\Phi\colon \mathcal{U} \to \mathcal{N}$ such that the barycentric coordinate of $\Phi(p)$ is $\phi_\alpha(p)$.

Note that Φ is locally Lipschitz. Clearly the Φ-preimage of any open simplex in $\mathcal{N}$ lies in Ω_α for some $\alpha \in \mathcal{A}$.

For each $\alpha \in \mathcal{A}$, choose $x_\alpha \in \Omega_\alpha$. Let us extend the map $\alpha \mapsto x_\alpha$ to a map $\Psi\colon \mathcal{N} \to \mathcal{U}$ that is κ-barycentric on each simplex. According to Theorem 14.7(a), this extension exists, Ψ is Lipschitz, and its Lipschitz constant depends only on ρ_0 and κ.

(a). Fix $x \in \mathcal{U}$. Denote by Δ the minimal simplex that contains $\Phi(x)$, and let $\alpha^0, \alpha^1, \dots, \alpha^n$ be the vertexes of Δ. Note that α is a vertex of Δ if and only if $\phi_\alpha(x) > 0$. Thus

$$|x - x_{\alpha^i}| < \tfrac{1}{3}\cdot\rho(x)$$

for any i. Therefore

$$\operatorname{diam}\Psi(\Delta) \leqslant \max_{i,j}\{|x_{\alpha^i} - x_{\alpha^j}|\} < \tfrac{2}{3}\cdot\rho(x).$$

In particular,

$$|x - \Psi\circ\Phi(x)| \leqslant |x - x_{\alpha^0}| + \operatorname{diam}\Psi(\Delta) < \rho(x).$$

(b). Assume the contrary; that is, $\Psi(\mathcal{N})$ is not included in the Ψ-image of the m-skeleton of $\mathcal{N}$. Then for some $k > m$, there is a k-simplex Δ^k in $\mathcal{N}$ such that the barycentric simplex $\sigma = \Psi|_{\Delta^k}$ is nondegenerate; that is,

$$W = \Psi(\Delta^k) \setminus \Psi(\partial\Delta^k) \neq \varnothing.$$

Applying Corollary 14.11 gives $\operatorname{LinDim}\mathcal{U} \geqslant k$, a contradiction. □

Proof of Theorem 14.13.

(b) ⇒ (c) ⇒ (d). The implication (b) ⇒ (c) follows from Lemma 14.12 and Theorem 14.7(c), and (c) ⇒ (d) is trivial.

(d) ⇒ (a). According to Theorem 7.7, there is a continuous map $f\colon K \to \mathbb{R}^m$ with a stable value. By the Tietze extension theorem, it is possible to extend f to a continuous map $F\colon \mathcal{U} \to \mathbb{R}^m$.

Fix $\varepsilon > 0$. Since F is continuous, there is a continuous positive function ρ defined on $\mathcal{U}$ such that

$$|x - y| < \rho(x) \quad\Rightarrow\quad |F(x) - F(y)| < \tfrac{1}{3}\cdot\varepsilon.$$

Apply Lemma 14.15 to ρ. For the resulting simplicial complex $\mathcal{N}$ and the maps $\Phi\colon \mathcal{U} \to \mathcal{N}$, $\Psi\colon \mathcal{N} \to \mathcal{U}$, we have

$$|F\circ\Psi\circ\Phi(x) - F(x)| < \tfrac{1}{3}\cdot\varepsilon$$

for any $x \in \mathcal{U}$.

According to Lemma 3.5, there is a locally Lipschitz map $F_\varepsilon\colon \mathcal{U} \to \mathbb{R}^{m+1}$ such that $|F_\varepsilon(x) - F(x)| < \tfrac{1}{3}\cdot\varepsilon$ for any $x \in \mathcal{U}$.

Note that $\Phi(K)$ is contained in a countable subcomplex of $\mathcal{N}$, say $\mathcal{N}'$. Indeed, since K is separable, there is a countable dense collection of points $\{x_n\}$ in K. Denote by Δ_n the minimal simplex of $\mathcal{N}$ that contains $\Phi(x_n)$. Then $\Phi(K) \subset \bigcup_i \Delta_n$.

Arguing by contradiction, assume $\operatorname{LinDim} \mathcal{U} < m$. By Lemma 14.15(b), the image $F_\varepsilon \circ \Psi \circ \Phi(K)$ lies in the F_ε-image of the $(m-1)$-skeleton of $\mathcal{N}'$. In particular, it can be covered by a countable collection of Lipschitz images of $(m-1)$-simplexes. Hence $0 \in \mathbb{R}^m$ is not a stable value of the restriction $F_\varepsilon \circ \Psi \circ \Phi|_K$. Since $\varepsilon > 0$ is arbitrary, then $0 \in \mathbb{R}^m$ is not a stable value of f—a contradiction.

(a) $\Rightarrow$ (b). Choose $q \in \mathcal{U}$ such that T_q contains a subcone E isometric to m-dimensional Euclidean space. Note that one can choose $\varepsilon > 0$ and a point array $(\dot{a}^0, \dot{a}^1, \dots, \dot{a}^m)$ in $E \subset \mathrm{T}_q$ such that $\bigcap_i \overline{\mathrm{B}}[\dot{a}^i, 1+\varepsilon] = \varnothing$ and $\bigcap_{i \neq j} \overline{\mathrm{B}}[\dot{a}^i, 1-\varepsilon] \neq \varnothing$ for each j.

For each i choose a geodesic γ^i from q that goes almost in the directions of $\dot{a}^i$. Choose small $\delta > 0$ and take the point a^i on γ^i at distance $\delta \cdot |\dot{a}^i|$ from q. We get a point array $(a^0, a^1, \dots, a^m)$ in $\mathcal{U}$ such that $\bigcap_i \overline{\mathrm{B}}[a^i, \delta] = \varnothing$ and $\bigcap_{i \neq j} \overline{\mathrm{B}}[a^i, \delta] \neq \varnothing$ for each j. Since $\delta > 0$ can be chosen arbitrarily small, (b) follows. □

D. Finite-dimensional spaces

Recall that a web embedding and its brace are defined in Definition 14.9.

14.16. Theorem. *Suppose $\mathcal{U}$ is a complete length* $\mathrm{CAT}(\kappa)$ *space such that* $\operatorname{LinDim} \mathcal{U} = m$, *and $\boldsymbol{a} = (a^0, a^1, \dots, a^m)$ is a point array in $\mathcal{U}$ with radius $< \frac{\varpi^\kappa}{2}$. Then the* $\operatorname{dist}_{\boldsymbol{a}}$*-web embedding* $\Phi\colon \mathbb{R}^m \hookrightarrow \mathcal{U}$ *with brace* $\operatorname{dist}_{a^0}$ *is locally Lipschitz.*

Note that if $\boldsymbol{a}$ is degenerate, that is, if $\operatorname{InWeb}(\operatorname{dist}_{\boldsymbol{a}}) = \varnothing$, then the domain of Φ is empty, and hence the conclusion of the theorem trivially holds.

14.17. Lemma. *Let $\mathcal{U}$ be a complete length* $\mathrm{CAT}(\kappa)$ *space, and let $\boldsymbol{a} = (a^0, a^1, \dots, a^k)$ be a point array with radius $< \frac{\varpi^\kappa}{2}$. Then for any $p \in \operatorname{InWeb}(\operatorname{dist}_{\boldsymbol{a}})$, there is $\varepsilon > 0$ such that if for some $q \in \operatorname{Web}(\operatorname{dist}_{\boldsymbol{a}})$ and $b \in \mathcal{U}$ we have*

$$(1) \qquad |p - q| < \varepsilon, \quad |p - b| < \varepsilon, \quad \textit{and} \quad \measuredangle\left[q\,{}^{b}_{a^i}\right] < \tfrac{\pi}{2} + \varepsilon$$

for each i, then the array $(b, a^0, a^1, \dots, a^m)$ is nondegenerate.

Proof. Without loss of generality, we may assume that $\mathcal{U}$ is geodesic and $\operatorname{diam} \mathcal{U} < \varpi^\kappa$. If not, consider instead of $\mathcal{U}$, a ball $\overline{\mathrm{B}}[z, r] \subset \mathcal{U}$ for some $z \in \mathcal{U}$ and $r < \frac{\varpi^\kappa}{2}$ such that $|z - a^i| \leqslant r$ for each i.

From the angle comparison (see Theorem 9.14(c)), it follows that $p \in$ $\in \mathrm{InWeb}\,\boldsymbol{a}$ if and only if both of the following conditions hold:

(1) $\max_i\{\measuredangle\left[p\,{}^{a^i}_{u}\right]\} \geqslant \frac{\pi}{2}$ for any $u \in \mathcal{U}$,

(2) for each i there is $u^i \in \mathcal{U}$ such that $\measuredangle\left[p\,{}^{a^j}_{u^i}\right] < \frac{\pi}{2}$ for all $j \neq i$.

Due to the semicontinuity of angles 9.34, there is $\varepsilon > 0$ such that for any $x \in \mathrm{B}(p, 10\cdot\varepsilon)$ we have

$$\measuredangle\left[x\,{}^{a^j}_{u^i}\right] < \tfrac{\pi}{2} - 10\cdot\varepsilon \quad \text{for all} \quad j \neq i. \tag{2}$$

Now assume that for sufficiently small $\varepsilon > 0$ there are points $b \in \mathcal{U}$ and $q \in \mathrm{Web}(\mathrm{dist}_{\boldsymbol{a}})$ such that (1) holds. According to Theorem 14.7(b), for all small $\varepsilon > 0$ we have

$$\mathrm{rad}\{b, a^0, a^1, \dots, a^k\} < \tfrac{\varpi^\kappa}{2}.$$

Fix a sufficiently small $\delta > 0$ and let

$$v^i = \mathrm{geod}_{[qu^i]}(\tfrac{1}{3}\cdot\delta) \quad \text{and} \quad w^i = \mathrm{geod}_{[v^i b]}(\tfrac{2}{3}\cdot\delta).$$

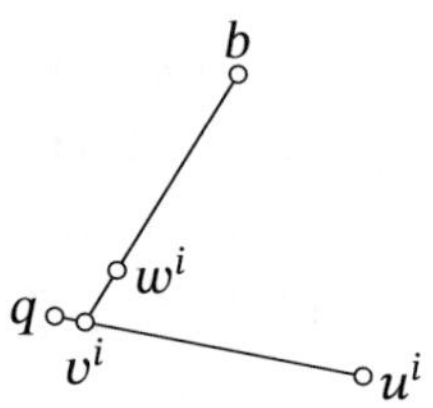

Clearly,

$$\begin{aligned}|b - w^i| &= |b - v^i| - \tfrac{2}{3}\cdot\delta \leqslant\\ &\leqslant |b - q| - \tfrac{1}{3}\cdot\delta.\end{aligned}$$

Further, inequalities (2) and (1) imply

$$\begin{aligned}|a^j - w^i| &< |a^j - v^i| + \tfrac{2}{3}\cdot\varepsilon\cdot\delta <\\ &< |a^i - q| - \varepsilon\cdot\delta <\\ &< |a^i - q|\end{aligned}$$

for all $i \neq j$.

Set $B^i = \overline{\mathrm{B}}[a^i, |a^i - q|]$ and $B^{m+1} = \overline{\mathrm{B}}[b, |a^i - q| - \frac{1}{3}\cdot\delta]$. Clearly,

$$\bigcap_{i \neq m+1} B^i = \{q\},$$

$$\bigcap_{i \neq j} B^i \ni w^j \quad \text{for } j \neq m+1, \text{ and}$$

$$\bigcap_{i} B^i = \{q\} \cap B^{m+1} = \varnothing.$$

Lemma 14.12 finishes the proof. □

Proof of Theorem 14.16. Suppose Φ is not locally Lipshitz.; that is, there are sequences $\boldsymbol{y}_n, \boldsymbol{z}_n \to \boldsymbol{x} \in \mathrm{Dom}\,\Phi$ such that

$$\frac{|\Phi(\boldsymbol{y}_n) - \Phi(\boldsymbol{z}_n|)}{|\boldsymbol{y}_n - \boldsymbol{z}_n|} \to \infty \quad \text{as} \quad n \to \infty. \tag{3}$$

Set $p = \Phi(\boldsymbol{x})$, $q_n = \Phi(\boldsymbol{y}_n)$, and $b_n = \Phi(\boldsymbol{z}_n)$. By 14.9, $p, q_n, b_n \in \mathrm{InWeb}(\mathrm{dist}_{\boldsymbol{a}})$ and $q_n, b_n \to p$ as $n \to \infty$. Choose $\varepsilon > 0$; note that (3) implies

$$\measuredangle\left[q_n\,{}^{a^i}_{b_n}\right] < \tfrac{\pi}{2} + \varepsilon$$

for all $i > 0$ and all large n. Further, according to Definition 14.9, the subfunction $(\mathrm{dist}_{\boldsymbol{a}^0}) \circ \Phi$ is locally Lipschitz. Therefore we also have

$$\measuredangle\left[q_n\,{}^{a^0}_{b_n}\right] < \tfrac{\pi}{2} + \varepsilon$$

for all large n. According to Lemma 14.17, the point array $b_n, a^0, \dots, a^k$ for large n is nondegenerate.

Applying Corollary 14.11, we have a contradiction. □

14.18. Exercise. Let $K_1, \dots, K_n$ be closed convex subsets in a complete length CAT(0) space $\mathcal{U}$. Suppose that TopDim $\mathcal{U} = m$ and that any $m+1$ subsets from $\{K_1, \dots K_n\}$ have a common point. Show that all the subsets $K_1, \dots, K_n$ have a common point.

This exercise generalizes Helly's theorem; compare it to 10.13, [**96**, Prop. 5.3] and [**87**].

E. Remarks

The following conjecture was formulated by Bruce Kleiner [**96**], see also [**74**, p. 133]. For separable spaces, it follows from Corollary 14.14.

14.19. Conjecture. *For any complete length* CAT *space $\mathcal{U}$, we have*

$$\mathrm{TopDim}\,\mathcal{U} = \mathrm{LinDim}\,\mathcal{U}.$$

Chapter 15

Dimension of CBB spaces

As the main dimension-like invariant, we will use the linear dimension LinDim; see Definition 7.9. In other words, by default "dimension" means "linear dimension".

A. Struts and rank

Our definitions of strut and distance charts differ from the one in [**44**]; it is closer to Perelman's definitions [**126, 127**].

The term "strut" seems to have the closest meaning to the original Russian term used by Yuriy Burago, Grigory Perelman, and Michael Gromov [**44**]. In the official translation, it appears as "burst", and in the authors' translation it was "strainer". Neither seems intuitive, so we decided to switch to "strut".

15.1. Definition of struts. *Let $\mathcal{L}$ be a complete length* CBB *space. We say that a point array $(a^0, a^1, \dots, a^k)$ in $\mathcal{L}$ is* κ-strutting *for a point $p \in \mathcal{L}$ if $\tilde\measuredangle^\kappa\left(p\,{}^{a^i}_{a^j}\right) > \frac{\pi}{2}$ for all $i \neq j$.*

Recall that the packing number is defined in Section 2.B. The following definition is motivated by the observation that $k = \operatorname{pack}_{\pi/2}(\mathbb{S}^{k-1}) - 1$ for any integer $k > 0$.

15.2. Definition. Let $\mathcal{L}$ be a complete length CBB space and let $p \in \mathcal{L}$. Let us define rank of $\mathcal{L}$ at p as

$$\operatorname{rank}_p = \operatorname{rank}_p \mathcal{L} := \operatorname{pack}_{\pi/2} \Sigma_p - 1.$$

Thus rank takes values in $\mathbb{Z}_{\geqslant 0} \cup \{\infty\}$.

15.3. Proposition. *Let $\mathcal{L}$ be a complete length* CBB(κ) *space and let $p \in \mathcal{L}$. Then the following conditions are equivalent:*

(a) $\operatorname{rank}_p \geqslant k$,

(b) *there is a point array $(a^0, a^1, \dots, a^k)$ that is κ-strutting at p.*

Proof of Proposition 15.3.
(b) $\Rightarrow$ (a). For each i, choose a point $\acute{a}^i \in \operatorname{Str}(p)$ sufficiently close to a^i (so $[p\acute{a}^i]$ exists for each i). One can choose $\acute{a}^i$ so that we still have $\tilde{\measuredangle}^\kappa\left(p\,{}^{\acute{a}^i}_{\acute{a}^j}\right) > \frac{\pi}{2}$ for all $i \neq j$.

From hinge comparison (see Theorem 8.14(c)),

$$\measuredangle(\uparrow_{[p\acute{a}^j]}, \uparrow_{[p\acute{a}^j]}) \geqslant \tilde{\measuredangle}^\kappa\left(p\,{}^{\acute{a}^i}_{\acute{a}^j}\right) > \tfrac{\pi}{2}$$

for all $i \neq j$. In particular, $\operatorname{pack}_{\pi/2} \Sigma_p \geqslant k + 1$.
(a) $\Rightarrow$ (b). Assume $(\xi^0, \xi^1, \dots, \xi^k)$ is an array of directions in Σ_p, such that $\measuredangle(\xi^i, \xi^j) > \frac{\pi}{2}$ if $i \neq j$.

Without loss of generality, we may assume that each direction ξ^i is geodesic; that is, for each i there is a geodesic γ^i in $\mathcal{L}$ such that $\gamma^i(0) = p$ and $\xi^i = (\gamma^i)^+(0)$. From the definition of angle, it follows that for sufficiently small $\varepsilon > 0$ the array of points $a^i = \gamma^i(\varepsilon)$ satisfies (b). ☐

15.4. Corollary. *Let $\mathcal{L}$ be a complete length* CBB *space and $k \in \mathbb{Z}_{\geqslant 0}$. Then the set of all points in $\mathcal{L}$ with rank $\geqslant k$ is open.*

Proof. Given an array of points $\boldsymbol{a} = (a^0, \dots, a^k)$ in $\mathcal{L}$, consider the set $\Omega_{\boldsymbol{a}}$ of all points $p \in \mathcal{L}$ such that array $\boldsymbol{a}$ is κ-strutting for a point p. Clearly $\Omega_{\boldsymbol{a}}$ is open.

According to Proposition 15.3, the set of points in $\mathcal{L}$ with rank $\geqslant k$ can be presented as

$$\bigcup_{\boldsymbol{a}} \Omega_{\boldsymbol{a}},$$

where the union is taken over all k-arrays $\boldsymbol{a}$ of points in $\mathcal{L}$. Hence the result. ☐

B. Right-inverse theorem

Suppose that $\boldsymbol{a} = (a^1, \dots, a^k)$ is a point array in a metric space $\mathcal{L}$. Recall that the map $\operatorname{dist}_{\boldsymbol{a}} : \mathcal{L} \to \mathbb{R}^n$ is defined by

$$\operatorname{dist}_{\boldsymbol{a}} p := (|a^1 - p|, \dots, |a^n - p|).$$

15.5. Right-inverse theorem. *Suppose $\mathcal{L}$ is a complete length* CBB(κ) *space, $p, b \in \mathcal{L}$, and $\boldsymbol{a} = (a^1, \dots, a^k)$ is a point array in $\mathcal{L}$.*

Assume that $(b, a^1, a^2, \dots, a^k)$ is κ-strutting for p. Then the distance map $\operatorname{dist}_{\boldsymbol{a}} : \mathcal{L} \to \mathbb{R}^k$ *has a right inverse defined in a neighborhood of* $\operatorname{dist}_{\boldsymbol{a}} p \in \mathbb{R}^k$*; that is, there is a submap* $\Phi : \mathbb{R}^k \multimap \mathcal{L}$ *such that* $\operatorname{Dom}\Phi \ni \operatorname{dist}_{\boldsymbol{a}} p$ *and* $\operatorname{dist}_{\boldsymbol{a}}[\Phi(\boldsymbol{x})] = \boldsymbol{x}$ *for any* $\boldsymbol{x} \in \operatorname{Dom}\Phi$*. Moreover,*

(a) *The map Φ can be chosen to be $C^{\frac12}$-continuous (that is, Hölder continuous with exponent $\frac12$) and such that*

$$\Phi(\operatorname{dist}_{\boldsymbol{a}} p) = p.$$

(b) *The distance map $\operatorname{dist}_{\boldsymbol{a}} : \mathcal{L} \to \mathbb{R}^k$ is locally co-Lipschitz (in particular, open) in a neighborhood of p.*

Part (b) of the theorem is closely related to [**44**, Theorem 5.4] by Yuriy Burago, Grigory Perelman, and Michael Gromov, but the proof presented in this work is different. Yet another proof can be built on [**111**, Proposition 4.3] by Alexander Lytchak.

Proof. Fix $\varepsilon, r, \lambda > 0$ such that the following conditions hold:

(i) Each distance function $\operatorname{dist}_{a^i}$ and dist_b is $\frac{\lambda}{2}$-concave in $\mathrm{B}(p, r)$.

(ii) For any $q \in \mathrm{B}(p,r)$, we have $\tilde\measuredangle^\kappa\left(q\,{}^{a^i}_{a^j}\right) > \frac{\pi}{2} + \varepsilon$ for all $i \neq j$ and $\tilde\measuredangle^\kappa\left(q\,{}^{b}_{a^i}\right) > \frac{\pi}{2} + \varepsilon$ for all i. In addition, $\varepsilon < \frac{1}{10}$.

Given $\boldsymbol{x} = (x^1, x^2, \dots, x^k) \in \mathbb{R}^k$, consider the function $f_{\boldsymbol{x}} : \mathcal{L} \to \mathbb{R}$ defined by

$$f_{\boldsymbol{x}} = \min_i\{h^i_{\boldsymbol{x}}\} + \varepsilon\cdot \operatorname{dist}_b,$$

where $h^i_{\boldsymbol{x}}(q) = \min\{0, |a^i - q| - x^i\}$. Note that for any $\boldsymbol{x} \in \mathbb{R}^k$, the function $f_{\boldsymbol{x}}$ is $(1+\varepsilon)$-Lipschitz and λ-concave in $\mathrm{B}(p,r)$. Denote by $\alpha_{\boldsymbol{x}}(t)$ the $f_{\boldsymbol{x}}$-gradient curve (see Chapter 16) that starts at p.

(1) *If for some $\boldsymbol{x} \in \mathbb{R}^k$ and $t_0 \leqslant \frac{r}{2}$ we have $|\operatorname{dist}_{\boldsymbol{a}} p - \boldsymbol{x}| \leqslant \frac{\varepsilon^2}{10}\cdot t_0$, then $\operatorname{dist}_{\boldsymbol{a}}[\alpha_{\boldsymbol{x}}(t_0)] = \boldsymbol{x}$.*

First note that (1) follows if for any $q \in \mathrm{B}(p,r)$, we have

(i) $(\boldsymbol{d}_q \operatorname{dist}_{a^i})(\nabla_q f_{\boldsymbol{x}}) < -\frac{1}{10}\cdot\varepsilon^2$ if $|a^i - q| > x^i$, and

(ii) $(\boldsymbol{d}_q \operatorname{dist}_{a^i})(\nabla_q f_{\boldsymbol{x}}) > \frac{1}{10}\cdot\varepsilon^2$ if

$$|a^i - q| - x^i = \min_j\left\{|a^j - q| - x^j\right\} < 0.$$

Indeed, since $t_0 \leqslant \frac{r}{2}$, then $\alpha_{\boldsymbol{x}}(t) \in \mathrm{B}(p,r)$ for all $t \in [0, t_0]$. Consider the following real-to-real functions:

$$(2)\qquad \begin{aligned} \phi(t) &:= \max_i\left\{|a^i - \alpha_{\boldsymbol{x}}(t)| - x^i\right\},\\ \psi(t) &:= \min_i\left\{|a^i - \alpha_{\boldsymbol{x}}(t)| - x^i\right\}. \end{aligned}$$

Then from ((i)), we have $\phi^+ < -\frac{1}{10}\cdot\varepsilon^2$ if $\phi > 0$ and $t \in [0, t_0]$. Similarly, from ((ii)), we have $\psi^+ > \frac{1}{10}\cdot\varepsilon^2$ if $\psi < 0$ and $t \in [0, t_0]$. Since $|\operatorname{dist}_{\boldsymbol{a}} p - \boldsymbol{x}| \leqslant \frac{\varepsilon^2}{10}\cdot t_0$, it follows that $\phi(0) \leqslant \frac{\varepsilon^2}{10}\cdot t_0$ and $\psi(0) \geqslant -\frac{\varepsilon^2}{10}\cdot t_0$. Thus $\phi(t_0) \leqslant 0$ and $\psi(t_0) \geqslant 0$. On the other hand, from (2) we have $\phi(t_0) \geqslant \psi(t_0)$. That is, $\phi(t_0) = \psi(t_0) = 0$; hence (1) follows.

Thus, to prove (1), it remains to prove ((i)) and ((ii)). First let us prove it assuming that $\mathcal{L}$ is geodesic.

Note that

$$(3) \qquad (\boldsymbol{d}_q \operatorname{dist}_b)(\uparrow_{[qa^i]}) \leqslant \cos\tilde{\measuredangle}^\kappa\left(q\,{}^{b}_{a^j}\right) < -\tfrac{\varepsilon}{2}$$

for all i, and

$$(4) \qquad (\boldsymbol{d}_q \operatorname{dist}_{a^j})(\uparrow_{[qa^i]}) \leqslant \cos\tilde{\measuredangle}^\kappa\left(q\,{}^{a^i}_{a^j}\right) < -\tfrac{\varepsilon}{2}$$

for all $j \neq i$. Further, (4) implies

$$(5) \qquad (\boldsymbol{d}_q h^j_{\boldsymbol{x}})(\uparrow_{[qa^i]}) \leqslant 0$$

for all $i \neq j$. The assumption in ((i)) implies

$$\boldsymbol{d}_q f_{\boldsymbol{x}} = \min_{j\neq i}\{\boldsymbol{d}_q h^j_{\boldsymbol{x}}\} + \varepsilon\cdot(\boldsymbol{d}_q \operatorname{dist}_b).$$

Thus

$$\begin{aligned} -(\boldsymbol{d}_q \operatorname{dist}_{a^i})(\nabla_q f_{\boldsymbol{x}}) &\geqslant \langle \uparrow_{[qa^i]}, \nabla_q f_{\boldsymbol{x}}\rangle \geqslant \\ &\geqslant (\boldsymbol{d}_q f_{\boldsymbol{x}})(\uparrow_{[qa^i]}) = \\ &= \min_{i\neq j}\{(\boldsymbol{d}_q h^i_{\boldsymbol{x}})(\uparrow_{[qa^i]})\} + \varepsilon\cdot(\boldsymbol{d}_q \operatorname{dist}_b)(\uparrow_{[qa^i]}). \end{aligned}$$

Therefore ((i)) follows from (3) and (5).

The assumption in ((ii)) implies that $f_{\boldsymbol{x}}(q) = h^i_{\boldsymbol{x}}(q) + \varepsilon\cdot \operatorname{dist}_b$ and

$$\boldsymbol{d}_q f_{\boldsymbol{x}} \leqslant \boldsymbol{d}_q \operatorname{dist}_{a^i} + \varepsilon\cdot(\boldsymbol{d}_p \operatorname{dist}_b).$$

Therefore,

$$\begin{aligned} (\boldsymbol{d}_q \operatorname{dist}_{a^i})(\nabla_q f_{\boldsymbol{x}}) &\geqslant \boldsymbol{d}_q f_{\boldsymbol{x}}(\nabla_q f_{\boldsymbol{x}}) \geqslant \\ &\geqslant \left[(\boldsymbol{d}_q f_{\boldsymbol{x}})(\uparrow_{[qb]})\right]^2 \geqslant \\ &\geqslant \left[\min_i\{\cos\tilde{\measuredangle}^\kappa\left(q\,{}^{b}_{a^i}\right)\} - \varepsilon^2\right]^2. \end{aligned}$$

Thus ((ii)) follows from (3), since $\varepsilon < \frac{1}{10}$.

Therefore (1) holds if $\mathcal{L}$ is geodesic. If $\mathcal{L}$ is not geodesic, perform the above estimate in $\mathcal{L}^\omega$, the ultrapower of $\mathcal{L}$. (Recall that according to Observation 4.9, $\mathcal{L}^\omega$ is geodesic.) This completes the proof of (1). △

Set $t_0(\boldsymbol{x}) = \frac{10}{\varepsilon^2}\cdot|\operatorname{dist}_{\boldsymbol{a}} p - \boldsymbol{x}|$, giving equality in (1). Define the submap Φ by

$$\Phi\colon \boldsymbol{x} \mapsto \alpha_{\boldsymbol{x}} \circ t_0(\boldsymbol{x}), \quad \operatorname{Dom}\Phi = \mathrm{B}(\operatorname{dist}_{\boldsymbol{a}} p, \tfrac{\varepsilon^2\cdot r}{20}) \subset \mathbb{R}^k.$$

It follows from (1) that $\operatorname{dist}_{\boldsymbol{a}}[\Phi(\boldsymbol{x})] = \boldsymbol{x}$ for any $\boldsymbol{x} \in \operatorname{Dom}\Phi$.

Clearly $t_0(p) = 0$; thus $\Phi(\operatorname{dist}_{\boldsymbol{a}} p) = p$. Further, by construction of $f_{\boldsymbol{x}}$,

$$|f_{\boldsymbol{x}}(q) - f_{\boldsymbol{y}}(q)| \leqslant |\boldsymbol{x} - \boldsymbol{y}|,$$

for any $q \in \mathcal{L}$. Therefore, according to Lemma 16.13, Φ is $C^{\frac{1}{2}}$-continuous. Thus (a).

Further, note that

$$|p - \Phi(\boldsymbol{x})| \leqslant (1+\varepsilon)\cdot t_0(\boldsymbol{x}) \leqslant \tfrac{11}{\varepsilon^2}\cdot|\operatorname{dist}_{\boldsymbol{a}} p - \boldsymbol{x}| \tag{6}$$

holds.

The above construction may be repeated for any $p' \in \mathrm{B}(p, \frac{r}{4})$, $\varepsilon' = \varepsilon$, and $r' = \frac{r}{2}$. Inequality (6) for the resulting map Φ' implies that for any $p', q \in \mathrm{B}(p, \frac{r}{4})$ there is $q' \in \mathcal{L}$ such that $\Phi'(q) = \Phi'(q')$ and

$$|p' - q'| \leqslant \tfrac{11}{\varepsilon^2}\cdot|\operatorname{dist}_{\boldsymbol{a}} p' - \boldsymbol{x}|.$$

That is, the distance map $\operatorname{dist}_{\boldsymbol{a}}$ is locally $\frac{11}{\varepsilon^2}$-co-Lipschitz in $\mathrm{B}(p, \frac{r}{4})$. □

C. Dimension theorem

The following theorem is the main result of this section.

15.6. Theorem. *Let $\mathcal{L}$ be a complete length* CBB(κ) *space, $q \in \mathcal{L}$, $R > 0$ and let $m \in \mathbb{Z}_{\geqslant 0}$. Then the following statements are equivalent.*

(A) LinDim $\mathcal{L} \geqslant m$.

(B) *There is a point $p \in \mathcal{L}$ that admits a κ-strutting array $(b, a^1, \dots, a^m) \in \mathcal{L}^{m+1}$.*

(C) *Let* Euk^m *be the set of all points $p \in \mathcal{L}$ such that there is a distance-preserving embedding $\mathbb{E}^m \hookrightarrow \mathrm{T}_p$ that preserves the cone structure (see* Section 6.E*). Then* Euk^m *contains a dense G-delta set in $\mathcal{L}$.*

(D) *There is a $C^{\frac{1}{2}}$-embedding*

$$\overline{\mathrm{B}}[1]_{\mathbb{E}^m} \hookrightarrow \mathrm{B}(q, R);$$

that is, a bi-Hölder embedding with exponent $\frac{1}{2}$.

(E)

$$\operatorname{pack}_\varepsilon \mathrm{B}(q, R) > \frac{c}{\varepsilon^m}$$

for fixed $c > 0$ and any $\varepsilon > 0$.

In particular:

(i) *If* $\operatorname{LinDim}\mathcal{L}=\infty$*, then all the statements* (C), (D), *and* (E) *are satisfied for all* $m\in\mathbb{Z}_{\geqslant 0}$.

(ii) *If the statement* (D) *or* (E) *is satisfied for some choice of* $q\in\mathcal{L}$ *and* $R>0$*, then it also is satisfied for any other choice of* q *and* R.

For finite-dimensional spaces, Theorem 15.13 gives a stronger version of the theorem above.

The above theorem, with the exception of statement (D) was proved by Conrad Plaut [**138**]. At that time, it was not known whether

$$\operatorname{LinDim}\mathcal{L}=\infty \quad\Rightarrow\quad \operatorname{TopDim}\mathcal{L}=\infty$$

for any complete length CBB(κ) space $\mathcal{L}$. The latter implication was proved by Grigory Perelman and the third author [**125**]; it was done by combining an idea of Conrad Plaut with the technique of gradient flow. Part (D) is somewhat stronger.

To prove Theorem 15.6 we will need the following three propositions.

15.7. Proposition. *Let p be a point in a a complete length* CBB(κ) *space $\mathcal{L}$. Assume there is a distance-preserving embedding* $\iota\colon\mathbb{E}^m\hookrightarrow \mathrm{T}_p\mathcal{L}$ *that preserves the cone structure. Then either*

(a) $\operatorname{Im}\iota=\mathrm{T}_p\mathcal{L}$*, or*

(b) *there is a point p' arbitrarily close to p such that there is a distance-preserving embedding* $\iota'\colon\mathbb{E}^{m+1}\hookrightarrow \mathrm{T}_{p'}\mathcal{L}$ *that preserves the cone structure.*

Proof. Assume $\iota(\mathbb{E}^m)$ is a proper subset of $\mathrm{T}_p\mathcal{L}$. Equivalently, there is a direction $\xi\in\Sigma_p\setminus\iota(\mathbb{S}^{m-1})$, where $\mathbb{S}^{m-1}\subset\mathbb{E}^m$ is the unit sphere.

Fix $\varepsilon>0$ so that $\measuredangle(\xi,\sigma)>\varepsilon$ for any $\sigma\in\iota(\mathbb{S}^{m-1})$. Choose a maximal ε-packing in $\iota(\mathbb{S}^{m-1})$; that is, an array $(\zeta^1,\zeta^2,\dots,\zeta^n)$ of directions in $\iota(\mathbb{S}^{m-1})$ such that $n=\operatorname{pack}_\varepsilon\mathbb{S}^{m-1}$ and $\measuredangle(\zeta^i,\zeta^j)>\varepsilon$ for any $i\neq j$.

Choose an array $(x,z^1,z^2,\dots,z^n)$ of points in $\mathcal{L}$ such that $\uparrow_{[px]}\approx\xi$, $\uparrow_{[pz^i]}\approx\approx\zeta^i$; here we write "$\approx$" for "sufficiently close". We can choose this array so that $\tilde\measuredangle^\kappa\left(p\,{}^{x}_{z^i}\right)>\varepsilon$ for all i and $\tilde\measuredangle^\kappa\left(p\,{}^{z^i}_{z^j}\right)>\varepsilon$ for all $i\neq j$. Applying Corollary 13.40, there is a point p' arbitrarily close to p such that all directions $\uparrow_{[p'x]}$, $\uparrow_{[p'z^1]}$, $\uparrow_{[p'z^2]},\dots,\uparrow_{[p'z^n]}$ belong to an isometric copy of $\mathbb{S}^{k-1}$ in $\Sigma_{p'}$. In addition, we may assume that $\tilde\measuredangle^\kappa\left(p'\,{}^{x}_{z^i}\right)>\varepsilon$ and $\tilde\measuredangle^\kappa\left(p'\,{}^{z^i}_{z^j}\right)>\varepsilon$. From the hinge comparison (see 8.14(c)), $\measuredangle(\uparrow_{[p'x]},\uparrow_{[p'z^i]})>\varepsilon$ and $\measuredangle(\uparrow_{[p'z^i]},\uparrow_{[p'z^j]})>\varepsilon$; that is,

$$\operatorname{pack}_\varepsilon\mathbb{S}^{k-1}\geqslant n+1>\operatorname{pack}_\varepsilon\mathbb{S}^{m-1}.$$

Hence $k>m$. □

15.8. Proposition. *Let $\mathcal{L}$ be a complete length* CBB(κ) *space. Then for any two points p, $\bar p \in \mathcal{L}$ and any R, $\bar R > 0$, there is a constant $\delta = \delta(\kappa, R, \bar R, |p - \bar p|) > 0$ such that*

$$\operatorname{pack}_{\delta\cdot\varepsilon} \mathrm{B}(\bar p, \bar R) \geqslant \operatorname{pack}_{\varepsilon} \mathrm{B}(p, R).$$

Proof. According to Corollary 8.33, we can assume that $\kappa \leqslant 0$.

Let $n = \operatorname{pack}_\varepsilon \mathrm{B}(p,R)$ and let $\{x^1, \dots, x^n\}$ be a maximal ε-packing in $\mathrm{B}(p,R)$; that is, $|x^i - x^j| > \varepsilon$ for all $i \neq j$. Without loss of generality, we may assume the x^i are in $\operatorname{Str}(\bar p)$. Thus, for each i there is a unique geodesic $[\bar p x^i]$ (see Theorem 8.11). Choose a factor $1 > s > 0$ so that $\bar R > s\cdot(|p - \bar p| + R)$. For each i, take $\bar x^i \in [\bar p x^i]$ so that $|\bar p - \bar x^i| = s\cdot(|p - x^i|)$. From 8.17(a),

$$\tilde\measuredangle^\kappa\left(\bar p\,{}^{\bar x^i}_{\bar x^j}\right) \geqslant \tilde\measuredangle^\kappa\left(\bar p\,{}^{x^i}_{x^j}\right).$$

The cosine law gives a constant $\delta =$ $= \delta(\kappa, R, \bar R, |p - \bar p|) > 0$ such that

$$|\bar x^i - \bar x^j| > \delta\cdot(|x^i - x^j|) > \delta\cdot\varepsilon$$

for all $i \neq j$. Hence the statement follows. ☐

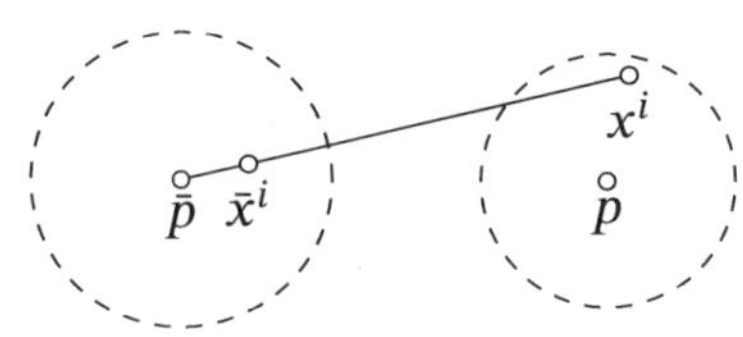

15.9. Proposition. *Let $\mathcal{L}$ be a complete length* CBB(κ) *space, $r < \varpi^\kappa$ and let $p \in \mathcal{L}$. Assume that*

$$\text{(1)} \qquad \operatorname{pack}_\varepsilon \mathrm{B}(p, r) > \operatorname{pack}_\varepsilon \overline{\mathrm{B}}[r]_{\mathbb{M}^m(\kappa)}$$

for $\varepsilon > 0$. Then there is a G-delta set $A \subset \mathcal{L}$ that is dense in a neighborhood of p and such that $\dim \operatorname{Lin}_q > m$ *for any $q \in A$.*

Proof. Choose a maximal ε-packing in $\mathrm{B}(p,r)$, that is, an array $(x^1, x^2, \dots, x^n)$ of points in $\mathrm{B}(p,r)$ such that $n = \operatorname{pack}_\varepsilon \mathrm{B}(p,r)$ and $|x^i - x^j| > \varepsilon$ for any $i \neq j$. Choose a neighborhood $\Omega \ni p$ such that $|q - x^i| < r$ for any $q \in \Omega$ and all i. Let

$$A = \Omega \cap \operatorname{Str}(x^1, x^2, \dots, x^n).$$

According to Theorem 8.11, A is a G-delta set that is dense in Ω.

Assume $k = \dim \operatorname{Lin}_q \leqslant m$ for $q \in A$. Consider an array $(v^1, v^2, \dots, v^n)$ of vectors in Lin_q, where $v^i = \log[q x^i]$. Clearly

$$|v^i| = |q - x^i| < r,$$

and from the hinge comparison (see 8.14(c)) we have

$$\tilde\gamma^\kappa\left[0\,{}^{v^i}_{v^j}\right] \geqslant |x^i - x^j| > \varepsilon.$$

Note that the ball $\mathrm{B}(0,r)_{\mathrm{Lin}_q}$ equipped with the metric $\rho(v,w) = \tilde{\gamma}^\kappa\,[0\,{}^v_w]$ is isometric to $\overline{\mathrm{B}}[r]_{\mathbb{M}^k(\kappa)}$. Thus

$$\operatorname{pack}_\varepsilon \overline{\mathrm{B}}[r]_{\mathbb{M}^k(\kappa)} \geqslant \operatorname{pack}_\varepsilon \mathrm{B}(p,r),$$

which contradicts $k \leqslant m$ and (1). □

The proof of Theorem 15.6 is essentially done in Propositions 15.7, 15.8, and 15.9, Theorem 15.10, and 15.5; now we assemble the proof from these parts.

Proof of Theorem 15.6. We will prove the implications

$$(\mathrm{C}) \Rightarrow (\mathrm{A}) \Rightarrow (\mathrm{B}) \Rightarrow (\mathrm{E}) \Rightarrow (\mathrm{C}) \Rightarrow (\mathrm{D}) \Rightarrow (\mathrm{E}).$$

The implication (C) ⇒ (A) is trivial.

(A) ⇒ (B). Choose a point $p \in \mathcal{L}$ such that $\dim \mathrm{Lin}_p \geqslant m$. Clearly one can choose an array $(\xi^0, \xi^1, \dots, \xi^m)$ of directions in Lin_p such that $\measuredangle(\xi^i,\xi^j) > \frac{\pi}{2}$ for all $i \neq j$. Choose an array $(x^0, x^1, \dots, x^m)$ of points in $\mathcal{L}$ such that each $\uparrow_{[px^i]}$ is sufficiently close to ξ^i; in particular, we have $\measuredangle\left[p\,{}^{x^i}_{x^j}\right] > \frac{\pi}{2}$. Choose points $a^i \in \,]px^i]$ sufficiently close to p. This can be done so that each $\tilde{\measuredangle}^\kappa\left(p\,{}^{a^i}_{a^j}\right)$ is arbitrarily close to $\measuredangle\left[p\,{}^{a^i}_{a^j}\right]$, in particular $\tilde{\measuredangle}^\kappa\left(p\,{}^{a^i}_{a^j}\right) > \frac{\pi}{2}$. Finally, set $b = a^0$.

(B) ⇒ (E). Let $p \in \mathcal{L}$ be a point that admits a κ-strutting array $(b, a^1, \dots, a^m)$ of points in $\mathcal{L}$. The right-inverse theorem (see 15.5(b)) implies that the distance map $\operatorname{dist}_{\boldsymbol{a}} : \mathcal{L} \to \mathbb{R}^m$,

$$\operatorname{dist}_{\boldsymbol{a}} : x \mapsto (|a^1 - x|, |a^2 - x|, \dots, |a^n - x|),$$

is open in a neighborhood of p. Since the distance map $\operatorname{dist}_{\boldsymbol{a}}$ is Lipschitz, for any $r > 0$, there is $c > 0$ such that

$$\operatorname{pack}_\varepsilon \mathrm{B}(p,r) > \frac{c}{\varepsilon^m}$$

for any $\varepsilon > 0$. Applying Proposition 15.8, we get a similar inequality for any other ball in $\mathcal{L}$; that is, for any $q \in \mathcal{L}$ and $R > 0$, there is $c' > 0$ such that

$$\operatorname{pack}_\varepsilon \mathrm{B}(q,R) > \frac{c'}{\varepsilon^m}.$$

(E) ⇒ (C). Note that for any $q' \in \mathcal{L}$ and $R' > |q - q'| + R$, we have

$$\begin{aligned}\operatorname{pack}_\varepsilon \mathrm{B}(q',R') &\geqslant \operatorname{pack}_\varepsilon \mathrm{B}(q,R) \geqslant \\ &\geqslant \frac{c}{\varepsilon^m} > \\ &> \operatorname{pack}_\varepsilon \overline{\mathrm{B}}[R']_{\mathbb{M}^{m-1}(\kappa)}\end{aligned}$$

for all sufficiently small $\varepsilon > 0$. Applying Proposition 15.9, $\overline{\mathrm{Euk}}^m$ contains a G-delta set that is dense in a neighborhood of any point $q' \in \mathcal{L}$.

(C) ⇒ (D). Since Euk^m contains a dense G-delta set in $\mathcal{L}$, we can choose $p \in$ $\in \mathrm{B}(q, R)$ with a distance-preserving cone embedding $\iota\colon \mathbb{E}^m \hookrightarrow \mathrm{T}_p$.

Repeating the construction in (A) ⇒ (B), we get a κ-strutting array $(p, a^1, \dots, a^m)$ for p.

Applying the right-inverse theorem (see 15.5), we obtain a $C^{\frac12}$-submap

$$\Phi\colon \mathbb{R}^m \multimap \mathrm{B}(q, R)$$

that is a right inverse for $\operatorname{dist}_{\boldsymbol{a}} : \mathcal{L} \to \mathbb{R}^m$ and such that $\Phi(\operatorname{dist}_{\boldsymbol{a}} p) = p$. In particular, Φ is a $C^{\frac12}$-embedding of $\operatorname{Dom}\Phi$.

(D) ⇒ (E). This proof is valid for general metric spaces; it is based on general relations between topological dimension, Hausdorff measure and $\mathrm{pack}_\varepsilon$.

Let $W \subset \mathrm{B}(q, R)$ be the image of Φ. Since $\operatorname{TopDim} W = m$, Szpilrajn's theorem (see 7.5) implies that

$$\operatorname{HausMes}_m W > 0.$$

Given $\varepsilon > 0$, consider a maximal ε-packing of W, that is, an array $(x^1, x^2, \dots, x^n)$ of points in W such that $n = \mathrm{pack}_\varepsilon W$ and $|x^i - x^j| > \varepsilon$ for all $i \neq j$. Note that W is covered by balls $\mathrm{B}(x^i, 2\cdot\varepsilon)$.

By the definition of Hausdorff measure,

$$\mathrm{pack}_\varepsilon W \geqslant \frac{c}{\varepsilon^m} \cdot \operatorname{HausMes}_m W$$

for a fixed constant $c > 0$ and all small $\varepsilon > 0$. Hence (E) follows. □

D. Inverse function theorem

15.10. Inverse function theorem. *Let $\mathcal{L}$ be an m-dimensional complete length CBB(κ) space and $p, b, a^1, a^2, \dots, a^m \in \mathcal{L}$.*

Assume that the point array $\boldsymbol{a} = (b, a^1, \dots, a^m)$ is κ-strutting for p. Then there are $R > 0$ and $\varepsilon > 0$ such that:

(a) *For all $i \neq j$ and any $q \in \mathrm{B}(p, R)$ we have*

$$\tilde\measuredangle^\kappa\left(q\,{}^{a^i}_{a^j}\right) > \tfrac{\pi}{2} + \varepsilon \quad \text{and} \quad \tilde\measuredangle^\kappa\left(q\,{}^{b}_{a^i}\right) > \tfrac{\pi}{2}.$$

(b) *The restriction of the distance map*

$$\operatorname{dist}_{\boldsymbol{a}} : x \mapsto (|a^1 - x|, \dots, |a^m - x|)$$

to the ball $\mathrm{B}(p, R)$ is an open $[\varepsilon, \sqrt{m}]$-bi-Lipschitz embedding $\mathrm{B}(p, R) \hookrightarrow$ $\hookrightarrow \mathbb{R}^m$.

(c) *The value R depends only on κ, $|p - a^i|$, $|a^i - a^j|$, and $|b - a^i|$ for all i and j.*

15.11. Definition. Suppose $\mathcal{L}$ is an m-dimensional complete length CBB(κ) space. If a point array $(b, a^1, a^2, \dots, a^m)$ and the value R satisfy the conditions in Theorem 15.10, then the restriction $\boldsymbol{x} = \operatorname{dist}_{\boldsymbol{a}} |_{B(p,R)}$ is called a *distance chart*, the restrictions $x^i = \operatorname{dist}_{a^i} |_{B(p,R)}$ are called *coordinates*, and the restriction $y = \operatorname{dist}_b |_{B(p,R)}$ is called the *strut* of the distance chart.

15.12. Lemma. *Let p be a point in an m-dimensional complete length* CBB(κ) *space $\mathcal{L}$. Assume for the directions $\xi, \zeta^1, \zeta^2, \dots, \zeta^k \in \Sigma_p$ the following conditions hold:*

(a) $\measuredangle(\xi, \zeta^i) > \frac{\pi}{2} - \varepsilon$ *for all i.*

(b) $\measuredangle(\zeta^i, \zeta^j) > \frac{\pi}{2} + \varepsilon$ *for all $i \neq j$. Then $k \leqslant m$.*

Proof. Without loss of generality, we can assume that all $\xi, \zeta^1, \zeta^2, \dots, \zeta^k$ are geodesic directions; let $\xi = \uparrow_{[px]}$ and $\zeta^i = \uparrow_{[pz^i]}$ for all i. Fix a small $r > 0$, and let $\bar{x} \in]px]$ and $\bar{z}^i \in]pz^i]$ be points such that

$$|p - \bar{x}| = |p - \bar{z}^1| = \dots = |p - \bar{z}^k| = r.$$

From the definition of angle, if r is sufficiently small we have

- $\tilde{\measuredangle}^\kappa\left(p\,{}^{\bar{x}}_{\bar{z}^i}\right) > \frac{\pi}{2} - \varepsilon$ for all i, and $\tilde{\measuredangle}^\kappa\left(p\,{}^{\bar{z}^i}_{\bar{z}^j}\right) > \frac{\pi}{2} + \varepsilon$ for all $i \neq j$.

Choose a point $p' \in \operatorname{Str}(\bar{x}, \bar{z}^1, \bar{z}^2, \dots, \bar{z}^k)$ sufficiently close to p that the above conditions still hold for p'; that is,

(1) $\tilde{\measuredangle}^\kappa\left(p'\,{}^{\bar{x}}_{\bar{z}^i}\right) > \frac{\pi}{2} - \varepsilon$ *for all i, and* $\tilde{\measuredangle}^\kappa\left(p'\,{}^{\bar{z}^i}_{\bar{z}^j}\right) > \frac{\pi}{2} + \varepsilon$ *for all $i \neq j$.*

Set $\acute{\xi} = \uparrow_{[p'\bar{x}]}$ and $\acute{\zeta}^i = \uparrow_{[p'\bar{z}^i]}$ for each i. By the hinge comparison (see (8.14(c))),

(2) $\measuredangle(\acute{\xi}, \acute{\zeta}^i) > \frac{\pi}{2} - \varepsilon$ *for all i, and* $\measuredangle(\acute{\zeta}^i, \acute{\zeta}^j) > \frac{\pi}{2} + \varepsilon$ *for all $i \neq j$.*

According to Corollary 13.40, all directions $\acute{\xi}, \acute{\zeta}^1, \acute{\zeta}^2, \dots, \acute{\zeta}^k$ lie in an isometric copy of the standard n-sphere in $\Sigma_{p'}$. Clearly $n \leqslant m - 1$. Thus it remains to prove the following claim, which is a partial case of the lemma.

(3) *If $\xi, \zeta^1, \zeta^2, \dots, \zeta^k \in \mathbb{S}^{m-1}$, $|\xi - \zeta^i| > \frac{\pi}{2} - \varepsilon$ for all i, and $|\zeta^i - \zeta^j| > \frac{\pi}{2} + \varepsilon$ for all $i \neq j$, then $k \leqslant m$.*

For each i, let $\bar{\zeta}^i$ be the closest point to ζ^i in $\Xi = \mathbb{S}^{m-1} \setminus B(\xi, \frac{\pi}{2}) \overset{\text{iso}}{=\!=} \mathbb{S}^{m-1}_+$ (if $\zeta \in \Xi$, then $\bar{\zeta}^i = \zeta^i$). By straightforward calculations, we have

$$|\bar{\zeta}^i - \bar{\zeta}^j| \geqslant |\zeta^i - \zeta^j| - \varepsilon > \tfrac{\pi}{2}.$$

Thus it is sufficient to show the claim

(4) $\operatorname{pack}_{\frac{\pi}{2}} \mathbb{S}^{m-1}_+ = m.$

Clearly, $\operatorname{pack}_{\frac{\pi}{2}} \mathbb{S}^{m-1}_+ \geqslant m.$

The opposite inequality is proved by induction on m. The base case $m = 1$ is obvious. Assume $(\bar{\zeta}^1, \bar{\zeta}^2, \dots, \bar{\zeta}^k)$ is an array of points in $\mathbb{S}^{m-1}_+$ with $|\bar{\zeta}^i - \bar{\zeta}^j| > \frac{\pi}{2}$. Without loss of generality, we can also assume that $\bar{\zeta}^k \in \partial\mathbb{S}^{m-1}_+$. For each $i < k$, let $\check{\zeta}^i = \uparrow_{[\bar{\zeta}^k\bar{\zeta}^i]} \in \Sigma_{\bar{\zeta}^k}\mathbb{S}^{m-1}_+ \overset{\text{iso}}{=\!=} \mathbb{S}^{m-2}_+$. By the hinge comparison (see 8.14(c)), $\measuredangle(\check{\zeta}^i, \check{\zeta}^j) > \frac{\pi}{2}$ for all $i < j < k$. Thus from the induction hypothesis we have $k - 1 \leqslant m - 1$. □

Proof of Theorem 15.10.

(a). Fix $\varepsilon > 0$ such that $\tilde{\measuredangle}^\kappa\left(p\,{}^{a^i}_{a^j}\right) > \frac{\pi}{2} + \varepsilon$ and $\tilde{\measuredangle}^\kappa\left(p\,{}^{b}_{a^i}\right) > \frac{\pi}{2} + \varepsilon$ for all $i \neq j$. Choose $R > 0$ sufficiently small that $\tilde{\measuredangle}^\kappa\left(q\,{}^{a^i}_{a^j}\right) > \frac{\pi}{2} + \varepsilon$ and $\tilde{\measuredangle}^\kappa\left(q\,{}^{b}_{a^i}\right) > \frac{\pi}{2} + \varepsilon$ for all $i \neq j$ and any $q \in \mathrm{B}(p, R)$. Clearly, (a) holds for $\mathrm{B}(p, R)$.

(b). Note that the distance map $\mathrm{dist}_{\boldsymbol{a}}$ is Lipschitz and its restriction $\mathrm{dist}_{\boldsymbol{a}}|_{\mathrm{B}(p,R)}$ is open; the latter follows from the right-inverse theorem (see 15.5(b)). Thus to prove (b), it is sufficient to show that

(5)
$$\max_i \left\{ \big||a^i - x| - |a^i - y|\big| \right\} > \tfrac{\varepsilon}{2}\cdot|x - y|$$

for any $x, y \in \mathrm{B}(p, R)$.

According to Lemma 15.12,

$$\measuredangle\left[x\,{}^{y}_{b}\right] \leqslant \tfrac{\pi}{2} - \varepsilon \quad \text{or} \quad \measuredangle\left[x\,{}^{y}_{a^i}\right] \leqslant \tfrac{\pi}{2} - \varepsilon \quad \text{for some } i.$$

In the latter case, since $|x - y| < 2\cdot R$ and R is small, the hinge comparison (see 8.14(c)) implies

(6)
$$|a^i - x| - |a^i - y| > \tfrac{\varepsilon}{2}\cdot|x - y| \quad \text{for some } i.$$

If $\measuredangle\left[x\,{}^{y}_{b}\right] \leqslant \frac{\pi}{2} - \varepsilon$, then switching x and y, we get

(7)
$$|a^j - y| - |a^j - x| > \tfrac{\varepsilon}{2}\cdot|x - y| \quad \text{for some } j.$$

Then (6) and (7) imply (5).

Finally, part (c) follows since the angle $\tilde{\measuredangle}^\kappa\left(q\,{}^{a^i}_{a^j}\right)$ depends continuously on κ, $|q - a^i|$, $|q - a^j|$, and $|a^i - a^j|$. □

E. Finite-dimensional spaces

The next theorem is a refinement of Theorem 15.6 for the finite-dimensional case; it was essentially proved by Yuriy Burago, Grigory Perelman, and Michael Gromov [**44**].

15.13. Theorem. *Suppose $\mathcal{L}$ is a complete length* CBB(κ) *space, m is a nonnegative integer, $0 < R \leqslant \varpi^\kappa$, and $q \in \mathcal{L}$. Then the following statements are equivalent:*

(a) $\operatorname{LinDim}\mathcal{L} = m$.

(b) *m is the maximal integer such that there is a point $p \in \mathcal{L}$ that admits a κ-strutting array $(b, a^1, \dots, a^m)$.*

(c) $\mathrm{T}_p \overset{\text{iso}}{=\!=} \mathbb{E}^m$ *for any point p in a dense G-delta set of $\mathcal{L}$.*

(d) *There is an open bi-Lipschitz embedding*

$$\overline{\mathrm{B}}[1]_{\mathbb{E}^m} \hookrightarrow \mathrm{B}(q, R) \subset \mathcal{L}.$$

(e) *For any $\varepsilon > 0$,*

$$\operatorname{pack}_\varepsilon \overline{\mathrm{B}}[R]_{\mathbb{M}^m(\kappa)} \geqslant \operatorname{pack}_\varepsilon \mathrm{B}(q, R).$$

Moreover, there is $c = c(q, R) > 0$ such that

$$\operatorname{pack}_\varepsilon \mathrm{B}(q, R) > \frac{c}{\varepsilon^m}.$$

Using Theorems 15.6 and 15.13, one can show that linear dimension is equal to many different types of dimension, such as *small* and *big inductive dimension* and *upper* and *lower box-counting dimension* (also known as *Minkowski dimension*), *homological dimension*, and so on.

The next two corollaries follow from (e).

15.14. Corollary. *Any finite-dimensional complete length* CBB *space is proper and geodesic.*

15.15. Corollary. *Let $(\mathcal{L}_n)$ be a sequence of length* CBB(κ) *spaces and $\mathcal{L}_n \to \mathcal{L}_\omega$ as $n \to \omega$. Assume* $\operatorname{LinDim} L_n \leqslant m$ *for all n. Then* $\operatorname{LinDim} L_\omega \leqslant m$.

15.16. Corollary. *Let $\mathcal{L}$ be a complete length* CBB(κ) *space. Then for any open $\Omega \subset \mathcal{L}$, we have*

$$\operatorname{LinDim} \mathcal{L} = \operatorname{LinDim} \Omega = \operatorname{TopDim} \Omega = \operatorname{HausDim} \Omega,$$

where TopDim *and* HausDim *denote topological* 7.2 *and Hausdorff dimension* 7.1, *respectively.*

In particular, $\mathcal{L}$ is dimension-homogeneous; that is, all open sets have the same linear dimension.

Proof of Corollary 15.16. The equality

$$\operatorname{LinDim} \mathcal{L} = \operatorname{LinDim} \Omega$$

follows from Theorems 15.6(A) and (C).

If $\operatorname{LinDim} \mathcal{L} = \infty$, then applying (see 15.6(D)) for $\mathrm{B}(q, R) \subset \Omega$, we find that there is a compact subset $K \subset \Omega$ having an arbitrarily large $\operatorname{TopDim} K$. Therefore

$$\operatorname{TopDim} \Omega = \infty.$$

By Szpilrajn's theorem (see 7.5), $\operatorname{HausDim} K \geqslant \operatorname{TopDim} K$. Thus we also have

$$\operatorname{HausDim} \Omega = \infty.$$

If $\operatorname{LinDim}\mathcal{L} = m < \infty$, then the first inequality in Theorem 15.13(e) implies that

$$\operatorname{HausDim} B(q, R) \leqslant m.$$

According to Corollary 15.14, $\mathcal{L}$ is proper and in particular has countable base. Thus applying Szpilrajn's theorem again, we have

$$\operatorname{TopDim}\Omega \leqslant \operatorname{HausDim}\Omega \leqslant m.$$

Finally, Theorem 15.13(d) implies that $m \leqslant \operatorname{TopDim}\Omega$. □

Proof of Theorem 15.13. The equivalence (a) ⇔ (b) follows from Theorem 15.6.

(a) ⇒ (c). If $\operatorname{LinDim}\mathcal{L} = m$, then by Theorem 15.6, Euk^m contains a dense G-delta set in $\mathcal{L}$. From Proposition 15.7, it follows that T_p is isometric to $\mathbb{E}^m$ for any $p \in \operatorname{Euk}^m$.

(c) ⇒ (d). This is proved in exactly the same way as implication (C) ⇒ (D) of Theorem 15.6, but applying the existence of a distance chart (see Theorem 15.10) instead of the right-inverse theorem (see 15.5).

(d) ⇒ (e). From (d), it follows that there is a point $p \in B(q, R)$ and $r > 0$ such that $B(p, r) \subset \mathcal{L}$ is bi-Lipschitz homeomorphic to a bounded open set of $\mathbb{E}^m$. Thus there is $c > 0$ such that

$$\operatorname{pack}_\varepsilon B(p, r) > \frac{c}{\varepsilon^m}. \tag{1}$$

Applying Proposition 15.8 shows that inequality (1), with different constants, holds for any other ball, in particular for $B(q, R)$.

Applying Proposition 15.9 gives the first inequality in (e).

(e) ⇒ (a). From Theorem 15.6, we have $\operatorname{LinDim}\mathcal{L} \geqslant m$. Applying Theorem 15.6 again, if $\operatorname{LinDim}\mathcal{L} \geqslant m + 1$, then for some $c > 0$ and any $\varepsilon > 0$,

$$\operatorname{pack}_\varepsilon B(q, R) \geqslant \frac{c}{\varepsilon^{m+1}}.$$

But

$$\frac{c'}{\varepsilon^m} \geqslant \operatorname{pack}_\varepsilon B(q, R)$$

for any $\varepsilon > 0$, a contradiction. □

15.17. Exercise. Suppose $\mathcal{L}$ is a complete length CBB space and $\Sigma_p\mathcal{L}$ is compact for any $p \in \mathcal{L}$. Prove that $\mathcal{L}$ is finite dimensional.

F. One-dimensional spaces

15.18. Theorem. *Let $\mathcal{L}$ be a one-dimensional complete length* CBB(κ) *space. Then $\mathcal{L}$ is isometric to a connected complete Riemannian one-dimensional manifold with possibly nonempty boundary.*

Proof. Clearly $\mathcal{L}$ is connected. It remains to show the following:

(1) *For any point $p \in \mathcal{L}$ there is $\varepsilon > 0$ such that* $\mathrm{B}(p,\varepsilon)$ *is isometric to either* $[0,\varepsilon)$ *or* $(-\varepsilon,\varepsilon)$.

First let us show

(2) *If $p \in \,]xy[$ for $x, y \in \mathcal{L}$ and $\varepsilon < \min\{|p-x|, |p-y|\}$, then* $\mathrm{B}(p,\varepsilon) \subset \,]xy[$. *In particular,* $\mathrm{B}(p,\varepsilon) \overset{\text{iso}}{=\!=} (-\varepsilon,\varepsilon)$.

Assume the contrary; that is, there is

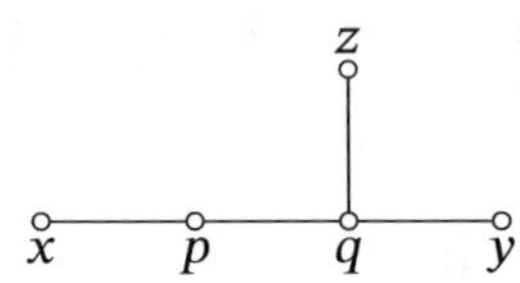

$$z \in \mathrm{B}(p,\varepsilon) \setminus \,]xy[.$$

Consider a geodesic $[pz]$, and let $q \in [pz] \cap [xy]$ be the point that maximizes the distance $|p-q|$. At q, we have three distinct directions: to x, y, and z. Moreover, $\measuredangle\left[q\,{}^{x}_{y}\right] = \pi$. Thus, according to Proposition 15.7, $\operatorname{LinDim}\mathcal{L} > 1$, a contradiction. △

Now we assume that no geodesic includes p as a nonendpoint. Since $\operatorname{LinDim}\mathcal{L}{=}1$ there is a point $y \neq p$.

Fix a positive value $\varepsilon < |p-y|$. Let us show

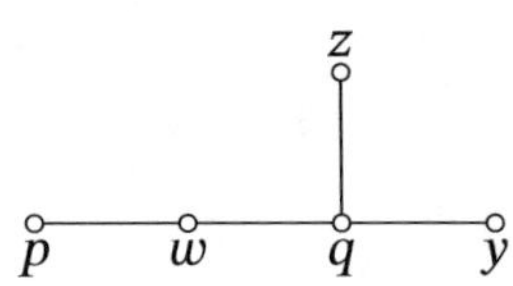

(3) $\mathrm{B}(p,\varepsilon) \subset [py]$; *in particular,* $\mathrm{B}(p,\varepsilon) \overset{\text{iso}}{=\!=} [0,\varepsilon)$.

Assume the contrary; let $z \in \mathrm{B}(p,\varepsilon) \setminus [py]$.

Choose a point $w \in \,]py[$ such that

$$|p-w| + |p-z| < \varepsilon.$$

Consider geodesic $[wz]$, and let $q \in [py] \cap [wz]$ be the point that maximizes the distance $|w-q|$. Since no geodesic includes p as a nonendpoint, we have $p \neq q$. As above, $\measuredangle\left[q\,{}^{p}_{y}\right] = \pi$ and $\uparrow_{[qz]}$ is distinct from $\uparrow_{[qp]}$ and $\uparrow_{[qp]}$. Thus, according to Proposition 15.7, $\operatorname{LinDim}\mathcal{L} > 1$, a contradiction. △

Clearly (2) + (3) ⇒ (1); hence the result. □

Chapter 16

Gradient flow

Gradient flow could be considered a nonsmooth version of first-order ordinary differential equations. It provides a universal tool in Alexandrov geometry with most significant applications to CBB spaces.

The theory of gradient flows of semiconvex functions on Hilbert spaces (which are of course both CBB(0) and CAT(0)) is classical, see for example [**33**].

The technique of gradient flows in the context of comparison geometry takes its roots in *Sharafutdinov's retraction*, introduced by Vladimir Sharafutdinov [**149**]. It has been used widely in comparison geometry since then. In CBB spaces, it was first used by Grigory Perelman and the third author [**125**, **131**]. A bit later, Jürgen Jost and Uwe Mayer [**89**, **115**] used the gradient flow in CAT space independently. Later, Alexander Lytchak unified and generalized these two approaches to a wide class of metric spaces [**111**]. It was developed yet further by Shin-ichi Ohta [**122**] and by Giuseppe Sevaré [**148**]. It is based on the more analytic approach suitable for the study of synthetic spaces with lower Ricci bounds was developed by Luigi Ambrosio, Nicola Gigli, and Giuseppe Sevaré in a general metric and metric measure setting [**19**].

The following exercise is a stripped-down version of Sharfutdinov's retraction; it gives the idea behind gradient flow.

16.1. Exercise. Assume that a one-parameter family of convex sets $K_t \subset \mathbb{E}^m$ is nested; that is, $K_{t_1} \supset K_{t_2}$ if $t_1 \leqslant t_2$. Show that there is a family of short maps $\phi_t \colon \mathbb{E}^m \to K_t$ such that $\phi_t|_{K_t} = \mathrm{id}$ for any t and $\phi_{t_2} \circ \phi_{t_1} = \phi_{t_2}$ if $t_1 \leqslant t_2$.

A. Gradient-like curves

Gradient-like curves will be used later in the construction of gradient curves. The latter are a special reparametrization of gradient-like curves.

16.2. Definition. Let $\mathcal{Z}$ be a complete length space, and let $f\colon \mathcal{Z} \looparrowright \mathbb{R}$ be locally Lipschitz semiconcave subfunction. Suppose that $\mathcal{Z}$ is either CBB or CAT.

A Lipschitz curve $\hat\alpha\colon [s_{\min}, s_{\max}) \to \operatorname{Dom} f$ will be called an *f-gradient-like curve* if

$$\hat\alpha^+ = \tfrac{1}{|\nabla_{\hat\alpha} f|}\cdot \nabla_{\hat\alpha} f;$$

that is, for any $s \in [s_{\min}, s_{\max})$, the right derivative $\hat\alpha^+(s)$ is defined and

$$\hat\alpha^+(s) = \tfrac{1}{|\nabla_{\hat\alpha(s)} f|}\cdot \nabla_{\hat\alpha(s)} f.$$

Note that this definition implies that $|\nabla_p f| > 0$ for any point p on $\hat\alpha$.

The following theorem gives a seemingly weaker condition that is equivalent to the definition of gradient-like curve.

16.3. Theorem. *Suppose $\mathcal{Z}$ is a complete length space, $f\colon \mathcal{Z} \looparrowright \mathbb{R}$ is a locally Lipschitz semiconcave subfunction, and $|\nabla_p f| > 0$ for any $p \in \operatorname{Dom} f$. Assume that $\mathcal{Z}$ is either* CBB *or* CAT.

A curve $\hat\alpha\colon [s_{\min}, s_{\max}) \to \operatorname{Dom} f$ is an f-gradient-like curve if and only if it is 1-Lipschitz and

(1) $$\varliminf_{s\to s_0+} \frac{f\circ\hat\alpha(s) - f\circ\hat\alpha(s_0)}{s - s_0} \geqslant |\nabla_{\hat\alpha(s_0)} f|$$

for almost all $s_0 \in [s_{\min}, s_{\max})$.

Proof. The only-if part follows directly from the definition. To prove the if part, note that for any $s_0 \in [s_{\min}, s_{\max})$ we have

$$\begin{aligned}\varliminf_{s\to s_0+} \frac{f\circ\hat\alpha(s) - f\circ\hat\alpha(s_0)}{s - s_0} &\geqslant \varliminf_{s\to s_0+} \frac{1}{s - s_0}\cdot \int\limits_{s_0}^{s} |\nabla_{\hat\alpha(\varsigma)} f|\cdot d\varsigma \geqslant \\ &\geqslant |\nabla_{\hat\alpha(s_0)} f|;\end{aligned}$$

the first inequality follows from (1) and the second from lower semicontinuity of the function $x \mapsto |\nabla_x f|$, see 13.31. From Lemma 13.22, we have

$$\hat\alpha^+(s_0) = \tfrac{1}{|\nabla_{\hat\alpha(s_0)} f|}\cdot \nabla_{\hat\alpha(s_0)} f.$$

Hence the result. □

Recall that second-order differential inequalities are understood in a barrier sense; see Section 3.E.

16.4. Theorem. *Let $\mathcal{Z}$ be a complete length space and let $f\colon \mathcal{Z} \multimap \mathbb{R}$ be locally Lipschitz and λ-concave. Suppose that $\mathcal{Z}$ is either* CBB *or* CAT. *Assume $\hat\alpha\colon [0, s_{\max}) \to \operatorname{Dom} f$ is an f-gradient-like curve. Then*

$$(f \circ \hat\alpha)'' \leqslant \lambda$$

everywhere on $[0, s_{\max})$.

Closely related statements were proved independently by Uwe Mayer [**115**, 2.36] and Shin-ichi Ohta [**122**, 5.7].

Before the proof, let us formulate and prove a corollary.

16.5. Corollary. *Let $\mathcal{Z}$ be a complete length space, let $f\colon \mathcal{Z} \multimap \mathbb{R}$ be a locally Lipschitz and semiconcave function, and let $\hat\alpha\colon [0, s_{\max}) \to \operatorname{Dom} f$ be an f-gradient-like curve. Suppose that $\mathcal{Z}$ is either* CBB *or* CAT. *Then the function $s \mapsto |\nabla_{\hat\alpha(s)} f|$ is right-continuous; that is, for any $s_0 \in [0, s_{\max})$ we have*

$$|\nabla_{\hat\alpha(s_0)} f| = \lim_{s \to s_0+} |\nabla_{\hat\alpha(s)} f|.$$

Proof. Applying Theorem 16.4 locally, we have that $f \circ \hat\alpha(s)$ is semiconcave. The statement follows since

$$(f \circ \hat\alpha)^+(s) = (\boldsymbol{d}_p f)\left(\tfrac{1}{|\nabla_{\hat\alpha(s)} f|} \cdot \nabla_{\hat\alpha(s)} f\right) = |\nabla_{\hat\alpha(s)} f|. \qquad \square$$

Proof of Theorem 16.4. For any $s > s_0$,

$$\begin{aligned}(f \circ \hat\alpha)^+(s_0) &= |\nabla_{\hat\alpha(s_0)} f| \geqslant \\ &\geqslant (d_{\hat\alpha(s_0)} f)(\uparrow_{[\hat\alpha(s_0)\hat\alpha(s)]}) \geqslant \\ &\geqslant \frac{f \circ \hat\alpha(s) - f \circ \hat\alpha(s_0)}{|\hat\alpha(s) - \hat\alpha(s_0)|} - \tfrac{\lambda}{2} \cdot |\hat\alpha(s) - \hat\alpha(s_0)|.\end{aligned}$$

Let $\lambda_+ = \max\{0, \lambda\}$. Since $s - s_0 \geqslant |\hat\alpha(s) - \hat\alpha(s_0)|$, for any $s > s_0$ we have

$$(f \circ \hat\alpha)^+(s_0) \geqslant \frac{f \circ \hat\alpha(s) - f \circ \hat\alpha(s_0)}{s - s_0} - \tfrac{\lambda_+}{2} \cdot (s - s_0). \tag{2}$$

Thus $f \circ \hat\alpha$ is λ_+-concave. That finishes the proof for $\lambda \geqslant 0$. For $\lambda < 0$ we get only that $f \circ \hat\alpha$ is 0-concave.

Note that $|\hat\alpha(s) - \hat\alpha(s_0)| = s - s_0 - o(s - s_0)$. Thus

$$(f \circ \hat\alpha)^+(s_0) \geqslant \frac{f \circ \hat\alpha(s) - f \circ \hat\alpha(s_0)}{s - s_0} - \tfrac{\lambda}{2} \cdot (s - s_0) + o(s - s_0). \tag{3}$$

Together, (2) and (3) imply that $f \circ \hat\alpha$ is λ-concave. $\square$

16.6. Proposition. *Let $\mathcal{L}$ be a complete length* CBB(κ) *space, $p, q \in \mathcal{L}$. Assume $\hat\alpha\colon [s_{\min}, s_{\max}) \to \mathcal{L}$ is a* dist_p*-gradient-like curve such that $\hat\alpha(s) \to z \in \,]pq[$ as $s \to s_{\max}+$. Then $\hat\alpha$ is a unit-speed geodesic that lies in $[pq]$.*

Proof. Clearly,

$$\tfrac{d^+}{dt}|q - \hat\alpha(t)| \geqslant -1. \tag{4}$$

On the other hand,

$$\begin{aligned}\tfrac{d^+}{dt}|p - \hat\alpha(t)| &\geqslant (\boldsymbol{d}_{\hat\alpha(t)} \operatorname{dist}_p)(\uparrow_{[\hat\alpha(t)q]}) \geqslant \\ &\geqslant -\cos \tilde\measuredangle^\kappa \left(\hat\alpha(t)\,{}^p_q\right).\end{aligned} \tag{5}$$

Inequalities (4) and (5) imply that the function $t \mapsto \tilde\measuredangle^\kappa \left(q\,{}^{\hat\alpha(t)}_p\right)$ is nondecreasing. Hence the result. $\square$

B. Gradient curves

In this section we define gradient curves and tie them tightly to gradient-like curves which were introduced in Section 16.A.

16.7. Definition. Let $\mathcal{Z}$ be a complete length space and let $f\colon \mathcal{Z} \looparrowright \mathbb{R}$ be a locally Lipschitz and semiconcave subfunction. Suppose that $\mathcal{Z}$ is either CBB or CAT.

A locally Lipschitz curve $\alpha\colon [t_{\min}, t_{\max}) \to \operatorname{Dom} f$ will be called an *f-gradient curve* if

$$\alpha^+ = \nabla_\alpha f;$$

that is, for any $t \in [t_{\min}, t_{\max})$, $\alpha^+(t)$ is defined and $\alpha^+(t) = \nabla_{\alpha(t)} f$.

The following exercise describes a global geometric property of a gradient curve without direct reference to its function. It uses the notion of *self-contracting curves* introduced by Aris Daniilidis, Olivier Ley, Stéphane Sabourau [**56**].

16.8. Exercise. Let $\mathcal{Z}$ be a complete length space, let $f\colon \mathcal{Z} \to \mathbb{R}$ a concave locally Lipschitz function, and let $\alpha\colon \mathbb{I} \to \mathcal{Z}$ an f-gradient curve. Suppose that $\mathcal{Z}$ is either CBB or CAT.

Show that α is *self-contracting*; that is,

$$t_1 \leqslant t_2 \leqslant t_3 \quad \Longrightarrow \quad |\alpha(t_1) - \alpha(t_3)|_{\mathcal{Z}} \geqslant |\alpha(t_2) - \alpha(t_3)|_{\mathcal{Z}}.$$

The next lemma states that gradient and gradient-like curves are special reparametrizations of each other.

16.9. Lemma. *Let $\mathcal{Z}$ be a complete length space and let $f\colon \mathcal{Z} \looparrowright \mathbb{R}$ be a locally Lipschitz semiconcave subfunction such that $|\nabla_p f| > 0$ for any $p \in \operatorname{Dom} f$. Suppose that $\mathcal{Z}$ is either* CBB *or* CAT.

Assume that $\alpha\colon [0, t_{\max}) \to \operatorname{Dom} f$ is a locally Lipschitz curve and that $\hat\alpha\colon [0, s_{\max}) \to \operatorname{Dom} f$ is its reparametrization by arc-length, so $\alpha = \hat\alpha \circ \varsigma$ for *a homeomorphism* $\varsigma\colon [0,t_{\max})\to[0,s_{\max})$. *Then*

$$\alpha^+ = \nabla_\alpha f$$
$$\Updownarrow$$
$$\hat\alpha^+ = \frac{1}{|\nabla_{\hat\alpha} f|}\cdot\nabla_{\hat\alpha} f \quad \text{and} \quad \varsigma^{-1}(s) = \int\limits_0^s \frac{\boldsymbol{d}\c{s}}{(f\circ\hat\alpha)'(\c{s})}.$$

Proof. ($\Downarrow$). According to Theorem 3.10,

$$\begin{aligned}\varsigma'(t) &\overset{\text{a.e.}}{=\!=} |\alpha^+(t)| = \\ &= |\nabla_{\alpha(t)} f|.\end{aligned} \tag{1}$$

Note that

$$\begin{aligned}(f\circ\alpha)'(t) &\overset{\text{a.e.}}{=\!=} (f\circ\alpha)^+(t) = \\ &= |\nabla_{\alpha(t)} f|^2.\end{aligned}$$

Setting $s=\varsigma(t)$, we have

$$\begin{aligned}(f\circ\hat\alpha)'(s) &\overset{\text{a.e.}}{=\!=} \frac{(f\circ\alpha)'(t)}{\varsigma'(t)} \overset{\text{a.e.}}{=\!=} \\ &\overset{\text{a.e.}}{=\!=} |\nabla_{\alpha(t)} f| = \\ &= |\nabla_{\hat\alpha(s)} f|.\end{aligned}$$

From Theorem 16.3, it follows that $\hat\alpha(t)$ is an f-gradient-like curve; that is,

$$\hat\alpha^+ = \frac{1}{|\nabla_{\hat\alpha} f|}\cdot\nabla_{\hat\alpha} f.$$

In particular, $(f\circ\hat\alpha)^+(s) = |\nabla_{\hat\alpha^+(s)} f|$, and by (1),

$$\begin{aligned}\varsigma^{-1}(s) &= \int\limits_0^s \frac{\boldsymbol{d}\c{s}}{|\nabla_{\hat\alpha(\c{s})} f|} = \\ &= \int\limits_0^s \frac{\boldsymbol{d}\c{s}}{(f\circ\hat\alpha)'(\c{s})}.\end{aligned}$$

(⇑). Clearly,

$$\begin{aligned}\varsigma(t) &= \int_0^t (f\circ\hat\alpha)^+(\varsigma(\xi))\cdot d\xi = \\ &= \int_0^t |\nabla_{\alpha(\xi)}f|\cdot d\xi.\end{aligned}$$

According to Corollary 16.5, the function $s \mapsto |\nabla_{\hat\alpha(s)}f|$ is right-continuous. Therefore so is the function $t \mapsto |\nabla_{\hat\alpha\circ\varsigma(t)}f| = |\nabla_{\alpha(t)}f|$. Hence, for any $t_0 \in$ $\in [0, t_{\max})$ we have

$$\begin{aligned}\varsigma^+(t_0) &= \lim_{t\to t_0+}\frac{1}{t-t_0}\cdot\int_{t_0}^t |\nabla_{\alpha(\xi)}f|\cdot d\xi = \\ &= |\nabla_{\alpha(t_0)}f|.\end{aligned}$$

Thus, we have

$$\begin{aligned}\alpha^+(t_0) &= \varsigma^+(t_0)\cdot\hat\alpha^+(\varsigma(t_0)) = \\ &= \nabla_{\alpha(t_0)}f.\end{aligned}$$

□

16.10. Exercise. Let $\mathcal{Z}$ be a complete length space, and let $f\colon \mathcal{Z}\to\mathbb{R}$ be a semiconcave locally Lipschitz function. Suppose that $\mathcal{Z}$ is either CBB or CAT. Assume $\alpha\colon \mathbb{I}\to\mathcal{Z}$ is a Lipschitz curve such that

$$\begin{aligned}\alpha^+(t) &\leqslant |\nabla_{\alpha(t)}f|, \\ (f\circ\alpha)^+(t) &\geqslant |\nabla_{\alpha(t)}f|^2\end{aligned}$$

for almost all t. Show that α is an f-gradient curve.

16.11. Exercise. Let $\mathcal{Z}$ be a complete length space and let $f\colon \mathcal{Z}\to\mathbb{R}$ be a concave locally Lipschitz function. Suppose that $\mathcal{Z}$ is either CBB or CAT. Show that $\alpha\colon \mathbb{R}\to\mathcal{Z}$ is an f-gradient curve if and only if

$$|x-\alpha(t_1)|^2_{\mathcal{Z}} - |x-\alpha(t_0)|^2_{\mathcal{Z}} \leqslant 2\cdot(t_1-t_0)\cdot(f\circ\alpha(t_1)-f(x))$$

for any $t_1 > t_0$ and $x\in\mathcal{Z}$.

C. Distance estimates

16.12. First distance estimate. *Let $\mathcal{Z}$ be a complete length space, and let $f\colon \mathcal{Z}\to\mathbb{R}$ be a locally Lipschitz λ-concave function. Suppose that $\mathcal{Z}$ is either* CBB *or* CAT. *Let $\alpha,\beta\colon [0,t_{\max})\to\mathcal{Z}$ be two f-gradient curves. Then*

$$|\alpha(t)-\beta(t)| \leqslant e^{\lambda\cdot t}\cdot|\alpha(0)-\beta(0)|$$

for any t.

Moreover, the statement holds for a locally Lipschitz λ-concave subfunction $f\colon \mathcal{Z} \multimap \mathbb{R}$ if there is a geodesic $[\alpha(t)\,\beta(t)]$ in $\operatorname{Dom} f$ *for any t.*

Proof. If $\mathcal{Z}$ is not geodesic, then pass to its ultrapower $\mathcal{Z}^\omega$.

Fix a choice of geodesic $[\alpha(t)\,\beta(t)]$ for each t.

Setting $\ell(t) = |\alpha(t) - \beta(t)|$, from the first variation inequality (see 6.7) and the estimate in 13.25 we get

$$\ell^+(t) \leqslant -\langle \uparrow_{[\alpha(t)\beta(t)]}, \nabla_{\alpha(t)} f\rangle - \langle \uparrow_{[\beta(t)\alpha(t)]}, \nabla_{\beta(t)} f\rangle \leqslant \lambda\cdot\ell(t).$$

Here one has to apply the first variation inequality for distance to the midpoint m of $[\alpha(t)\,\beta(t)]$, and then apply the triangle inequality. Hence the result. □

16.13. Second distance estimate. *Let $\mathcal{Z}$ be a complete length space, $\varepsilon > 0$, and let $f, g\colon \mathcal{Z} \to \mathbb{R}$ be two λ-concave locally Lipschitz functions such that $|f - g| < \varepsilon$. Suppose that $\mathcal{Z}$ is either* CBB *or* CAT. *Assume $\alpha, \beta\colon [0, t_{\max}) \to \mathcal{Z}$ are, respectively, f- and g-gradient curves. Let $\ell\colon t \mapsto |\alpha(t) - \beta(t)|$. Then*

$$\ell^+ \leqslant \lambda\cdot\ell + \tfrac{2\cdot\varepsilon}{\ell}.$$

In particular, if $\alpha(0) = \beta(0)$ and $t_{\max} < \infty$, then

$$|\alpha(t) - \beta(t)| \leqslant c\cdot\sqrt{\varepsilon\cdot t}$$

for a constant $c = c(t_{\max}, \lambda)$.

Moreover, the same conclusion holds for locally Lipschitz λ-concave subfunctions $f, g\colon \mathcal{Z} \multimap \mathbb{R}$ if for any $t \in [0, t_{\max})$ there is a geodesic $[\alpha(t)\,\beta(t)]$ in $\operatorname{Dom} f \cap \operatorname{Dom} g$.

Proof. Set $\ell = \ell(t) = |\alpha(t) - \beta(t)|$. Fix t, and let $p = \alpha(t)$ and $q = \beta(t)$. From the first variation formula and Lemma 13.24,

$$\begin{aligned}
\ell^+ &\leqslant -\langle \uparrow_{[pq]}, \nabla_p f\rangle - \langle \uparrow_{[qp]}, \nabla_q g\rangle \leqslant \\
&\leqslant -\left(f(q) - f(p) - \lambda\cdot\tfrac{\ell^2}{2}\right)/\ell - \left(g(p) - g(q) - \lambda\cdot\tfrac{\ell^2}{2}\right)/\ell \leqslant \\
&\leqslant \lambda\cdot\ell + \tfrac{2\cdot\varepsilon}{\ell}.
\end{aligned}$$

By integrating, we get the second statement. □

D. Existence and uniqueness

In general, the "past" of gradient curves can not be determined by the "present". For example, consider the concave function $f\colon \mathbb{R} \to \mathbb{R}$, $f(x) = -|x|$. The two curves $\alpha(t) = \min\{0, t\}$ with $\beta(t) = 0$ are f-gradient with $\alpha(t) = \beta(t) = 0$ for all $t \geqslant 0$; however $\alpha(t) \neq \beta(t)$ for all $t < 0$. Another example can be given as follows.

16.14. Example. Let f be as in Example 13.18; that is, $f\colon (x,y) \mapsto -|x| - |y|$ be the concave function on the (x,y)-plane; its gradient field is sketched on the figure.

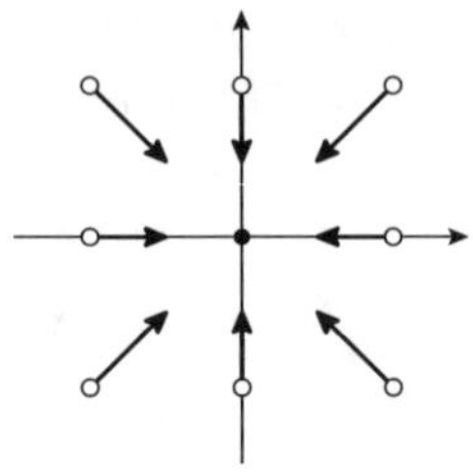

Let α be an f-gradient curve that starts at $p = (x,y)$ for $x > y > 0$. Then

$$\alpha(t) = \begin{cases} (x-t, y-t) & \text{for } 0 \leqslant t \leqslant x-y, \\ (x-t, 0) & \text{for } x-y \leqslant t \leqslant x, \\ (0,0) & \text{for } x \leqslant t. \end{cases}$$

In particular, gradient curves can merge even in the region where $|\nabla f| \neq 0$. Hence their *past* cannot be uniquely determined from their *present*.

The next theorem shows that the future gradient curve is determined by its present.

16.15. Picard's theorem. *Let $\mathcal{Z}$ be a complete length space, let $f\colon \mathcal{Z} \hookrightarrow \mathbb{R}$ be a semiconcave subfunction. Suppose that $\mathcal{Z}$ is either* CBB *or* CAT. *Assume $\alpha, \beta\colon [0, t_{\max}) \to \operatorname{Dom} f$ are two f-gradient curves such that $\alpha(0) = \beta(0)$. Then $\alpha(t) = \beta(t)$ for any $t \in [0, t_{\max})$.*

Proof. The proof follows from the first distance estimate (see 16.12). □

16.16. Local existence. *Let $\mathcal{Z}$ be a complete length space and let $f\colon \mathcal{Z} \hookrightarrow \mathbb{R}$ be locally Lipschitz λ-concave subfunction. Suppose that $\mathcal{Z}$ is either* CBB *or* CAT. *Then for any $p \in \operatorname{Dom} f$,*

(a) *if $|\nabla_p f| > 0$, then for some $\varepsilon > 0$, there is an f-gradient-like curve $\hat\alpha\colon [0,\varepsilon) \to \mathcal{Z}$ that starts at p (that is, $\hat\alpha(0) = p$);*

(b) *for some $\delta > 0$, there is an f-gradient curve $\alpha\colon [0,\delta) \to \mathcal{Z}$ that starts at p (that is $\alpha(0) = p$).*

This theorem was proved by Grigory Perelman and the third author [**125**]; we present a simplified proof given by Alexander Lytchak [**111**].

Proof. If $|\nabla_p f| = 0$, then the constant curve $\alpha(t) = p$ is f-gradient.

Otherwise, choose $\varepsilon > 0$ such that $\mathrm{B}(p,\varepsilon) \subset \operatorname{Dom} f$, the restriction $f|_{\mathrm{B}(p,\varepsilon)}$ is Lipschitz, and $|\nabla_x f| > \varepsilon$ for all $x \in \mathrm{B}(p,\varepsilon)$; the latter is possible due to semicontinuity of |gradient| (see 13.31).

The curves $\hat\alpha$ and α will be constructed in the following three steps. First we construct an f^ω-gradient-like curve $\hat\alpha_\omega\colon [0,\varepsilon) \to \mathcal{Z}^\omega$ as an ω-limit of a certain sequence of broken geodesics in $\mathcal{Z}$. Second, we parametrize $\hat\alpha_\omega$ as in Lemma 16.9, to obtain an f^ω-gradient curve α_ω in $\mathcal{Z}^\omega$. Third, applying Picard's theorem (see 16.15) together with Lemma 4.5, we obtain that α_ω lies in $\mathcal{Z} \subset \mathcal{Z}^\omega$ and therefore one can take $\alpha = \alpha_\omega$ and $\hat\alpha = \hat\alpha_\omega$.

Note that if $\mathcal{Z}$ is proper, then $\mathcal{Z}$ is a metric component of $\mathcal{Z}^\omega$ and $f = f^\omega|_{\mathcal{Z}}$. Thus, in this case, the third step is not necessary.

Step 1. Given $n \in \mathbb{N}$, by an open-closed argument, we can construct a unit-speed curve $\hat{\alpha}_n : [0,\varepsilon] \to \mathcal{Z}$ starting at p, with a partition of $[0,\varepsilon)$ into a countable number of half-open intervals $[\varsigma_i, \bar{\varsigma}_i)$ such that for each i we have

(i) $\hat{\alpha}_n([\varsigma_i, \bar{\varsigma}_i])$ is a geodesic and $\bar{\varsigma}_i - \varsigma_i < \frac{1}{n}$,

(ii) $f \circ \hat{\alpha}_n(\bar{\varsigma}_i) - f \circ \hat{\alpha}_n(\varsigma_i) > (\bar{\varsigma}_i - \varsigma_i)\cdot(|\nabla_{\hat{\alpha}_n(\varsigma_i)} f| - \frac{1}{n})$.

Passing to a subsequence of $\hat{\alpha}_n$ such that $f \circ \hat{\alpha}_n$ uniformly converges, let

$$h(s) = \lim_{n\to\infty} f \circ \hat{\alpha}_n(s).$$

Let $\hat{\alpha}_\omega = \lim_{n\to\omega} \hat{\alpha}_n$; it is a curve in $\mathcal{Z}^\omega$ that starts at $p \in \mathcal{Z} \subset \mathcal{Z}^\omega$.

Clearly $\hat{\alpha}_\omega$ is 1-Lipschitz. From (ii) and 13.28, we have

$$(f^\omega \circ \hat{\alpha}_\omega)^+(\varsigma) \geqslant |\nabla_{\hat{\alpha}_\omega(\varsigma)} f^\omega|.$$

According to Theorem 16.3, $\hat{\alpha}_\omega : [0,\varepsilon) \to \mathcal{Z}^\omega$ is an f^ω-gradient-like curve.

Step 2. Clearly $h(s) = f^\omega \circ \alpha_\omega$. Therefore, according to Theorem 16.4, h is λ-concave. Thus we can define a homeomorphism $\varsigma : [0,\delta] \to [0,\varepsilon]$ by

$$\varsigma^{-1}(s) = \int_0^s \frac{d\c{s}}{h'(\c{s})}, \tag{1}$$

According to Lemma 16.9, $\alpha(t) = \hat{\alpha} \circ \varsigma(t)$ is an f^ω-gradient curve in $\mathcal{Z}^\omega$.

Step 3. Clearly, $\nabla_p f = \nabla_p f^\omega$ for any $p \in \mathcal{Z} \subset \mathcal{Z}^\omega$; more formally, if $\iota : \mathcal{Z} \hookrightarrow \mathcal{Z}^\omega$ is the natural embedding, then $(\boldsymbol{d}_p\iota)(\nabla_p f) = \nabla_p f^\omega$. Thus it is sufficient to show that α_ω lies in $\mathcal{Z}$. Assume the contrary; then according to Lemma 4.5, there is a subsequence $\hat{\alpha}_{n_k}$ such that

$$\hat{\alpha}_\omega \neq \hat{\alpha}'_\omega := \lim_{k\to\omega} \hat{\alpha}_{n_k}.$$

Clearly $h(s) = f^\omega \circ \hat{\alpha}_\omega = f^\omega \circ \hat{\alpha}'_\omega$. Thus for $\varsigma : [0,\delta] \to [0,\varepsilon]$ defined by (1), we have that both curves $\hat{\alpha}_\omega \circ \varsigma$ and $\hat{\alpha}'_\omega \circ \varsigma$ are f^ω-gradient. From Picard's theorem (see 16.15), we have $\hat{\alpha}_\omega \circ \varsigma = \hat{\alpha}'_\omega \circ \varsigma$. Therefore $\hat{\alpha}_\omega = \hat{\alpha}'_\omega$, a contradiction. □

E. Convergence

16.17. Ultralimit of gradient curves. *Assume*

- $\mathcal{Z}_n$ *is a sequence of complete spaces,* $\mathcal{Z}_n \to \mathcal{Z}_\omega$ *as* $n \to \omega$, *and* $p_n \to p_\omega$ *for a sequence of points* $p_n \in \mathcal{Z}_n$;
- *all spaces* $\mathcal{Z}_n$ *are either* CBB(κ) *or* CAT(κ);

- *$f_n \colon \mathcal{Z}_n \multimap \mathbb{R}$ are ℓ-Lipschitz and λ-concave, $f_n \to f_\omega$ as $n \to \omega$, and $p_\omega \in \operatorname{Dom} f_\omega$.*

Then:

(a) *f_ω is λ-concave.*

(b) *If $|\nabla_{p_\omega} f_\omega| > 0$, then there is $\varepsilon > 0$ such that, the f_n-gradient-like curves $\hat\alpha_n \colon [0, \varepsilon) \to \mathcal{Z}_n$ are defined for ω-almost all n. Moreover, a curve $\hat\alpha_\omega \colon [0, \varepsilon) \to \mathcal{Z}_\omega$ is a gradient-like curve that starts at p_ω if and only if $\hat\alpha_n(s) \to \hat\alpha_\omega(s)$ as $n \to \omega$ for all $s \in [0, \varepsilon)$.*

(c) *For some $\delta > 0$, the f_n-gradient curves $\alpha_n \colon [0, \delta) \to \mathcal{Z}_n$ are defined for ω-almost all n. Moreover, a curve $\alpha_\omega \colon [0, \delta) \to \mathcal{Z}_\omega$ is a gradient curve that starts at p_ω if and only if $\alpha_n(t) \to \alpha_\omega(t)$ as $n \to \omega$ for all $t \in [0, \delta)$.*

Note that according to Exercise 4.17, part (a) does not hold for general metric spaces. The idea of the proof is the same as in the proof of local existence (see 16.16).

Proof of 16.17.

(a). Fix a geodesic $\gamma_\omega \colon \mathbb{I} \to \operatorname{Dom} f_\omega$; we need to show that the function

(1)
$$t \mapsto f_\omega \circ \gamma_\omega(t) - \tfrac{\lambda}{2} \cdot t^2$$

is concave.

Since the f_n are ℓ-Lipschitz, so is f_ω. Therefore it is sufficient to prove concavity in the interior of $\mathbb{I}$. In particular, we can assume that γ_ω is sufficiently short and can be extended behind its ends p_ω and q_ω as a minimizing geodesic. If $\mathcal{Z}$ is CBB, then by Theorem 8.11, γ_ω is the unique geodesic connecting p_ω to q_ω. The same holds true if $\mathcal{Z}$ is CAT by the uniqueness of geodesics (see 9.8).

Construct two sequences of points $p_n, q_n \in \mathcal{Z}_n$ such that $p_n \to p_\omega$ and $q_n \to q_\omega$ as $n \to \omega$. Applying either Theorem 8.11 or 9.8, we can assume that for each n there is a geodesic γ_n from p_n to q_n in $\mathcal{Z}_n$.

Since f_n is λ-concave, the function

$$t \mapsto f_n \circ \gamma_n(t) - \tfrac{\lambda}{2} \cdot t^2$$

is concave.

The ω-limit of the sequence γ_n is a geodesic in $\mathcal{Z}_\omega$ from p_ω to q_ω. By uniqueness of such geodesics, we have that $\gamma_n \to \gamma_\omega$ as $n \to \omega$. Passing to the limit, we have (1).

If part of (b). Take $\varepsilon > 0$ so small that $\mathrm{B}(p_\omega, \varepsilon) \subset \operatorname{Dom} f_\omega$ and $|\nabla_{x_\omega} f_\omega| > 0$ for any $x_\omega \in \mathrm{B}(p_\omega, \varepsilon)$ (this is possible by 13.31).

Clearly $\hat\alpha_\omega$ is 1-Lipschitz. From 13.28, we get

$$(f_\omega \circ \hat\alpha_\omega)^+(s) \geqslant |\nabla_{\hat\alpha_\omega(s)} f^\omega|.$$

According to Theorem 16.3, $\hat{\alpha}_\omega : [0,\varepsilon) \to \mathcal{Z}^\omega$ is an f_ω-gradient-like curve.

If part of (c). Assume first that $|\nabla_{p_\omega} f_\omega| > 0$, so we can apply the if part of (b). Let $h_n = f_n \circ \hat{\alpha}_n : [0,\varepsilon) \to \mathbb{R}$ and $h_\omega = f_\omega \circ \hat{\alpha}_\omega$. From Theorem 16.4, the h_n are λ-concave, and clearly $h_n \to h_\omega$ as $n \to \omega$. Let us define reparametrizations

$$\varsigma_n^{-1}(s) = \int\limits_0^s \frac{d\c{s}}{h_n'(\c{s})}, \qquad \varsigma_\omega^{-1}(s) = \int\limits_0^s \frac{d\c{s}}{h_\omega'(\c{s})}.$$

The λ-convexity of the h_n implies that $\sigma_n \to \sigma_\omega$ as $n \to \omega$. By Lemma 16.9, $\alpha_n = \hat{\alpha}_n \circ \varsigma_n$. Applying the if part of (b) together with Lemma 16.9, we get that $\alpha_\omega = \hat{\alpha}_\omega \circ \varsigma_\omega$ is gradient curve.

The remaining case $|\nabla_{p_\omega} f_\omega| = 0$ can be reduced to the one above using the following trick. Consider the sequence of spaces $\mathcal{Z}_n^\times = \mathcal{Z}_n \times \mathbb{R}$, with the sequence of subfunctions $f_n^\times : \mathcal{Z}_n^\times \to \mathbb{R}$ defined by

$$f_n^\times(p,t) = f_n(p) + t.$$

Applying either Theorem 11.7(b) or Theorem 11.6(b), we have that $\mathcal{Z}_n^\times$ is a CBB(κ_-) space for $\kappa_- = \min\{\kappa, 0\}$, or CAT($\kappa_+$) space for $\kappa_+ = \max\{\kappa, 0\}$. Note that the $f_n^\times$ are λ_+-concave for $\lambda_+ = \max\{\lambda, 0\}$. Now let $\mathcal{Z}_\omega^\times = \mathcal{Z}_\omega \times \mathbb{R}$, and $f_\omega^\times(p,t) = f_\omega(p) + t$.

Clearly $\mathcal{Z}_n^\times \to \mathcal{Z}_\omega^\times$, $f_n^\times \to f_\omega^\times$ as $n \to \omega$, and $|\nabla_x f_\omega^\times| > 0$ for any $x \in \operatorname{Dom} f_\omega^\times$. Thus for the sequence $f_n^\times : \mathcal{Z}_n^\times \multimap \mathbb{R}$, we can apply the if part of (b). It remains to note that the curve $\alpha_\omega^\times(t) = (\alpha_\omega(t), t)$ is an $f_\omega^\times$-gradient curve in $\mathcal{Z}_\omega^\times$ if and only if $\alpha_\omega(t)$ is an f_ω-gradient curve.

Only-if part of (c) *and* (b). The only-if part of (c) follows from the if part of (c) and Picard's theorem (see 16.15). Applying Lemma 16.9, we get the only-if part of (b). □

From local existence (see 16.16) and the distance estimates (see 16.12), we obtain the following.

16.18. Global existence. *Let $f : \mathcal{Z} \multimap \mathbb{R}$ be a locally Lipschitz and λ-concave subfunction on a complete length space $\mathcal{Z}$. Suppose that $\mathcal{Z}$ is either* CBB *or* CAT. *Then for any $p \in \operatorname{Dom} f$, there is $t_{\max} \in (0,\infty]$ such that there is an f-gradient curve $\alpha : [0, t_{\max}) \to \mathcal{Z}$ with $\alpha(0) = p$. Moreover, for any sequence $t_n \to t_{\max}-$, the sequence $\alpha(t_n)$ does not have a limit point in* Dom f.

The following theorem guarantees the existence of gradient curves for all times for the special type of semiconcave functions that play important role in the theory. It follows from 16.18, Theorem 16.4, and Lemma 16.9.

16.19. Theorem. *Let $\mathcal{Z}$ be a complete length space and let $f : \mathcal{Z} \to \mathbb{R}$ satisfies*

$$f'' + \kappa \cdot f \leqslant \lambda$$

for real constants κ and λ. Suppose that $\mathcal{Z}$ is either CBB *or* CAT. *Then f has* complete gradient; *that is, for any $x \in \mathcal{Z}$ there is a f-gradient curve $\alpha\colon [0,\infty) \to$ $\to \mathcal{Z}$ that starts at x.*

F. Gradient flow

In this section we define gradient flow for semiconcave subfunctions and reformulate theorems obtained earlier in this chapter using this new terminology.

Let $\mathcal{Z}$ be a complete length space and let $f\colon \mathcal{Z} \hookrightarrow \mathbb{R}$ be a locally Lipschitz semiconcave subfunction. Suppose that $\mathcal{Z}$ is either CBB or CAT. For any $t \geqslant 0$, we write $\mathrm{Flow}_f^t(x) = y$ if there is an f-gradient curve α such that $\alpha(0) = x$ and $\alpha(t) = y$. The partially defined map Flow_f^t from $\mathcal{Z}$ to itself is called the *f-gradient flow* for time t.

From 16.13, it follows that for any $t \geqslant 0$, the domain of definition of Flow_f^t is an open subset of $\mathcal{Z}$; that is, Flow_f^t is a submap. Moreover, if f is defined on all of $\mathcal{Z}$ and $f'' + \mathrm{K}\cdot f \leqslant \lambda$ for constants $\mathrm{K}, \lambda \in \mathbb{R}$, then according to Theorem 16.19, $\mathrm{Flow}_f^t(x)$ is defined for all pairs $(x,t) \in \mathcal{Z} \times \mathbb{R}_{\geqslant 0}$.

Clearly $\mathrm{Flow}_f^{t_1+t_2} = \mathrm{Flow}_f^{t_1} \circ \mathrm{Flow}_f^{t_2}$; in other words, gradient flow is given by an action of the semigroup $(\mathbb{R}_{\geqslant 0}, +)$.

From the first distance estimate (see 16.12), we have the following:

16.20. Proposition. *Let $\mathcal{Z}$ be a complete length* CBB *or* CAT *space and let $f\colon \mathcal{Z} \to \mathbb{R}$ be a semiconcave function. Then the map $x \mapsto \mathrm{Flow}_f^t(x)$ is locally Lipschitz.*

Moreover, if f is λ-concave, then Flow_f^t is $e^{\lambda\cdot t}$-Lipschitz.

The next proposition states that gradient flow is stable under Gromov–Hausdorff convergence. The proposition follows directly from the proposition on ultralimit of gradient curves 16.17.

16.21. Proposition. *Supose $\mathcal{Z}_\infty, \mathcal{Z}_1, \mathcal{Z}_2, \dots$ are complete length* $\mathrm{CBB}(\kappa)$ *space, $\mathcal{Z}_n \xrightarrow{\tau} \mathcal{Z}_\infty$, and $f_n\colon \mathcal{Z}_n \to \mathbb{R}$ is a sequence of λ-concave functions that converges to $f_\infty\colon \mathcal{Z}_\infty \to \mathbb{R}$. Then $\mathrm{Flow}_{f_n}^t\colon \mathcal{Z}_n \to \mathcal{Z}_n$ converges to $\mathrm{Flow}_{f_\infty}^t\colon \mathcal{Z}_\infty \to \mathcal{Z}_\infty$.*

G. Line splitting theorem

Let $\mathcal{X}$ be a metric space and let $A, B \subset \mathcal{X}$. We say that $\mathcal{X}$ is a *direct sum* of A and B, briefly

$$\mathcal{X} = A \oplus B,$$

if there are projections $\mathrm{proj}_A\colon \mathcal{X} \to A$ and $\mathrm{proj}_B\colon \mathcal{X} \to B$ such that

$$|x-y|^2 = |\mathrm{proj}_A(x) - \mathrm{proj}_A(y)|^2 + |\mathrm{proj}_B(x) - \mathrm{proj}_B(y)|^2$$

for any two points $x, y \in \mathcal{X}$.

Note that if

$$\mathcal{X} = A \oplus B,$$

then

- A intersects B at a single point,
- both sets A and B are convex sets in $\mathcal{X}$.

Recall that a line in a metric space is a both-sided infinite geodesic; thus it minimizes the length on each segment.

16.22. Line splitting theorem. *Let $\mathcal{L}$ be a complete length* CBB(0) *space and γ be a line in $\mathcal{L}$. Then*

$$\mathcal{L} = \mathcal{L}' \oplus \gamma(\mathbb{R})$$

for a subset $\mathcal{L}' \subset \mathcal{L}$.

For smooth two-dimensional surfaces, this theorem was proved by Stefan Cohn-Vossen [**55**]. For Riemannian manifolds of higher dimensions it was proved by Victor Toponogov [**156**]. Then it was generalized by Anatoliy Milka [**118**] to Alexandrov spaces; nearly the same proof is used in [**37**, 1.5].

Further generalizations of the splitting theorem for Riemannian manifolds with nonnegative Ricci curvature were obtained by Jeff Cheeger and Detlef Gromoll [**53**]. This was further generalized by Jeff Cheeger and Toby Colding for limits of Riemannian manifolds with almost nonnegative Ricci curvature [**51**] and to their synthetic generalizations, so-called *RCD spaces*, by Nicola Gigli [**68, 69**]. Jost-Hinrich Eschenburg obtained an analogous result for Lorentzian manifolds [**64**], that is, pseudo-Riemannian manifolds of signature $(1, n)$.

We present a proof that uses gradient flow for Busemann functions. It is close in spirit to the proof given in [**53**].

Before going into the proof, let us state a few corollaries of the theorem.

16.23. Corollary. *Let $\mathcal{L}$ be a complete length* CBB(0) *space. Then there is an isometric splitting*

$$\mathcal{L} = \mathcal{L}' \oplus H,$$

where $H \subset \mathcal{L}$ is a subset isometric to a Hilbert space, and $\mathcal{L}' \subset \mathcal{L}$ is a convex subset that contains no line.

16.24. Corollary. *Let $\mathcal{K}$ be a finite-dimensional complete length* CBB(0) *cone and let $v_+, v_- \in \mathcal{K}$ be a pair of opposite vectors (that is, $v_+ + v_- = 0$, see Definiton* 13.36*). Then there is an isometry $\iota\colon K \to K' \times \mathbb{R}$ such that $\iota : v_\pm \mapsto$ $\mapsto (0', \pm|v_\pm|)$, where K' is a complete length* CBB(0) *space having a cone structure with tip $0'$.*

16.25. Corollary. *Let $\mathcal{L}$ be an m-dimensional complete length* CBB(1) *space,* $2 \leqslant m < \infty$, *and* $\operatorname{rad} \mathcal{L} = \pi$. *Then*

$$\mathcal{L} \stackrel{\text{iso}}{=\!=} \mathbb{S}^m.$$

The following lemma is closely relevant to the first distance estimate (see 16.12); its proof goes along the same lines.

16.26. Lemma. *Let $\mathcal{L}$ be a complete length* CBB(0) *space. Suppose $f : \mathcal{L} \to \mathbb{R}$ be a concave* 1*-Lipschitz function. Consider two f-gradient curves α and β. Then for any $t, s \geqslant 0$ we have*

$$|\alpha(s) - \beta(t)|^2 \leqslant |p - q|^2 + 2\cdot(f(p) - f(q))\cdot(s - t) + (s - t)^2,$$

where $p = \alpha(0)$ and $q = \beta(0)$.

Proof. If $\mathcal{L}$ is not geodesic, then pass to its ultrapower $\mathcal{L}^\omega$.

Since f is 1-Lipschitz, $|\nabla f| \leqslant 1$. Therefore

$$f \circ \beta(t) \leqslant f(q) + t$$

for any $t \geqslant 0$.

Set $\ell(t) = |p - \beta(t)|$. Applying Lemma 13.24(a) and the first variation inequality (see 6.7), we get

$$\begin{aligned} \ell^2(t)^+ &\leqslant 2\cdot (f \circ \beta(t) - f(p)) \leqslant \\ &\leqslant 2\cdot (f(q) + t - f(p)). \end{aligned}$$

Therefore

$$\ell^2(t) - \ell^2(0) \leqslant 2\cdot (f(q) - f(p)) \cdot t + t^2.$$

It proves the needed inequality in case $s = 0$. Combining it with the first distance estimate (see 16.12), we get the result in case $s \leqslant t$. The case $s \geqslant t$ follows by switching the roles of s and t. □

Proof of 16.22. Consider two Busemann functions, bus_+ and bus_-, asociated with half-lines $\gamma : [0, \infty) \to \mathcal{L}$ and $\gamma : (-\infty, 0] \to \mathcal{L}$, respectively; that is,

$$\operatorname{bus}_\pm(x) = \lim_{t\to\infty} |\gamma(\pm t) - x| - t.$$

According to Exercise 8.25, both functions $\operatorname{bus}_\pm$ are concave.

Fix $x \in \mathcal{L}$. Note that since γ is a line, we have

$$\operatorname{bus}_+(x) + \operatorname{bus}_-(x) \geqslant 0.$$

On the other hand, by Exercise 8.23(b), $f(t) = \operatorname{dist}_x^2 \circ \gamma(t)$ is 2-concave. In particular, $f(t) \leqslant t^2 + at + b$ for some constants $a, b \in \mathbb{R}$. Passing to the limit as $t \to \pm\infty$, we have

$$\operatorname{bus}_+(x) + \operatorname{bus}_-(x) \leqslant 0.$$

Hence

$$\mathrm{bus}_+(x) + \mathrm{bus}_-(x) = 0$$

for any $x \in \mathcal{L}$. In particular, the functions $\mathrm{bus}_\pm$ are *affine*; that is, they are convex and concave at the same time.

It follows that for any x,

$$\begin{aligned} |\nabla_x \mathrm{bus}_\pm| &= \sup\{\boldsymbol{d}_x \mathrm{bus}_\pm(\xi) : \xi \in \Sigma_x\} = \\ &= \sup\{-\boldsymbol{d}_x \mathrm{bus}_\mp(\xi) : \xi \in \Sigma_x\} \equiv \\ &\equiv 1. \end{aligned}$$

By Exercise 16.10, a 1-Lipschitz curve α such that $\mathrm{bus}_\pm(\alpha(t)) = t + c$ is a $\mathrm{bus}_\pm$-gradient curve. In particular, $\alpha(t)$ is a bus_+-gradient curve if and only if $\alpha(-t)$ is a bus_--gradient curve. It follows that for any $t > 0$, the $\mathrm{bus}_\pm$-gradient flows commute; that is,

$$\mathrm{Flow}^t_{\mathrm{bus}_+} \circ \mathrm{Flow}^t_{\mathrm{bus}_-} = \mathrm{id}_{\mathcal{L}}.$$

Setting

$$\mathrm{Flow}^t = \begin{cases} \mathrm{Flow}^t_{\mathrm{bus}_+} & \text{if } t \geqslant 0 \\ \mathrm{Flow}^t_{\mathrm{bus}_-} & \text{if } t \leqslant 0 \end{cases}$$

defines an $\mathbb{R}$-action on $\mathcal{L}$.

Consider the level set $\mathcal{L}' = \mathrm{bus}_+^{-1}(0) = \mathrm{bus}_-^{-1}(0)$; it is a closed convex subset of $\mathcal{L}$, and therefore forms an Alexandrov space. Consider the map $h : \mathcal{L}' \times \mathbb{R} \to \mathcal{L}$ defined by $h : (x, t) \mapsto \mathrm{Flow}^t(x)$. Note that h is onto. Applying Lemma 16.26 for $\mathrm{Flow}^t_{\mathrm{bus}_+}$ and $\mathrm{Flow}^t_{\mathrm{bus}_-}$ shows that h is short and noncontracting at the same time; that is, h is an isometry. □

H. Radial curves

The radial curves are specially reparametrized gradient curves for distance functions. This parametrization makes them behave like unit-speed geodesics in a natural comparison sense (see 16.I).

16.27. Definition. Assume $\mathcal{L}$ is a complete length CBB space, $\kappa \in \mathbb{R}$, and $p \in \mathcal{L}$. A curve

$$\sigma : [s_{\min}, s_{\max}) \to \mathcal{L}$$

is called a (p, κ)-*radial curve* if

$$s_{\min} = |p - \sigma(s_{\min})| \in (0, \tfrac{\varpi^\kappa}{2})$$

and σ satisfies the differential equation

(1) $$\sigma^+(s) = \frac{\mathrm{tg}^\kappa |p - \sigma(s)|}{\mathrm{tg}^\kappa s} \cdot \nabla_{\sigma(s)} \mathrm{dist}_p$$

for any $s \in [s_{\min}, s_{\max})$, where $\mathrm{tg}^\kappa x := \frac{\mathrm{sn}^\kappa x}{\mathrm{cs}^\kappa x}$.

If $x = \sigma(s_{\min})$, we say that σ *starts* at x.

Note that according to the definition, $s_{\max} \leqslant \frac{\varpi^\kappa}{2}$.

In the the next section, we will see that (p,κ)-radial curves work best for CBB(κ) spaces.

16.28. Definition. Let $\mathcal{L}$ be a complete length CBB space and let $p \in \mathcal{L}$. A unit-speed geodesic $\gamma\colon \mathbb{I} \to \mathcal{L}$ is called a *p-radial geodesic* if $|p - \gamma(s)| \equiv s$.

The proofs of the following two propositions follow directly from the definitions.

16.29. Proposition. *Let $\mathcal{L}$ be a complete length* CBB *space and let $p \in \mathcal{L}$. Assume $\frac{\varpi^\kappa}{2} \geqslant s_{\max}$. Then any p-radial geodesic $\gamma\colon [s_{\min}, s_{\max}) \to \mathcal{L}$ is a (p,κ)-radial curve.*

16.30. Proposition. *Suppose $\mathcal{L}$ is a complete length* CBB *space, $p \in \mathcal{L}$. Then for any (p,κ)-radial curve $\sigma\colon [s_{\min}, s_{\max}) \to \mathcal{L}$ and $s \in [s_{\min}, s_{\max})$, we have $|p - \sigma(s)| \leqslant s$, and therefore, σ is* 1*-Lipschitz.*

Moreover, if for some s_0 we have $|p - \sigma(s_0)| = s_0$, then the restriction $\sigma|_{[s_{\min}, s_0]}$ is a p-radial geodesic.

16.31. Existence and uniqueness. *Let $\mathcal{L}$ be a complete length* CBB *space, $\kappa \in \mathbb{R}$, $p \in \mathcal{L}$, and $x \in \mathcal{L}$. Assume $0 < |p - x| < \frac{\varpi^\kappa}{2}$. Then there is a unique (p,κ)-radial curve $\sigma\colon [|p - x|, \frac{\varpi^\kappa}{2}) \to \mathcal{L}$ that starts at x; that is, $\sigma(|p - x|) = x$.*

Proof.

Existence. Let us define *integral tangent*

$$\operatorname{itg}^\kappa\colon [0, \tfrac{\varpi^\kappa}{2}) \to \mathbb{R}, \quad \operatorname{itg}^\kappa(t) = \int_0^t \operatorname{tg}^\kappa \mathfrak{t}\cdot \mathbf{d}\mathfrak{t}.$$

Clearly $\operatorname{itg}^\kappa$ is smooth and increasing. From Proposition 3.18 it follows that the composition

$$f = \operatorname{itg}^\kappa \circ \operatorname{dist}_p$$

is semiconcave in $\mathrm{B}(p, \frac{\varpi^\kappa}{2})$.

According to 16.16, there is an f-gradient curve $\alpha\colon [0, t_{\max}) \to \mathcal{L}$ defined on the maximal interval such that $\alpha(0) = x$.

Now consider the solution $\tau(t)$ for the initial value problem $\tau' = \operatorname{tg}^\kappa \tau$, $\tau(0) = r$. Note that $\tau(t)$ is also a gradient curve for the function $\operatorname{itg}^\kappa$ defined on $[0, \frac{\varpi^\kappa}{2})$. Direct calculations show that the composition $\alpha \circ \tau^{-1}$ is a (p,κ)-radial curve.

Uniqueness. Assume σ^1, σ^2 are two $(p,\varkappa)$-radial curves that start at x. Then the compositions $\sigma^i \circ \tau$ both give f-gradient curves. By Picard's theorem (see 16.15), we have $\sigma^1 \circ \tau \equiv \sigma^2 \circ \tau$. Therefore $\sigma^1(s) = \sigma^2(s)$ for any $s \geqslant r$ such that both sides are defined. □

I. Radial comparisons

In this section we show that radial curves behave in a comparison sense like unit-speed geodesics. Recall that notation $\tilde{\measuredangle}^\varkappa\{a;b,c\}$ is introduced in Section 1.A.

16.32. Radial monotonicity. *Let $\mathcal{L}$ be a complete length* CBB($\varkappa$) *space and let p, q be distinct points in $\mathcal{L}$. Let $\sigma\colon [s_{\min}, \frac{\varpi^\varkappa}{2}) \to \mathcal{L}$ be a $(p,\varkappa)$-radial curve. Then the function*

$$\psi\colon s \mapsto \tilde{\measuredangle}^\varkappa\{|q-\sigma(s)|;|p-q|,s\}$$

is nonincreasing in its domain of definition. Moreover, if $\psi(s)$ is undefined, then $|q-\sigma(s)| < s - |p-q|$.

If one extends the definition of $\tilde{\measuredangle}^\varkappa\{a;b,c\}$ by stating $\tilde{\measuredangle}^\varkappa\{a;b,c\} = 0$ if $a < |b-c|$ and $\tilde{\measuredangle}^\varkappa\{a;b,c\} = \pi$ if $a > b+c$. Then the radial monotonicity implies that ψ is a nonincreasing function defined in $[s_{\min}, \frac{\varpi^\varkappa}{2})$.

Radial monotonicity implies the following by straightforward calculations.

16.33. Corollary. *Let $\varkappa \leqslant 0$, let $\mathcal{L}$ be a complete* CBB($\varkappa$) *space, and let $p, q \in \mathcal{L}$. Let $\sigma\colon [s_{\min}, \infty) \to \mathcal{L}$ be a $(p,\varkappa)$-radial curve. Then for any $w \geqslant 1$, the function*

$$s \mapsto \tilde{\measuredangle}^\varkappa\{|q-\sigma(s)|;|p-q|, w\cdot s\}$$

is nonincreasing in its domain of definition.

16.34. Radial comparison. *Let $\mathcal{L}$ be a complete length* CBB($\varkappa$) *space and let $p \in \mathcal{L}$. Let $\rho\colon [r_{\min}, \frac{\varpi^\varkappa}{2}) \to \mathcal{L}$ and $\sigma\colon [s_{\min}, \frac{\varpi^\varkappa}{2}) \to \mathcal{L}$ be two $(p,\varkappa)$-radial curves. Let*

$$\phi_{\min} = \tilde{\measuredangle}^\varkappa\left(p\,{}^{\rho(r_{\min})}_{\sigma(s_{\min})}\right).$$

Then for any $r \in [r_{\min}, \frac{\varpi^\varkappa}{2})$ and $s \in [s_{\min}, \frac{\varpi^\varkappa}{2})$, we have

$$\tilde{\measuredangle}^\varkappa\{|\rho(r)-\sigma(s)|;r,s\} \leqslant \phi_{\min}, \tag{1}$$

if the left-hand side is defined. Moreover,

$$|\rho(r)-\sigma(s)| \leqslant \tilde{\gamma}^\varkappa\{\phi_{\min};r,s\}. \tag{2}$$

for all r, s.

We prove Theorems 16.32 and 16.34 simultaneously. The proof is an application of 13.24 plus trigonometric manipulations. We give a proof first in the simplest case $\varkappa = 0$, and then in the harder case $\varkappa \neq 0$. The arguments for both cases are nearly the same, but the case $\varkappa \neq 0$ requires an extra twist.

Proof of Theorems 16.32 and 16.34 in case $\varkappa = 0$. Set

$$\begin{aligned} R &= R(r) = |p - \rho(r)|, \\ S &= S(s) = |p - \sigma(s)|, \\ \ell &= \ell(r,s) = |\rho(r) - \sigma(s)|, \\ \phi &= \phi(r,s) = \tilde{\measuredangle}^0\{\ell(r,s); r, s\}. \end{aligned}$$

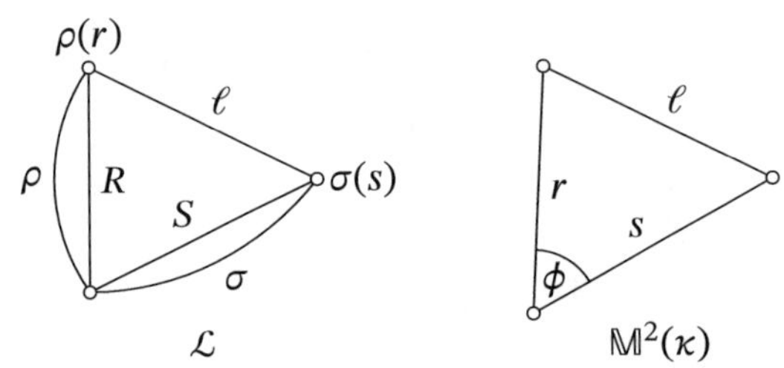

Therefore it will be sufficient to prove the following inequalities:

$(*)^0_\phi$ $$\frac{\partial^+}{\partial r}\phi(s_{\min}, r) \leqslant 0, \qquad \frac{\partial^+}{\partial s}\phi(s, r_{\min}) \leqslant 0$$

$(**)^0_\phi$ $$s \cdot \frac{\partial^+}{\partial s}\phi + r \cdot \frac{\partial^+}{\partial r}\phi \leqslant 0.$$

Indeed, from $((*)^0_\phi)$, we get that ψ is locally nonincreasing. If ψ is defined in $[s_{\min}, \frac{\varpi^\varkappa}{2})$, then it implies the radial monotonicity. Otherwise, note that ψ is continuous, and it is defined on a closed subset of $[s_{\min}, \frac{\varpi^\varkappa}{2})$. Suppose $[s_{\min}, s_{\max}]$ is a maximal interval where ψ is defined. Note that $\phi(s_{\max}) = 0$, therefore $|q - \sigma(s_{\max})| = |s_{\max} - |p - q||$. Note that $|q - \sigma(s_n)| < |s_n - |p - q||$ for some sequence $s_n \to s_{\max}+$. The triangle inequality and Proposition 16.30 imply that $|q - \sigma(s_n)| \geqslant -s_n + |p - q|$. Therefore $|q - \sigma(s_n)| < s_n - |p - q|$. By the triangle inequality $|q - \sigma(s)| < s - |p - q|$ for any $s > s_n$. Hence $\phi(s)$ is undefined for all $s > s_{\max}$, and the radial monotonicity follows.

Similarly, (1) follows from $((*)^0_\phi)$ and $((**)^0_\phi)$. Indeed, one can connect $(s_{\min}, r_{\min})$ and (s_0, r_0) in $[s_{\min}, \infty) \times [r_{\min}, \infty)$ by a concatenation of a coordinate segment and a segment defined by $r/s = r_0/s_0$ as in the figure. By $((*)^0_\phi)$ and $((**)^0_\phi)$, we have that ϕ does not increase while the pair (r, s) moves along this concatenation with nondecreasing r and s. Thus

$$\phi(r_0, s_0) \leqslant \phi(r_{\min}, s_{\min}) = \phi_{\min}.$$

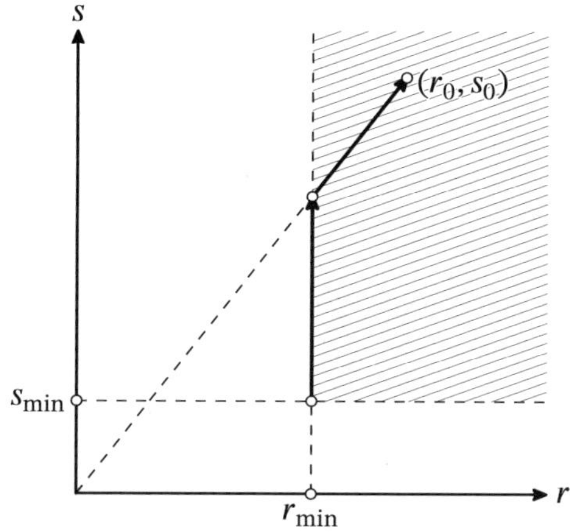

Finally, if $\phi(r_0, s_0)$ is defined, then (1) implies (2); otherwise, $\ell(r_0, s_0) < |r_0 - s_0|$, and (2) trivially holds.

Let us rewrite the inequalities $((*)^0_\phi)$ and $((**)^0_\phi)$ in an equivalent form:

$$(*)^0_\ell \qquad \begin{aligned} &\tfrac{\partial^+}{\partial s}\ell(s, r_{\min}) \leqslant \cos\tilde{\measuredangle}^0\{r_{\min}; s, \ell\},\\ &\tfrac{\partial^+}{\partial r}\ell(s_{\min}, r) \leqslant \cos\tilde{\measuredangle}^0\{s_{\min}; r, \ell\},\end{aligned}$$

$$(**)^0_\ell \qquad s\cdot\tfrac{\partial^+}{\partial s}\ell + r\cdot\tfrac{\partial^+}{\partial r}\ell \leqslant s\cdot\cos\tilde{\measuredangle}^0\{r; s, \ell\} + r\cdot\cos\tilde{\measuredangle}^0\{s; r, \ell\} = \ell.$$

Let

$$(A)^0 \qquad f = \tfrac{1}{2}\cdot\operatorname{dist}^2_p.$$

Clearly f is 1-concave, and

$$(B)^0 \qquad \rho^+(r) = \tfrac{1}{r}\cdot\nabla_{\rho(r)}f \quad\text{and}\quad \sigma^+(s) = \tfrac{1}{s}\cdot\nabla_{\sigma(s)}f.$$

Thus from Corollary 13.24, we have

$$(C)^0 \qquad \tfrac{\partial^+}{\partial r}\ell = -\tfrac{1}{r}\cdot\langle\nabla_{\rho(r)}f, \uparrow_{[\rho(r)\sigma(s)]}\rangle \leqslant \frac{\ell^2 + R^2 - S^2}{2\cdot\ell\cdot r}.$$

Since $R(r) \leqslant r$ and $S(s_{\min}) = s_{\min}$, we have

$$(D)^0 \qquad \begin{aligned}\tfrac{\partial^+}{\partial r}\ell(r, s_{\min}) &\leqslant \frac{\ell^2 + r^2 - s^2_{\min}}{2\cdot\ell\cdot r} =\\ &= \cos\tilde{\measuredangle}^0\{s_{\min}; r, \ell\},\end{aligned}$$

which is the first inequality in $((*)^0_\ell)$. By swapping ρ and σ, we obtain the second inequality in $((*)^0_\ell)$. Further, adding $((C)^0)$ and its mirror-inequality for $\tfrac{\partial^+}{\partial s}\ell$, we have

$$(E)^0 \qquad r\cdot\tfrac{\partial^+}{\partial r}\ell + s\cdot\tfrac{\partial^+}{\partial s}\ell \leqslant \frac{\ell^2 + R^2 - S^2}{2\cdot\ell} + \frac{\ell^2 + S^2 - R^2}{2\cdot\ell} = \ell,$$

which is $((**)^0_\ell)$. □

Remark. Alternatively, we could redefine $\cos\psi$ and $\cos\phi$ via the cosine law even if the corresponding comparison triangles do not exist. That is, set $(\cos\tilde{\measuredangle}^0)\{\ell;r,s\} := \frac{r^2+s^2-\ell^2}{2rs}$; it is always defined but might have absolute value bigger than 1. Then inequality $((*)^0_\ell)$ implies that $\cos\psi$ is nondecreasing on $[s_{\min},\frac{\varpi^\kappa}{2})$. This in particular shows that if $\cos\psi(s_0) > 1$ for some s_0, then $\cos\psi(s) > 1$ for $s \geqslant s_0$. Or, equivalently, if $|q-\sigma(s_0)| < s_0 - |p-q|$, then $|q-\sigma(s)| < s - |p-q|$ for $s \geqslant s_0$.

Similarly, inequalities $((*)^0_\ell)$ and $((**)^0_\ell)$ imply that $\cos\phi(r,s)$ is nondecreasing while the point (r,s) moves along this concatenation in the figure with nondecreasing r and s. Likewise, if at some point along this concatenation the triangle inequality fails, then it also fails for all later times.

Proof of 16.32 and 16.34 in case $\kappa \neq 0$. As before, let

$$R = R(r) = |p-\rho(r)|, \qquad \ell = \ell(r,s) = |\rho(r)-\sigma(s)|,$$
$$S = S(s) = |p-\sigma(s)|, \qquad \phi = \phi(r,s) = \tilde{\measuredangle}^\kappa\{\ell(r,s);r,s\}.$$

It suffices to prove the following three inequalities:

$$(*)^\pm_\phi \qquad \frac{\partial^+}{\partial r}\phi(s_{\min},r) \leqslant 0, \qquad \frac{\partial^+}{\partial s}\phi(s,r_{\min}) \leqslant 0,$$

$$(**)^\pm_\phi \qquad \operatorname{sn}^\kappa s\cdot \operatorname{cs}^\kappa S\cdot\frac{\partial^+}{\partial s}\phi + \operatorname{sn}^\kappa r\cdot \operatorname{cs}^\kappa R\cdot\frac{\partial^+}{\partial r}\phi \leqslant 0..$$

Then radial monotonicity follows from $((*)^\pm_\phi)$ the same way as in the $\kappa = 0$ case. The radial comparison follows from $((*)^0_\phi)$ and $((**)^\pm_\phi)$. Indeed, the functions $s \mapsto \operatorname{sn}^\kappa s\cdot\operatorname{cs}^\kappa S$ and $r \mapsto \operatorname{sn}^\kappa r\cdot\operatorname{cs}^\kappa R$ are Lipschitz. Thus there is a solution for the differential equation

$$(r',s') = (\operatorname{sn}^\kappa s\cdot\operatorname{cs}^\kappa S, \operatorname{sn}^\kappa r\cdot\operatorname{cs}^\kappa R)$$

with any initial data $(r_0,s_0) \in \left[r_{\min},\frac{\varpi^\kappa}{2}\right)\times\left[s_{\min},\frac{\varpi^\kappa}{2}\right)$. (Unlike the case $\kappa = 0$, the solution cannot be written explicitly.) Since $\operatorname{sn}^\kappa s\cdot\operatorname{cs}^\kappa S, \operatorname{sn}^\kappa r\cdot\operatorname{cs}^\kappa R > 0$, this solution $t \mapsto (r(t),s(t))$ must meet one of the coordinate rays $\{r_{\min}\}\times\left[s_{\min},\frac{\varpi^\kappa}{2}\right)$ or $\left[r_{\min},\frac{\varpi^\kappa}{2}\right)\times\{s_{\min}\}$. That is, one can connect the pair $(s_{\min},r_{\min})$ to (s_0,r_0) by a concatenation of a coordinate segment (vertical or horizontal) and part of the solution $(r(t),s(t))$. According to $((*)^\pm_\phi)$ and $((**)^\pm_\phi)$, the value of ϕ does not increase while the pair (r,s) moves along this concatenation in direction of increasing r and s. So, the same argument as in the $\kappa = 0$ case, implies (1) and (2).

As before, we rewrite the inequalities ($(*)^{\pm}_{\phi}$) and ($(**)^{\pm}_{\phi}$) in terms of ℓ:

$$(*)^{\pm}_{\ell} \qquad \begin{aligned} \tfrac{\partial^+}{\partial s}\ell(s, r_{\min}) &\leqslant \cos\tilde{\measuredangle}^{\kappa}\{r_{\min}; s, \ell\}, \\ \tfrac{\partial^+}{\partial r}\ell(s_{\min}, r) &\leqslant \cos\tilde{\measuredangle}^{\kappa}\{s_{\min}; r, \ell\}, \end{aligned}$$

$$(**)^{\pm}_{\ell} \qquad \begin{aligned} &\operatorname{sn}^{\kappa} s\cdot \operatorname{cs}^{\kappa} S\cdot \tfrac{\partial^+}{\partial s}\ell + \operatorname{sn}^{\kappa} r\cdot \operatorname{cs}^{\kappa} R\cdot \tfrac{\partial^+}{\partial r}\ell \leqslant \\ &\quad\leqslant \operatorname{sn}^{\kappa} s\cdot \operatorname{cs}^{\kappa} S\cdot \cos\tilde{\measuredangle}^{\kappa}\{r; s, \ell\} + \operatorname{sn}^{\kappa} r\cdot \operatorname{cs}^{\kappa} R\cdot \cos\tilde{\measuredangle}^{\kappa}\{s; r, \ell\}. \end{aligned}$$

Let

$$(A)^{\pm} \qquad f = -\tfrac{1}{\kappa}\cdot \operatorname{cs}^{\kappa}\circ \operatorname{dist}_p = \operatorname{md}^{\kappa}\circ \operatorname{dist}_p - \tfrac{1}{\kappa}.$$

Clearly $f'' + \kappa\cdot f \leqslant 0$ and

$$(B)^{\pm} \qquad \begin{aligned} \rho^+(r) &= \frac{1}{\operatorname{tg}^{\kappa} r\cdot \operatorname{cs}^{\kappa} R}\cdot \nabla_{\rho(r)} f, \\ \sigma^+(s) &= \frac{1}{\operatorname{tg}^{\kappa} s\cdot \operatorname{cs}^{\kappa} S}\cdot \nabla_{\sigma(s)} f. \end{aligned}$$

Thus from Corollary 13.24, we have

$$(C)^{\pm} \qquad \begin{aligned} \tfrac{\partial^+}{\partial r}\ell &= -\frac{1}{\operatorname{tg}^{\kappa} r\cdot \operatorname{cs}^{\kappa} R}\cdot \langle \nabla_{\rho(r)} f, \uparrow_{[\rho(r)\sigma(s)]}\rangle \leqslant \\ &\leqslant \frac{1}{\operatorname{tg}^{\kappa} r\cdot \operatorname{cs}^{\kappa} R}\cdot \frac{\operatorname{cs}^{\kappa} S - \operatorname{cs}^{\kappa} R\cdot \operatorname{cs}^{\kappa}\ell}{\kappa\cdot \operatorname{sn}^{\kappa}\ell} = \\ &= \frac{\frac{\operatorname{cs}^{\kappa} S}{\operatorname{cs}^{\kappa} R} - \operatorname{cs}^{\kappa}\ell}{\kappa\cdot \operatorname{tg}^{\kappa} r\cdot \operatorname{sn}^{\kappa}\ell}. \end{aligned}$$

Note that for all $\kappa \neq 0$, the function $x \mapsto \frac{1}{\kappa\cdot \operatorname{cs}^{\kappa} x}$ is increasing. Thus, since $R(r) \leqslant r$ and $S(s_{\min}) = s_{\min}$, we have

$$(D)^{\pm} \qquad \begin{aligned} \tfrac{\partial^+}{\partial r}\ell(r, s_{\min}) &\leqslant \frac{\frac{\operatorname{cs}^{\kappa} s_{\min}}{\operatorname{cs}^{\kappa} r} - \operatorname{cs}^{\kappa}\ell}{\kappa\cdot \operatorname{tg}^{\kappa} r\cdot \operatorname{sn}^{\kappa}\ell} = \\ &= \frac{\operatorname{cs}^{\kappa} s_{\min} - \operatorname{cs}^{\kappa}\ell\cdot \operatorname{cs}^{\kappa} r}{\kappa\cdot \operatorname{sn}^{\kappa} r\cdot \operatorname{sn}^{\kappa}\ell} = \\ &= \cos\tilde{\measuredangle}^{\kappa}\{s_{\min}; r, \ell\}, \end{aligned}$$

which is the first inequality in $((*)^{\pm}_{\ell})$ for $\kappa \neq 0$. By swapping ρ and σ, we obtain the second inequality in $((*)^{\pm}_{\ell})$. Further, adding $((C)^{\pm})$ and its mirror-inequality for $\frac{\partial^+}{\partial s}\ell$, we have

$(E)^{\pm}$

$$\begin{aligned}\operatorname{sn}^\kappa r\cdot \operatorname{cs}^\kappa R\cdot \tfrac{\partial^+}{\partial r}\ell + \operatorname{sn}^\kappa s\cdot \operatorname{cs}^\kappa S\cdot \tfrac{\partial^+}{\partial s}\ell &\leqslant \\ \leqslant \frac{\operatorname{cs}^\kappa S\cdot \operatorname{cs}^\kappa r - \operatorname{cs}^\kappa \ell\cdot \operatorname{cs}^\kappa R\cdot \operatorname{cs}^\kappa r}{\kappa\cdot \operatorname{sn}^\kappa \ell} &+ \frac{\operatorname{cs}^\kappa R\cdot \operatorname{cs}^\kappa s - \operatorname{cs}^\kappa \ell\cdot \operatorname{cs}^\kappa S\cdot \operatorname{cs}^\kappa s}{\kappa\cdot \operatorname{sn}^\kappa \ell} = \\ = \operatorname{sn}^\kappa r\cdot \operatorname{cs}^\kappa R\cdot \frac{\operatorname{cs}^\kappa s - \operatorname{cs}^\kappa \ell\cdot \operatorname{cs}^\kappa r}{\kappa\cdot \operatorname{sn}^\kappa r\cdot \operatorname{sn}^\kappa \ell} &+ \operatorname{sn}^\kappa s\cdot \operatorname{cs}^\kappa S\cdot \frac{\operatorname{cs}^\kappa r - \operatorname{cs}^\kappa \ell\cdot \operatorname{cs}^\kappa s}{\kappa\cdot \operatorname{sn}^\kappa s\cdot \operatorname{sn}^\kappa \ell} = \\ = \operatorname{sn}^\kappa r\cdot \operatorname{cs}^\kappa R\cdot \cos \tilde\measuredangle^\kappa\{r; s, \ell\} &+ \operatorname{sn}^\kappa s\cdot \operatorname{cs}^\kappa S\cdot \cos \tilde\measuredangle^\kappa\{s; r, \ell\},\end{aligned}$$

which is $((**)^{\pm}_{\ell})$. □

16.35. Exercise. Suppose $\mathcal{L}$ is a complete length CBB(κ) space and $x, y, z \in \mathcal{L}$. Assume $\tilde\measuredangle^\kappa\left(z\,{}^x_y\right) = \pi$. Show that there is a geodesic $[xy]$ that contains z. In particular, x can be connected to y by a minimizing geodesic.

J. Gradient exponent

Let $\mathcal{L}$ be a complete length CBB(κ) space, let $p \in \mathcal{L}$, and let $\xi \in \Sigma_p$. Consider a sequence of points $x_n \in \mathcal{L}$ such that $\uparrow_{[px_n]} \to \xi$. Let $r_n = |p - x_n|$, and let $\sigma_n : [r_n, \frac{\varpi^\kappa}{2}) \to \mathcal{L}$ be the (p,κ)-radial curve that starts at x_n.

By the radial comparison (see 16.34), the curves $\sigma_n : [r_n, \frac{\varpi^\kappa}{2}) \to \mathcal{L}$ converge to a curve $\sigma_\xi : (0, \frac{\varpi^\kappa}{2}) \to \mathcal{L}$, and this limit is independent of the choice of the sequence x_n. Let $\sigma_\xi(0) = p$, and if $\kappa > 0$ define

$$\sigma_\xi(\tfrac{\varpi^\kappa}{2}) = \lim_{t\to \frac{\varpi^\kappa}{2}} \sigma_\xi(t).$$

The resulting curve σ_ξ will be called the (p,κ)-*radial curve* in direction ξ.

The *gradient exponential map* $\operatorname{gexp}^\kappa_p : \overline{\mathrm{B}}[0, \frac{\varpi^\kappa}{2}]_{\mathrm{T}_p} \to \mathcal{L}$ is defined by

$$\operatorname{gexp}^\kappa_p : r\cdot\xi \mapsto \sigma_\xi(r).$$

Here are properties of radial curves reformulated in terms of the gradient exponential map:

16.36. Theorem. *Let $\mathcal{L}$ be a complete length* CBB(κ) *space. Then:*

(a) *If $p, q \in \mathcal{L}$ are points such that $|p - q| \leqslant \frac{\varpi^\kappa}{2}$, then for any geodesic $[pq]$ in $\mathcal{L}$ we have*

$$\operatorname{gexp}^\kappa_p(\log[pq]) = q.$$

(b) *For any* $v, w \in \overline{\mathrm{B}}[0, \frac{\varpi^\kappa}{2}]_{\mathrm{T}_p}$,

$$|\mathrm{gexp}_p^\kappa v - \mathrm{gexp}_p^\kappa w| \leqslant \tilde{\gamma}^\kappa\,[0\,{}^{v}_{w}]\,.$$

In other words, if we denote by $\mathcal{T}_p^\kappa$ *the set* $\overline{\mathrm{B}}[0, \frac{\varpi^\kappa}{2}]_{\mathrm{T}_p}$ *equipped with the metric* $|v - w|_{\mathcal{T}_p^\kappa} = \tilde{\gamma}^\kappa\,[0\,{}^{v}_{w}]$, *then*

$$\mathrm{gexp}_p^\kappa : \mathcal{T}_p^\kappa \to \mathcal{L}$$

is a short map.

(c) *Suppose* $p, q \in \mathcal{L}$ *and* $|p - q| \leqslant \frac{\varpi^\kappa}{2}$. *If* $v \in \mathrm{T}_p$, $|v| \leqslant 1$, *and*

$$\sigma(t) = \mathrm{gexp}_p^\kappa(t \cdot v),$$

then the function

$$s \mapsto \tilde{\measuredangle}^\kappa(\sigma|_0^s, q) := \tilde{\measuredangle}^\kappa\{|q - \sigma(s)|; |q - \sigma(0)|, s\}$$

is nonincreasing in its domain of definition.

Proof. Follows directly from the construction of gexp_p^κ and the radial comparison (see 16.34). □

Applying the theorem above together with Theorem 15.13(c), we obtain the following.

16.37. Corollary. *Let* p *be a point in an* m*-dimensional complete length* CBB(κ) *space* $\mathcal{L}$, $m < \infty$, *and* $0 < R \leqslant \frac{\varpi^\kappa}{2}$. *Then there is a short map* $f : \overline{\mathrm{B}}[R]_{\mathbb{M}^m(\kappa)} \to \mathcal{L}$ *such that* $\mathrm{Im}\, f = \overline{\mathrm{B}}[p, R] \subset \mathcal{L}$.

16.38. Exercise. Let $\mathcal{L} \subset \mathbb{E}^2$ be the Euclidean half-plane. Clearly $\mathcal{L}$ is a two-dimensional complete length CBB(0) space. Given a point $x \in \mathbb{E}^2$, denote by $\mathrm{proj}(x)$ the closest point to x on $\mathcal{L}$.

Apply the radial comparison (see 16.34) to show that for any interior point $p \in \mathcal{L}$ and any $v \in \mathbb{R}^2$ we have

$$\mathrm{gexp}_p v = \mathrm{proj}(p + v).$$

16.39. Exercise. Suppose x, p, and q are points in a complete length CBB(κ) space, and $x \in [pq[$. Show that there is a unique vector $v \in \mathrm{T}_p$ such that $\mathrm{gexp}_p v = x$.

K. Remarks

Gradient flow on Riemannian manifolds. The gradient flow for general semiconcave functions on smooth Riemannian manifolds can be introduced with much less effort. To do this note that the distance estimates proved in the Section 16.C can be proved in the same way for gradient curves of smooth semiconcave subfunctions. By the Greene–Wu lemma [**70**], given a λ-concave function f, a compact set $K \subset \operatorname{Dom} f$, and $\varepsilon > 0$ there is a smooth $(\lambda - \varepsilon)$-concave function that is ε-close to f on K. Hence one can apply smoothing and pass to the limit as $\varepsilon \to 0$. Note that by the second distance estimate 16.13, the limit curve obtained does not depend on the smoothing.

Gradient curves of a family of functions. Gradient flow can be extended to a family of functions. This type of flow was studied by Chanyoung Jun [**91, 92**], by Lucas Ferreira and Julio Valencia-Guevara [**65**], and by Alexander Mielke, Riccarda Rossi, and Giuseppe Savaré [**117**]. We will follow the simplified and generalized approach given by Alexander Lytchak and the third author [**107**], where an application related to this type of flow is given. The original motivation of Chanyoung Jun came from the study of pursuit-evasion problems. Another application of this type of flow comes from the fact that the optimal transport plan, or equivalently geodesics in the Wasserstein metric, can be described as gradient flow for a family of semiconcave functions. This observation was used by the third author to prove that Alexandrov spaces with nonnegative curvature have nonnegative Ricci curvature in the sense of Lott–Villani–Sturm [**130**].

Suppose that $\mathcal{Z}$ is either CBB or CAT. Let f_t be a family of functions defined on open subsets $\operatorname{Dom} f_t$ of $\mathcal{Z}$. More precisely, we assume that the parameter t lies in a real interval $\mathbb{I}$ and

$$\Omega = \{ (x,t) \in \mathcal{Z} \times \mathbb{I} : x \in \operatorname{Dom} f_t \}$$

is an open subset in $\mathcal{Z} \times \mathbb{I}$.

A family of functions f_t is called *Lipschitz* if the function $(x,t) \mapsto f_t(x)$ is L-Lipschitz for some constant L.

A family of functions f_t will be called *semiconcave* if the function $x \mapsto f_t(x)$ is λ-concave for each t. A family f_t is called *locally semiconcave* if for each $(p_0, t_0) \in \Omega$ there is a neighborhood Ω' and $\lambda \in \mathbb{R}$ such that the restriction of f_t to Ω' is semiconcave.

One cannot expect that a direct generalization of Definition 16.7 holds for every family of functions f_t; that is, gradient curves of a family f_t cannot be defined as curves satisfying the equation $\alpha^+ = \nabla_\alpha f$.

For example, consider a 1-Lipschitz curve α in the real line. It is reasonable to assume that α is an f_t-gradient curve for the family $f_t(x) = -|x - \alpha(t)|$. (Indeed α can be realized as a limit of gradient curves for a family of functions obtained by smoothing f_t.) On the other hand, $\alpha^+(t)$ might be undefined, and even if it is defined, in general $\alpha^+(t) \neq 0$, while $\nabla_{\alpha(t)} f_t \equiv 0$.

Instead we define an f_t-*gradient curve* as a Lipschitz curve α that satisfies the following inequality for any point p, time t, and small $\varepsilon > 0$:

$$\text{(1)} \qquad \operatorname{dist}_p \circ \alpha(t+\varepsilon) \leqslant \operatorname{dist}_p \circ \alpha(t) - \varepsilon \cdot \boldsymbol{d}_{\alpha(t)} f_t(\uparrow_{[\alpha(t)p]}) + o(\varepsilon).$$

If there is no geodesic $[\alpha(t)\,p]$, then we impose no condition.

If $\alpha^+(t) = \nabla_{\alpha(t)} f_t$ for all t, then (1) holds by the definition of gradient 13.17. On the other hand, the example above shows that the converse does not hold; that is, (1) generalizes Definition 16.7. The defining inequality (1) is closely related to the so-called *evolution variational inequality* [**19**, Theorem 4.0.4(iii)].

16.40. Distance estimate. *Let f_t and h_t be two Lipschitz families of λ-concave functions on a complete length space $\mathcal{Z}$, and $s \geqslant 0$. Suppose that $\mathcal{Z}$ is either* CBB *or* CAT. *Assume f_t and h_t have common domain $\Omega \subset \mathcal{Z} \times \mathbb{R}$, and $|f_t(x) - h_t(x)| \leqslant s$ for any $(x,t) \in \Omega$. Assume $t \mapsto \alpha(t)$ and $t \mapsto \beta(t)$ are f_t- and h_t-gradient curves, respectively, defined on a common interval $t \in [a,b)$, and let $\ell(t) = |\alpha(t) - \beta(t)|_{\mathcal{Z}}$. If for all t, a minimizing geodesic $[\alpha(t)\,\beta(t)]$ lies in $\{\, x \in \mathcal{Z} : (x,t) \in \Omega \,\}$, then*

$$\ell'(t) \leqslant \lambda \cdot \ell(t) + 2 \cdot s / \ell(t),$$

whenever the left-hand side is defined. Moreover,

$$\ell(t)^2 + \tfrac{2 \cdot s}{\lambda} \leqslant \big(\ell(a)^2 + \tfrac{2 \cdot s}{\lambda}\big) \cdot e^{2 \cdot \lambda \cdot (t-a)}.$$

In particular, these inequalities hold for any $t \in \mathbb{I}$ if $\Omega \supset B(p, 2 \cdot r) \times \mathbb{I}$ and $\alpha(t), \beta(t) \in B(p,r)$ for any $t \in \mathbb{I}$.

Note that if $f_t = h_t$, then $s = 0$; in this case the second inequality can be written as

$$\text{(2)} \qquad \ell(t) \leqslant \ell(a) \cdot e^{\lambda \cdot (t-a)}.$$

In particular, it implies uniqueness of the future of gradient curves with given initial data. This inequality also makes it possible to estimate the distance between two gradient curves for close functions. In particular, it implies convergence for f_t^n-gradient curves if a sequence of ℓ-Lipschitz and λ-concave families f_t^n converges uniformly as $n \to \infty$.

Proof of 16.40. Fix a time moment t and set $f = f_t$ and $h = h_t$. Let m be the midpoint of the geodesic $[\alpha(t)\beta(t)]$. Let $\gamma\colon [0, \ell(t)] \to \mathcal{Z}$ be an arclength parametrization of $[\alpha(t)\beta(t)]$. Note that $\boldsymbol{d}_{\alpha(t)} f(\uparrow_{[\alpha(t)m]})$ is the right derivative of $f \circ \gamma$ at 0 and $-\boldsymbol{d}_{\alpha(t)} h(\uparrow_{[\beta(t)m]})$ is the left derivative of $h \circ \gamma$ at $\ell(t)$. Since f and h are λ-concave,

$$f \circ \beta(t) \leqslant f \circ \alpha(t) + \ell(t) \cdot \boldsymbol{d}_{\alpha(t)} f(\uparrow_{[\alpha(t)m]}) + \tfrac{1}{2} \cdot \lambda \cdot \ell(t)^2,$$
$$h \circ \alpha(t) \leqslant h \circ \beta(t) + \ell(t) \cdot \boldsymbol{d}_{\alpha(t)} h(\uparrow_{[\beta(t)m]}) + \tfrac{1}{2} \cdot \lambda \cdot \ell(t)^2.$$

Adding these inequalities and taking into account $|f(x) - h(x)| < s$ for any x, we conclude that

$$\boldsymbol{d}_{\alpha(t)} f(\uparrow_{[\alpha(t)m]}) + \boldsymbol{d}_{\alpha(t)} h(\uparrow_{[\beta(t)m]}) \geqslant \lambda \cdot \ell(t) + 2 \cdot s/\ell(t).$$

Applying the triangle inequality and (1) at m, we obtain

$$\begin{aligned}\ell(t+\varepsilon) &= |\alpha(t+\varepsilon) - \beta(t+\varepsilon)| \leqslant \\ &\leqslant |\alpha(t+\varepsilon) - m| + |\beta(t+\varepsilon) - m| \leqslant \\ &\leqslant |\alpha(t) - m| - \varepsilon \cdot \boldsymbol{d}_{\alpha(t)} f(\uparrow_{[\alpha(t)m]}) + \\ &\quad + |\beta(t+\varepsilon) - m| - \varepsilon \cdot \boldsymbol{d}_{\beta(t)} h(\uparrow_{[\beta(t)m]}) + o(\varepsilon) = \\ &= \ell(t) - \varepsilon \cdot (\lambda \cdot \ell(t) + 2 \cdot s/\ell(t)) + o(\varepsilon)\end{aligned}$$

for $\varepsilon > 0$. The first inequality follows.

Since α and β are Lipschitz, $t \mapsto \ell(t)$ is a Lipschitz function. By Rademacher's theorem, its derivative ℓ' is defined almost everywhere and satisfies the fundamental theorem of calculus. Therefore the first inequality implies the second. ☐

16.41. Proposition. *Suppose $\mathcal{Z}$ is a complete length space that is either* CBB *or* CAT. *Let f_t be a family of λ-concave functions for $t \in [a, b)$, where* $\operatorname{Dom} f_t \supset$ $\supset B(z, 2 \cdot r)$ *for fixed $z \in \mathcal{Z}$, $r > 0$ and any t.*

Let $\alpha\colon [a, b) \to B(z, r)$ be Lipschitz. Then α is an f_t-gradient curve if and only if

$$\text{(3)}\quad \operatorname{dist}_p \circ \alpha(t+\varepsilon) \leqslant \operatorname{dist}_p \circ \alpha(t) - \varepsilon \cdot \left[\frac{f_t(p) - f_t \circ \alpha(t)}{|p - \alpha(t)|} - \tfrac{\lambda}{2} \cdot |p - \alpha(t)|\right] + o(\varepsilon)$$

for any $t \in [a, b)$ and $p \in B(z, r) \setminus \{\alpha(t)\}$.

Proof. Note that the geodesics $[\alpha(t)p]$ lie in $\operatorname{Dom} f_t$ for any t.

Since f_t is λ-concave, we have

$$\boldsymbol{d}_{\alpha(t)} f_t(\uparrow_{[\alpha(t)p]}) \geqslant \frac{f(p) - f \circ \alpha(t)}{|p - \alpha(t)|} - \tfrac{\lambda}{2} \cdot |p - \alpha(t)|.$$

Hence the only-if part follows.

Given $p \in \mathcal{Z}$ and t, consider a point $\bar{p} \in [\alpha(t)p]$. Applying (3) for $\bar{p}$, and the triangle inequality, we have

$$\operatorname{dist}_p \circ \alpha(t+\varepsilon) \leqslant \operatorname{dist}_p \circ \alpha(t) - \varepsilon \cdot \left[\frac{f(\bar{p}) - f \circ \alpha(t)}{|\bar{p} - \alpha(t)|} - \tfrac{\lambda}{2} \cdot |\bar{p} - \alpha(t)|\right] + o(\varepsilon).$$

By taking $\bar{p}$ close to $\alpha(t)$, the value $\frac{f(\bar{p}) - f \circ \alpha(t)}{|\bar{p} - \alpha(t)|} - \frac{\lambda}{2} \cdot |\bar{p} - \alpha(t)|$ can be made arbitrarily close to $\boldsymbol{d}_{\alpha(t)} f_t(\uparrow_{[\alpha(t)p]})$. Therefore, given $\delta > 0$, the inequality

$$\operatorname{dist}_p \circ \alpha(t+\varepsilon) \leqslant \operatorname{dist}_p \circ \alpha(t) - \varepsilon \cdot \boldsymbol{d}_{\alpha(t)} f_t(\uparrow_{[\alpha(t)p]}) + \varepsilon \cdot \delta$$

holds for all sufficiently small $\varepsilon > 0$. Therefore (1) holds. ☐

Now we are ready to formulate and prove global existence of gradient curves for time-dependent families—an analogue of 16.18.

16.42. Theorem. *Suppose $\mathcal{Z}$ is a complete length space that is either* CBB *or* CAT. *Let $\{f_t\}$ be a family of functions defined on an open set*

$$\Omega = \{\,(x,t) \in \mathcal{Z} \times \mathbb{R} \,:\, x \in \operatorname{Dom} f_t\,\}.$$

Suppose that f_t is Lipschitz and locally semiconcave. Then for any time a and initial point $p \in \operatorname{Dom} f_a$, there is a unique f_t-gradient curve $t \mapsto \alpha(t)$ defined on a maximal semiopen interval $[a,b)$. Moreover, if $b < \infty$, then $(\alpha(t),t)$ escapes from any closed set $K \subset \Omega$.

Proof. Let L be a Lipschitz constant of f_t. Fix $b > a$ sufficiently small that $\operatorname{Dom} f_t \supset B(p,\varepsilon\cdot L)$ for any $t \in [a,b)$. Consider a sequence $a = t_0 < t_1 < \dots$ $\dots < t_n = b$, and a piecewise constant family of functions on $B(p,\varepsilon\cdot L)$ defined by $\hat f_t = f_{t_i}$ if $t_i \leqslant t < t_{i+1}$.

Note that $\hat f_t$ is time-independent on each interval $[t_i,t_{i+1})$. By 16.18 applied recursively on each interval $[t_i,t_{i+1})$, the proposition holds for $\hat f_t$. That is, there is a unique $\hat f_t$-gradient curve $\hat\alpha$ that starts at p and is defined on the interval $[a,b)$.

The distance estimates 16.40 show that as the partition gets finer, the gradient curves $\hat\alpha$ form a Cauchy sequence; denote its limit by α. Then

$$\begin{aligned}\operatorname{dist}_p \circ \hat\alpha(t+\varepsilon) &\leqslant \operatorname{dist}_p \circ \hat\alpha(t) - \varepsilon\cdot\left[\frac{\hat f_t(p) - \hat f_t\circ\hat\alpha(t)}{|p-\alpha(t)|} - \tfrac{\lambda}{2}\cdot|p-\hat\alpha(t)|\right] + o(\varepsilon) \leqslant \\ &\leqslant \operatorname{dist}_p \circ \hat\alpha(t) - \\ &\quad - \varepsilon\cdot\left[\frac{f_t(p) - f_t\circ\hat\alpha(t) - 2\cdot s}{|p-\alpha(t)|} - \tfrac{\lambda}{2}\cdot|p-\hat\alpha(t)|\right] + o(\varepsilon),\end{aligned}$$

where

$$s = \sup_{t,x}\{|f_t(x) - \hat f_t(x)|\}.$$

Since $s \to 0$ as $\hat\alpha \to \alpha$, then (3) holds for α; that is, α is an f_t-gradient curve.

This proves short time existence. Applying this argument recursively, we can find a gradient curve defined on a maximal interval $[a,b)$. Uniqueness of this curve follows from the distance estimate (2).

Note that α is L-Lipschitz. In particular, if $b < \infty$, then $\alpha(t) \to p'$ as $t \to b$. If $(p',b) \in \Omega$, then we can repeat the procedure; otherwise α escapes from any closed set in Ω. ☐

Gradient curves for non-Lipschitz functions. In this book, we only consider gradient curves for locally Lipschitz semiconcave subfunctions; this turns out to be sufficient for all our needs. However, instead of Lipschitz semiconcave subfunctions, it is more natural to consider upper semicontinuous semiconcave functions with target in $[-\infty, \infty)$, and to assume in addition that the functions take finite values at a dense set in the domain of definition. Suppose that $\mathcal{Z}$ is a complete length space that is either CBB or CAT. The set of such subfunctions on $\mathcal{Z}$ will be denoted by $\mathrm{LCC}(\mathcal{Z})$ (for **l**ower semi**c**ontinous and semi**c**oncave).

In this section we describe the adjustments needed to construct gradient curves for the subfunctions in $\mathrm{LCC}(\mathcal{Z})$.

When $\mathcal{Z} = H$ is a Hilbert space, the theory we develop is equivalent to the classical theory of gradient flows on Hilbert space mentioned earlier [**33**].

Further examples of such functions include the entropy and other closely related functionals on the Wasserstein space over a CBB(0) space. Another important example is given by the Cheeger energy on metric measure spaces, its gradient flow leads to the notion of the heat flow on such spaces. The gradient flow for these functions plays an important role in the theory of optimal transport; see [**158**] and references therein.

Differential. First we need to extend the definition of differential (Definition 6.15) to LCC subfunctions.

Let $\mathcal{Z}$ be a complete length space and let $f \in \mathrm{LCC}(\mathcal{Z})$. Suppose that $\mathcal{Z}$ is either CBB or CAT. Given a point $p \in \mathrm{Dom}\, f$ and a geodesic direction $\xi = \uparrow_{[pq]}$, let $\hat{\boldsymbol{d}}_p f(\xi) = (f \circ \mathrm{geod}_{[pq]})^+(0)$. Since f is semiconcave, the value $\hat{\boldsymbol{d}}_p f(\xi)$ is defined if $f \circ \mathrm{geod}_{[pq]}(t)$ is finite at all sufficiently small values $t > 0$, but $\hat{\boldsymbol{d}}_p f(\xi)$ may take value ∞. Note that $\hat{\boldsymbol{d}}_p f$ is defined on a dense subset of Σ_p.

Let

$$\boldsymbol{d}_p f(\zeta) = \varlimsup_{\xi \to \zeta} \hat{\boldsymbol{d}}_p f(\xi),$$

and let $\boldsymbol{d}_p f(v) = |v| \cdot \boldsymbol{d}_p f(\xi)$ if $v = |v| \cdot \xi$ for some $\xi \in \Sigma_p$.

In other words, we define differential as the smallest upper semicontinuous positive-homogeneous function $\boldsymbol{d}_p f \colon \mathrm{T}_p \to \mathbb{R}$ such that if $\hat{\boldsymbol{d}}_p f(\xi)$ is defined, then $\boldsymbol{d}_p f(\xi) \geqslant \hat{\boldsymbol{d}}_p f(\xi)$.

Existence and uniqueness of the gradient. Note that in the proof of 13.20, we used the Lipschitz condition just once, to show that

$$s = \sup\{(\boldsymbol{d}_p f)(\xi) : \xi \in \Sigma_p\} = \varlimsup_{x \to p} \frac{f(x) - f(p)}{|x - p|} < \infty.$$

The value s above will be denoted by $|\nabla|_p f$. Note that if the gradient $\nabla_p f$ is defined, then $|\nabla|_p f = |\nabla_p f|$, and otherwise $|\nabla|_p f = \infty$.

Summarizing the discussion above, we have the following.

13.20′ Existence and uniqueness of the gradient. *Assume $\mathcal{Z}$ is a complete space and $f \in \mathrm{LCC}(\mathcal{Z})$. Suppose that $\mathcal{Z}$ is either* CBB *or* CAT. *Then for any point $p \in \operatorname{Dom} f$, either there is a unique gradient $\nabla_p f \in \mathrm{T}_p$ or $|\nabla|_p f = \infty$.*

Further, in all the results of Section 13.F we may assume only that both f and the gradient of f are defined at the points under consideration. The proofs are the same.

Sections 13.G–16.D require almost no changes. Mainly, where appropriate one needs to change $|\nabla_p f|$ to $|\nabla|_p f$ and/or assume that the gradient is defined at the points of interest. Also, (1) in Theorem 16.3 is taken as the definition of gradient-like curve. Then the theorem states that any gradient-like curve $\alpha\colon \mathbb{I} \to \mathcal{Z}$ satisfies Definition 16.2 at $t \in \mathbb{I}$ if $\nabla_{\hat{\alpha}(s)} f$ is defined. Further, Definition 16.7, should be changed to the following.

16.7′. Definition. *Let $\mathcal{Z}$ be a complete length space and $f \in \mathrm{LCC}(\mathcal{Z})$. Suppose that $\mathcal{Z}$ is either* CBB *or* CAT.

*A curve $\alpha\colon [t_{\min}, t_{\max}) \to \operatorname{Dom} f$ will be called an f-*gradient curve *if*

$$\alpha^+(t) = \nabla_{\alpha(t)} f$$

when $\nabla_{\alpha(t)} f$ is defined and

$$(f \circ \alpha)^+(t) = \infty$$

otherwise.

In the proof of local existence 16.16, condition (ii) should be changed to the following condition:

(ii)′ $f \circ \hat{\alpha}_n(\bar{\varsigma}_i) - f \circ \hat{\alpha}_n(\varsigma_i) > (\bar{\varsigma}_i - \varsigma_i) \cdot \max\{n, |\nabla|_{\hat{\alpha}_n(\varsigma_i)} f - \tfrac{1}{n})\}$.

Any gradient curve $\alpha[0, \ell) \to \mathcal{Z}$ for a subfunction $f \in \mathrm{LCC}(\mathcal{Z})$ satisfies the equation

$$\alpha^+(t) = \nabla_{\alpha(t)} f$$

at all values t, with the possible exception of $t = 0$. In particular, the gradient of f is defined at all points of any f-gradient curve, with the possible exception of the initial point.

16.43. Example. Let $\mathcal{X} = L^2(\mathbb{R}^n)$, and let $F\colon \mathcal{X} \to \mathbb{R}$ be given by $F(f) = -\int |\nabla f|^2 d\mathcal{L}$. Then F is a concave but not locally Lipschitz functional on $\mathcal{X}$ and it is finite precisely at functions $f \in W^{1,2}(\mathbb{R}^n) \subset L^2(\mathbb{R}^n)$. Integration by parts shows that for any smooth $f \in W^{1,2}(\mathbb{R}^n)$ it holds that $\nabla_f F = \Delta f$. The gradient flow of F is given by the heat flow f_t starting at f and f_t is smooth for all positive t.

Slower radial curves. Let $\kappa \geqslant 0$. Assume that for some function ψ, the curves defined by the equation

$$\sigma^+(s) = \psi(s, |p - \sigma(s)|) \cdot \nabla_{\sigma(s)} \operatorname{dist}_p$$

satisfy radial comparison Theorem 16.34. Then in fact the $\sigma(s)$ are radial curves; that is,

$$\psi(s, |p - \sigma(s)|) = \frac{\operatorname{tg}^\kappa |p - \sigma(s)|)}{\operatorname{tg}^\kappa s},$$

see Exercise 16.38.

In case $\kappa < 0$, such a function ψ is not unique. In particular, one can take curves defined by the simpler equation

$$\sigma^+(s) = \frac{\operatorname{sn}^\kappa |p - \sigma(s)|}{\operatorname{sn}^\kappa s} \cdot \nabla_{\sigma(s)} \operatorname{dist}_p = \frac{1}{\operatorname{sn}^\kappa s} \cdot \nabla_{\sigma(s)} (\operatorname{md}^\kappa \circ \operatorname{dist}_p).$$

Among all curves of that type, the radial curves for curvature κ as defined in Definition 16.27 maximize the growth of $|p - \sigma(s)|$.

Semisolutions

1.3. Suppose α is a closed spherical curve. By Crofton's formula, the length of α is $\pi\cdot n_\alpha$, where n_α denotes the average number of crossings of α with equators.

Since α is closed, almost all equators cross it at an even number of points (we assume that ∞ is an even number). If length $\alpha < 2\cdot\pi$, then $n_\alpha < 2$. Therefore there is an equator that does not cross α—hence the result.

2.9. (a). Note that any Cauchy sequence x_n in $(\mathcal{X}, \|*-*\|)$ is also Cauchy in $\mathcal{X}$. Since $\mathcal{X}$ is complete, x_n converges; denote its limit by x_∞.

Passing to a subsequence, we may assume that $\|x_{n-1}-x_n\| < \frac{1}{2^n}$. It follows that there is a 1-Lipschitz curve $\alpha\colon [0,1] \to (\mathcal{X}, \|*-*\|)$ such that $x_n = \alpha(\frac{1}{2^n})$ and $x_\infty = \alpha(0)$. In particular, $\|x_n - x_\infty\| \to 0$ and $n \to \infty$.

(b). Fix two points $x, y \in \mathcal{X}$ such that $\ell = \|x-y\| < \infty$. Let α_n be a sequence of paths from x to y such that length$(\alpha_n) \to \ell$ as $n \to \infty$. Without loss of generality, we may assume that each α_n is $(\ell+1)$-Lipschitz.

Since $\mathcal{X}$ is compact, there is a partial limit α_∞ of α_n as $n \to \infty$. By semicontinuity of length, length $\alpha_\infty \leqslant \ell$; that is; α is a shortest path in $\mathcal{X}$.

Source. Part (a) appears as a Corollary in [**84**]; see also [**128**, Lemma 2.3].

2.10. The following example was suggested by Fedor Nazarov [**121**].

Consider the unit ball (B, ρ_0) in the space c_0 of all sequences converging to zero equipped with the sup-norm.

Consider another metric ρ_1 which is different from ρ_0 by the conformal factor

$$\phi(\boldsymbol{x}) = 2 + \tfrac{1}{2}\cdot x_1 + \tfrac{1}{4}\cdot x_2 + \tfrac{1}{8}\cdot x_3 + \cdots,$$

where $\boldsymbol{x} = (x_1, x_2, \dots) \in B$. That is, if $\boldsymbol{x}(t)$, $t \in [0, \ell]$, is a curve parametrized by ρ_0-length, then its ρ_1-length is

$$\operatorname{length}_{\rho_1} \boldsymbol{x} = \int_0^\ell \phi \circ \boldsymbol{x}.$$

Note that the metric ρ_1 is bi-Lipschitz equivalent to ρ_0.

Assume $\boldsymbol{x}(t)$ and $\boldsymbol{x}'(t)$ are two curves parametrized by ρ_0-length that differ only in the mth coordinate; denote them by $x_m(t)$ and $x_m'(t)$, respectively. Note that if $x_m'(t) \leqslant x_m(t)$ for any t and the function $x_m'(t)$ is locally 1-Lipschitz at all t such that $x_m'(t) < x_m(t)$, then

$$\operatorname{length}_{\rho_1} \boldsymbol{x}' \leqslant \operatorname{length}_{\rho_1} \boldsymbol{x}.$$

Moreover this inequality is strict if $x_m'(t) < x_m(t)$ for some t.

Fix a curve $\boldsymbol{x}(t)$, $t \in [0, \ell]$, parametrized by ρ_0-length. We can choose m large so that $x_m(t)$ is sufficiently close to 0 for any t. In particular, for some values t, we have $y_m(t) < x_m(t)$, where

$$y_m(t) = (1 - \tfrac{t}{\ell})\cdot x_m(0) + \tfrac{t}{\ell}\cdot x_m(\ell) - \tfrac{1}{100}\cdot \min\{t, \ell - t\}.$$

Consider the curve $\boldsymbol{x}'(t)$ as above with

$$x_m'(t) = \min\{x_m(t), y_m(t)\}.$$

Note that $\boldsymbol{x}'(t)$ and $\boldsymbol{x}(t)$ have the same endpoints, and by the above

$$\operatorname{length}_{\rho_1} \boldsymbol{x}' < \operatorname{length}_{\rho_1} \boldsymbol{x}.$$

That is, for any curve $\boldsymbol{x}(t)$ in (B, ρ_1), we can find a shorter curve $\boldsymbol{x}'(t)$ with the same endpoints. In particular, (B, ρ_1) has no geodesics.

2.11. Choose a sequence of positive numbers $\varepsilon_n \to 0$ and an ε_n-net N_n of K for each n. Assume N_0 is a one-point set, so $\varepsilon_0 > \operatorname{diam} K$. Connect each point $x \in N_{k+1}$ to a point $y \in N_k$ by a curve of length at most ε_k.

Consider the union K' of all these curves with K; observe that K' is compact and path-connected.

Source: This problem was suggested by Eugene Bilokopytov [**28**].

2.14. Choose two points $x, y \in \mathcal{X}$; let $\ell = |x - y|$. Suppose $f\colon E \to \mathcal{X}$ is a distance-preserving map such that $0, \ell \in E \subset [0, \ell]$, $f(0) = x$, and $f(\ell) = y$.

Show that we can choose f so that E is maximal; that is, f cannot be extended to a distance-preserving map on a larger subset of $[0, \ell]$.

Show that there is no open interval (a, b) in the complement of E such that $a, b \in E$.

Apply the completeness of $\mathcal{X}$ to show that E is closed. Conclude that $E = [0,\ell]$.

2.17. Consider the following subset of $\mathbb{R}^2$ equipped with the induced length metric

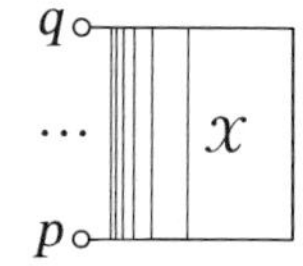

$$\mathcal{X} = \big((0,1]\times\{0,1\}\big)\cup\big(\{1,\tfrac{1}{2},\tfrac{1}{3},\dots\}\times[0,1]\big).$$

Note that $\mathcal{X}$ is locally compact and geodesic.

Its completion $\bar{\mathcal{X}}$ is isometric to the closure of $\mathcal{X}$ equipped with the induced length metric; $\bar{\mathcal{X}}$ is obtained from $\mathcal{X}$ by adding two points $p = (0,0)$ and $q = (0,1)$.

The point p admits no compact neighborhood in $\bar{\mathcal{X}}$ and there is no geodesic connecting p to q in $\bar{\mathcal{X}}$.

Source: This example is taken from [**34**].

2.23. Let $\mathcal{X}$ be a compact metric space. Let us identify $\mathcal{X}$ with its image in $\mathrm{Bnd}(\mathcal{X},\mathbb{R})$ under the Kuratowsky embedding (Section 2.G). Denote by $\mathcal{K}$ the *linear* convex hull of $\mathcal{X}$ in the space of bounded functions on $\mathcal{X}$; that is, $x\in\mathcal{K}$ if and only if x cannot be separated from $\mathcal{X}$ by a hyperplane.

Since $\mathcal{X}$ is compact, so is $\mathcal{K}$. It remains to observe that $\mathcal{K}$ is a length space since it is convex.

Remark. Alternatively, one can use the embedding of $\mathcal{X}$ into its injective hull; see [**86**].

4.2. Let $F = \{\,n\in\mathbb{N} : f(n) = n\,\}$; we need to show that $\omega(F) = 1$.

Consider an oriented graph Γ with vertex set $\mathbb{N}\setminus F$ such that m is connected to n if $f(m) = n$. Show that each connected component of Γ has at most one cycle. Use it to subdivide vertices of Γ into three sets S_1, S_2, and S_3 such that $f(S_i)\cap S_i = \varnothing$ for each i.

Conclude that $\omega(S_1) = \omega(S_2) = \omega(S_3) = 0$ and hence

$$\omega(F) = \omega(\mathbb{N}\setminus(S_1\cup S_2\cup S_3)) = 1.$$

Source: The presented proof was given by Robert Solovay [**151**], but the key statement is due to Miroslav Katětov [**94**].

4.6. Choose a nonprincipal ultrafilter ω and set $L(\boldsymbol{s}) = s_\omega$. It remains to observe that L is linear.

Remark. By this exercise, ω corresponds to a vector in $(\ell^\infty)^*\setminus\ell^1$.

4.7. Use Exercise 4.2.

4.10. (a). Show that there is $\delta > 0$ such that sides of any geodesic triangle in $\mathbb{M}^2(1)$ intersect a disk of radius δ. Observe that $\mathbb{M}^2(n) = \frac{1}{\sqrt{n}}\cdot\mathbb{M}^2(1)$, and use it to show that any geodesic triangle in $\mathcal{T}$ is a tripod.

(b). Observe and use that $\mathbb{M}^2(n)$ are homogeneous.

(c). Choose $p_1 \in \mathbb{M}^2(1)$, denote by p_n the corresponding point in $\mathbb{M}^2(n) = \frac{1}{\sqrt{n}}\cdot\mathbb{M}^2(1)$. Suppose $p_n \to p_\omega$ as $n \to \omega$; we can assume that $p_\omega \in \mathcal{T}$. By (b), it is sufficient to show that p_ω has a continuum degree.

Choose distinct geodesics $\gamma_1, \gamma_2 \colon [0, \infty) \to \mathbb{M}^2(1)$ that start at a point p_1. Show that the limits of γ_1 and γ_2 run in the different connected components of $\mathcal{T} \setminus \{p_\omega\}$. Since there is a continuum of distinct geodesics starting at p, we get that the degree of p_ω is at least continuum.

On the other hand, the set of sequences of points $q_n \in \mathbb{M}^2(n)$ has cardinality continuum. In particular, the set of points in $\mathcal{T}$ has cardinality at most continuum. It follows that the degree of any vertex is at most continuum.

Remark. The properties (b) and (c) describe the tree $\mathcal{T}$ up to isometry [**61**]. In particular, $\mathcal{T}$ does not depend on the choice of the ultrafilter.

4.13. Show and use that the spaces $\mathcal{X}^\omega$ and $(\mathcal{X}^\omega)^\omega$ have discrete metrics and both have cardinality of the continuum.

4.14. Choose a bijection $\iota \colon \mathbb{N} \to \mathbb{N} \times \mathbb{N}$. Given a set $S \subset \mathbb{N}$, consider the sequence $S_1, S_2, \ldots$ of subsets in $\mathbb{N}$ defined by $m \in S_n$ if $(m, n) = \iota(k)$ for some $k \in S$. Set $\omega_1(S) = 1$ if and only if $\omega(S_n) = 1$ for ω-almost all n. It remains to check that ω_1 meets the conditions of the exercise.

Comment. It turns out that $\omega_1 \neq \omega$ for any ι; see the post of Andreas Blass [**30**].

4.15. Consider the infinite metric $\mathcal{T}$ tree with unit edges shown on the diagram. Observe that $\mathcal{T}$ is proper.

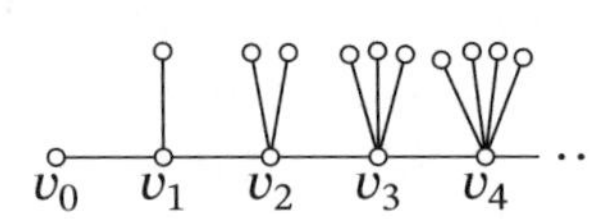

Consider the vertex $v_\omega = \lim_{n\to\omega} v_n$ in the ultrapower $\mathcal{T}^\omega$. Observe that ω has an infinite degree. Conclude that $\mathcal{T}^\omega$ is not locally compact.

4.17. Let $\mathcal{X}_n$ be the square $\{(x, y) \in \mathbb{R}^2, |x| \leqslant 1, |y| \leqslant 1\}$ with the metric induced by the ℓ^n-norm and let $f_n(x, y) = x$ for all n. Observe that $\mathcal{X}_\omega$ is the square with the metric induced by the ℓ^∞-norm where the limit function $f_\omega(x, y) = x$ is not concave.

5.13. (a). Suppose $\mathcal{X}_n \xrightarrow[\text{GH}]{} \mathcal{X}$ and $\mathcal{X}_n$ are simply connected length metric spaces. It is sufficient to show that any nontrivial covering map $f \colon \tilde{\mathcal{X}} \to \mathcal{X}$ corresponds to a nontrivial covering map $f_n \colon \tilde{\mathcal{X}}_n \to \mathcal{X}_n$ for large n.

The latter can be constructed by covering $\mathcal{X}_n$ by small balls that lie close to sets in $\mathcal{X}$ evenly covered by f, preparing a few copies of these sets and gluing them in the same way as the inverse images of the evenly covered sets in $\mathcal{X}$ are glued to obtain $\tilde{\mathcal{X}}$.

(b). Let $\mathcal{V}$ be a cone over Hawaiian earrings. Consider the *doubled cone* $\mathcal{W}$—two copies of $\mathcal{V}$ with their base points glued (see the diagram).

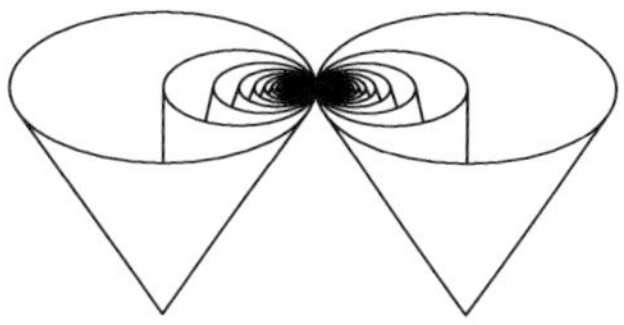

The space $\mathcal{W}$ can be equipped with a length metric (for example, the induced length metric from the embedding shown).

Show that $\mathcal{V}$ is simply connected, but $\mathcal{W}$ is not; the latter is a good exercise in topology.

If we delete from the earrings all small circles, then the obtained double cone becomes simply connected and it remains close to $\mathcal{W}$. That is, $\mathcal{W}$ is a Gromov–Hausdorff limit of simply connected spaces.

Remark. Note that from part (b), the limit does not admit a nontrivial covering. So, if we define the fundamental group as the inverse image of groups of deck transformations for all the coverings of the given space, then one may say that a Gromov–Hausdorff limit of simply connected length spaces is simply connected.

5.14. (a). Suppose that a metric on $\mathbb{S}^2$ is close to the disk $\mathbb{D}^2$. Note that $\mathbb{S}^2$ contains a circle γ that is close to the boundary curve of $\mathbb{D}^2$. By the Jordan curve theorem, γ divides $\mathbb{S}^2$ into two disks, say D_1 and D_2.

By Exercise 5.13(a), the Gromov–Hausdorff limits of D_1 and D_2 have to contain the whole $\mathbb{D}^2$, otherwise the limit would admit a nontrivial covering.

Consider points $p_1 \in D_1$ and $p_2 \in D_2$ that are close to the center of $\mathbb{D}^2$. If n is large, the distance $|p_1 - p_2|_n$ has to be very small. On the other hand, any curve from p_1 to p_2 must cross γ, so it has length about 2—a contradiction.

(b). Make holes in the unit 3-disc, that do not change its topology and do not change its length metric much and pass to its doubling in the boundary.

Source: The exercise is taken from [**37**].

5.15. Modify proof of Theorem 5.12, or apply Theorem 5.16(b).

6.6. If $\measuredangle\left[p\,{}^x_z\right] + \measuredangle\left[p\,{}^y_z\right] < \pi$, then by the triangle inequality for angles 6.5 we have $\measuredangle\left[p\,{}^x_y\right] < \pi$. The latter implies that $[xy]$ fails to be minimizing near p.

6.10. By the definition of a right derivative, there is a geodesic γ such that both limits

$$\overline{\lim_{\varepsilon\to 0+}}\ \frac{|\alpha(\varepsilon)-\gamma(\varepsilon)|_{\mathcal{X}}}{\varepsilon} \quad\text{and}\quad \overline{\lim_{\varepsilon\to 0+}}\ \frac{|\beta(\varepsilon)-\gamma(\varepsilon)|_{\mathcal{X}}}{\varepsilon}$$

are arbitrarily small. By the triangle inequality, we get

$$\varlimsup_{\varepsilon\to 0+} \frac{|\alpha(\varepsilon)-\beta(\varepsilon)|_{\mathcal{X}}}{\varepsilon} = 0.$$

6.13. Apply the definition.

6.14. Observe that

$$\operatorname{speed}_t \alpha = |\alpha^+(t)| = |\alpha^-(t)|.$$

Apply Theorem 3.10 to show that

$$|\alpha^+(t) - \alpha^-(t)|_{\mathrm{T}_{\alpha(t)}} = 2\cdot \operatorname{speed}_t \alpha.$$

7.10. Choose two non-Euclidean norms $\|{*}\|_{\mathcal{X}}$ and $\|{*}\|_{\mathcal{Y}}$ on $\mathbb{R}^{10}$ such that the sum $\|{*}\|_{\mathcal{X}} + \|{*}\|_{\mathcal{Y}}$ is Euclidean. See **[147]** for more details.

8.3. Assume $|p - x^i| = |q - y^i|$ for each i. Observe and use that

$$|x^i - x^j| \leqslant |y^i - y^j| \quad\Longleftrightarrow\quad \tilde\measuredangle^\kappa\left(p\,{}^{x^i}_{x^j}\right) \leqslant \tilde\measuredangle^\kappa\left(q\,{}^{y^i}_{y^j}\right).$$

8.3. Apply the four-point comparison 8.1.

8.8. Modify the induced length metric on the unit sphere in an infinite-dimensional Hilbert space in small neighborhoods of a countable collection of points. To prove that the obtained space is CBB(0), you may need to use the technique from Halbeisen's example 13.6.

8.12. Mimic the proof of Theorem 8.11.

8.13. On the plane, any nonnegatively curved metric having an everywhere dense set of singular points will do the job, where by singular point we mean a point having total angle around it strictly smaller than $2\cdot\pi$.

Indeed, if x_i is a singular point, then there is $0 < \varepsilon_i < 1/20$ such that no geodesic with ends outside of $\mathrm{B}(x_i, r)$ can meet the ball $\mathrm{B}(x_i, \varepsilon_i\cdot r)$. The set

$$\Omega_n = \bigcup_i \mathrm{B}(x_i, \tfrac{\varepsilon_i}{n})$$

is open and everywhere dense. Note that Ω_n may intersect a geodesic of length $1/n$ only within $\frac{1}{10n}$ of its endpoints. The intersection of the Ω_n is a G-delta dense set that does not intersect the interior of any geodesic.

8.15. Note that rescaling does not change the space. Therefore if the space is CBB(κ), then it is CBB($\lambda\cdot\kappa$) for any $\lambda > 0$. Passing to the limit as $\lambda \to 0$, we may assume that the space is CBB(0).

The point-on-side comparison (see 8.14(b)) for $p = v$, $x = w$, $y = -w$, and $z = 0$ implies that

$$\|v + w\|^2 + \|v - w\|^2 \leqslant 2\cdot\|v\|^2 + 2\cdot\|w\|^2.$$

Applying the comparison for $p = v + w$, $x = w - v$, $y = v - w$ and $z = 0$ gives the opposite inequality. That is, the parallelogram identity

$$\|v + w\|^2 + \|v - w\|^2 = 2\cdot\|v\|^2 + 2\cdot\|w\|^2$$

holds for any vectors v and w. Whence the statement follows.

8.16. Apply the hinge comparison (see 8.14(c)).

8.18. Without loss of generality, we may assume that the points x, v, w, y appear on the geodesic $[xy]$ in that order. By the point-on-side comparison (see 8.14(b)) we have

$$\begin{aligned}\tilde\measuredangle^\kappa(x\,{}^{y}_{p}) \leqslant \tilde\measuredangle^\kappa(x\,{}^{w}_{p}) &\leqslant \tilde\measuredangle^\kappa(x\,{}^{v}_{p}),\\ \tilde\measuredangle^\kappa(y\,{}^{w}_{p}) \geqslant \tilde\measuredangle^\kappa(y\,{}^{v}_{p}) &\geqslant \tilde\measuredangle^\kappa(y\,{}^{x}_{p}).\end{aligned}$$

Therefore

$$\begin{aligned}\tilde\measuredangle^\kappa(x\,{}^{y}_{p}) < \tilde\measuredangle^\kappa(x\,{}^{w}_{p}) &\implies \tilde\measuredangle^\kappa(x\,{}^{y}_{p}) < \tilde\measuredangle^\kappa(x\,{}^{v}_{p}),\\ \tilde\measuredangle^\kappa(y\,{}^{x}_{p}) < \tilde\measuredangle^\kappa(y\,{}^{w}_{p}) &\impliedby \tilde\measuredangle^\kappa(y\,{}^{x}_{p}) < \tilde\measuredangle^\kappa(y\,{}^{v}_{p}).\end{aligned}$$

By Alexandrov's lemma (Lemma 6.3), we have

$$\begin{aligned}\tilde\measuredangle^\kappa(x\,{}^{y}_{p}) < \tilde\measuredangle^\kappa(x\,{}^{v}_{p}) &\iff \tilde\measuredangle^\kappa(y\,{}^{x}_{p}) < \tilde\measuredangle^\kappa(y\,{}^{v}_{p}),\\ \tilde\measuredangle^\kappa(x\,{}^{y}_{p}) < \tilde\measuredangle^\kappa(x\,{}^{w}_{p}) &\iff \tilde\measuredangle^\kappa(y\,{}^{x}_{p}) < \tilde\measuredangle^\kappa(y\,{}^{w}_{p}).\end{aligned}$$

Hence the statement follows.

8.19. See the construction of Urysohn's space [**71**, $3.11\frac{3}{2}{}_{+}$] or [**134**].

8.20. Read [**101**].

8.21. Apply the angle-sidelength monotonicity (see 8.17) twice.

8.22. The first part follows from the angle-sidelength monotonicity 8.17. An example for the second part can be found among metrics on $\mathbb{R}^2$ induced by a norm. (Compare to Exercise 8.15.)

Remark. This exercise is inspired by Busemann's definition [**45**].

8.25 and 9.28. (a) By the function comparison definitions of CBB(0) space (see 8.23(b)), for any $p \in \mathcal{L}$ and $\varepsilon > 0$ the function dist_p is ε-concave everywhere sufficiently far from p. Applying the definition of Busemann function, we get the result.

The CAT(0) case is analogous; we have to apply Theorem 9.25(b) and use ε-convexity.

(b). By the definition of Busemann function (see 6.1),

$$\begin{aligned}\exp(\operatorname{bus}_\gamma) &= \lim_{t\to\infty} \exp(\operatorname{dist}_{\gamma(t)} - t) = \\ &= \lim_{t\to\infty} \left[\exp(\operatorname{dist}_{\gamma(t)} - t) + \exp(-\operatorname{dist}_{\gamma(t)} - t)\right] = \\ &= \lim_{t\to\infty} \left(2\cdot \exp(-t)\cdot \cosh\circ \operatorname{dist}_{\gamma(t)}\right).\end{aligned}$$

By the function comparison definitions of CAT(κ) space (see 9.25(b)) or CBB(κ) space (see 8.23(b)), for any $p \in \mathcal{U}$ the function $f = \cosh\circ\operatorname{dist}_p$ satisfies $f'' + \kappa\cdot f \geqslant 1$ (respectively, $f'' + \kappa\cdot f \leqslant 1$). The result follows.

8.32. Read [**129**].

8.46. If $\operatorname{diam}(\mathcal{L}/G) > \frac{\pi}{2}$, then for some $x \in \mathcal{L}$ we have

$$\sup\{\operatorname{dist}_{G\cdot x}(y) : y \in \mathcal{L}\} > \tfrac{\pi}{2}.$$

Use comparison to show that there is a unique point y^* that lies at maximal distance from the orbit $G\cdot x$. Observe that y^* is a fixed point.

8.47. Assume there are four such points x_1, x_2, x_3, x_4. Since the space $\mathcal{L}$ is CBB(1), it is also CBB(0). By the angle comparison, the sum of the angles in any geodesic triangle in an CBB(0) space is $\geqslant \pi$. Therefore the average of the $\measuredangle\left[x_i\,{}^{x_j}_{x_k}\right]$ is larger than $\frac{\pi}{3}$. On the other hand, since each x_i has space of directions $\leqslant \frac{1}{2}\cdot\mathbb{S}^n$ and the perimeter of any triangle in $\frac{1}{2}\cdot\mathbb{S}^n$ is at most π, the average of $\measuredangle\left[x_i\,{}^{x_j}_{x_k}\right]$ is at most $\frac{\pi}{3}$—a contradiction.

Source: Based on the main idea in [**83**].

9.4. Suppose that

$$\tilde\measuredangle^\kappa\left(x^0\,{}^{x^1}_{x^2}\right) + \tilde\measuredangle^\kappa\left(x^0\,{}^{x^2}_{x^3}\right) < \tilde\measuredangle^\kappa\left(x^0\,{}^{x^1}_{x^3}\right).$$

Show that

$$\tilde\measuredangle^\kappa\left(x^2\,{}^{x^0}_{x^1}\right) + \tilde\measuredangle^\kappa\left(x^2\,{}^{x^1}_{x^3}\right) + \tilde\measuredangle^\kappa\left(x^2\,{}^{x^3}_{x^0}\right) > 2\cdot\pi.$$

Conclude that one can take $p = x^2$.

9.5. Modify the configuration in Reformulation 9.2.

9.6. Read [**146**]; the original proof [**25**] is harder to follow.

An example for the second part of the problem can be found among 4-point metric spaces. It is sufficient to take four vertices of a generic convex quadrangle and increase one of its diagonals slightly; it will still satisfy the inequality for all relabeling but will fail to meet 9.2.

9.9. Suppose that a geodesic $[px]$ is not extendable beyond x. We may assume that $|p-x| < \varpi^\kappa$; otherwise move p along the geodesic toward x.

By the uniqueness of geodesics (see 9.8), any point y in a neighborhood $\Omega \ni x$ is connected to p by a unique geodesic path; denote it by γ_y. Note that $h_t(y) = \gamma_y(t)$ defines a homotopy, called the *geodesic homotopy*, between the identity map of Ω and the constant map with value p.

Since $[px]$ is not extendable, $x \notin h_t(\Omega)$ for any $t < 1$. In particular, the local homology groups vanish at x—a contradiction.

9.10. Choose a sequence of directions ξ_n at p; by $\gamma_n : \mathbb{R} \to \mathcal{U}$ the corresponding local geodesics. Since the space $\mathcal{U}$ is locally compact, we may pass to a converging subsequence of (γ_n); its limit is a local geodesic by Corollary 9.22. Denote the limit by γ_∞ and its direction by ξ_∞. By comparison, ξ_∞ is a limit of (ξ_n).

9.16. Follow the solution in the Exercise 8.15, reversing all the inequalities.

9.17. It is sufficient to show that if v and y are midpoints of geodesics $[uw]$ and $[xz]$ in $\mathcal{U}$, then

$$|v-y| \leqslant \tfrac{1}{2}\cdot(|u-x| + |w-z|).$$

Denote by p the midpoint of $[uz]$. Applying the angle-sidelength monotonicity 9.15 twice, we have

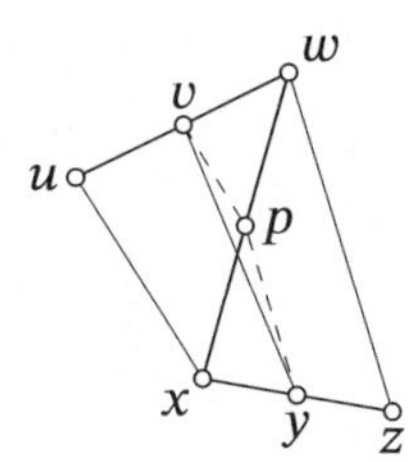

$$|v-p| \leqslant \tfrac{1}{2}\cdot|w-z|.$$

Similarly we have

$$|y-p| \leqslant \tfrac{1}{2}\cdot|u-x|.$$

It remains to add these two inequalities and apply the triangle inequality.

Remark. This inequality also follows directly from the majorization theorem (see 9.56).

9.20. The only-if part is evident. Use Theorem 9.14 to show that (c) ⇒ (b) ⇒ ⇒ (a). By 9.14, condition (a) implies that the natural map is distance-preserving on the sides $[\tilde{x}\tilde{y}]$ and $[\tilde{x}\tilde{z}]$. Applying it again, we have that condition (a) holds for all permutations of the labels x, y, z. Whence the natural map is distance-preserving on all three sides.

Remark. These conditions imply that the natural map can be extended to a distance-preserving map to the solid model triangle. In fact the image of the line-of-sight map (Definition 9.32) is isometric to the model triangle.

9.28. See the solution of Exercise 8.25.

9.33. (a) Suppose that $x_n \to x_\infty$, $y_n \to y_\infty$ as $n \to \infty$, but $[x_n y_n]$ does not converge to $[x_\infty y_\infty]$. Since the space is proper, we can pass to a subsequence such that $[x_n y_n]$ converges to another geodesic. That is, we have at least two geodesics between x_∞ and y_∞.
(b). Let Δ_n be a sequence of solid spherical triangles with angle $\frac{\pi}{4}$ and adjacent sides $\pi - \frac{1}{n}$. Let us glue each Δ_n to $[0, \pi]$ along an isometry of one of the longer sides. It remains to show that the obtained space $\mathcal{X}$ is a needed example.
Source: The example (b) is taken from [**34**, Chapter I, Exercise 3.14].

9.41. Subdivide Q into a a half-plane A bounded by the extension of γ_1 and the remaining solid angle B; it has angle measure measure $\pi - \alpha$. First glue B along γ_2, and then glue A. Each time apply the Reshetnyak gluing theorem 9.39, to show that the obltained space is CAT(0).

9.42. Suppose that A is not convex. Then there is a geodesic $[xy]$ with ends in A that does not lie in A completely. Note that $[xy]$ can be lifted to two different geodesics with the same ends in the doubling, and apply uniqueness of geodesics (Corollary 9.13).

9.43. Since K is π-convex, it is CAT(1). By Theorem 11.7, the spherical suspension $\operatorname{Susp} K$ is CAT(1) as well. Let us glue $\operatorname{Susp} K$ to $\mathcal{U}$ along K; according to the Reshetnyak gluing theorem, the resulting space, say $\mathcal{U}'$, is CAT(1).

Consider the geodesic path $\gamma\colon [0,1]$ from p to a pole of the suspension in $\mathcal{U}'$. Set $K_t = \mathcal{U} \cap \overline{\mathrm{B}}[\gamma(t), \frac{\pi}{2}]$. By Corollary 9.27, K_t is π-convex for any t; monotonicity and continuity of the family should be evident.
Source: This construction was used in [**107**]. Applying it together with Sharafutdinov retraction leads to another solution of Exercise 9.76.

9.44. Apply the Reshetnyak gluing theorem or its reformulation 9.40.

9.59. By Theorem 9.56, there is a majorization $F : D \to \mathcal{U}$ of the polygonal line β. Show and use that D is a convex plane polygon and its external angles cannot exceed the corresponding external angle of β.

9.60. This exercise generalizes the so-called Fáry–Milnor theorem. An elementary proof is given by the first author and Richard Bishop [**4**]; another proof is given by Stephan Stadler [**152**].

9.61. *Easier way.* Let $(t,s) \mapsto \gamma_t(s)$ be the line-of-sight map for α with respect to $\alpha(0)$, and let $(t,s) \mapsto \tilde{\gamma}_t(s)$ be the line-of-sight map for $\tilde{\alpha}$ with respect to $\tilde{\alpha}(0)$. Consider the map $F\colon \operatorname{Conv}\tilde{\alpha} \to \mathcal{U}$ such that $F\colon \tilde{\gamma}_t(s) \mapsto \gamma_t(s)$.

Show that F majorizes α and conclude that F is distance-preserving.

Harder way. Prove and apply the following statement together with the majorization theorem.

- Let α and β be two convex curves in $\mathbb{M}^2(\kappa)$. Assume

$$\operatorname{length}\alpha = \operatorname{length}\beta < 2\cdot\varpi^\kappa$$

and there is a short bijecction $f\colon \alpha \to \beta$. Then f is an isometry.

9.62. Suppose that points p, x, q, y appear on the curve in that cyclic order. Assume that the geodesics $[pq]$ and $[xy]$ do not intersect. Use the argument in the proof of the majorization theorem 9.56 to show that in this case there are nonequivalent majorization maps.

Now we can assume that pairs of geodesics $[pq]$ and $[xy]$ intersect for all choices of points p, x, q, y on the curve in that cyclic order. Show that in this case the convex hull K of the curve is isometric to a convex figure.

Note that the composition of a majorization map and closest point projection to K is a majorization. Show and use that the boundary of a convex figure in the plane admits a unique majorization up to equivalence.

Remark. A typical rectifiable closed curve in a CAT(0) space can be majorized by more than one convex figure. There are two exceptions: (1) the majorization map is distance-preserving, and (2) the curve is geodesic triangle. It is expected that there are no other exceptions; this question was asked by Richard Bishop in a private conversation.

9.64. Show that quadrangle $[x^1x^2x^3x^4]$ is majorized by the solid quadrangle $[\tilde{x}^1\tilde{x}^2\tilde{x}^3\tilde{x}^4]$. Further show that the majorization is isometric; argue as in Exercise 8.18.

9.70. If ℓ and m do not intersect, then the double cover $\mathcal{X}$ is not simply connected. In particular, by the Hadamard–Cartan theorem, $\mathcal{X}$ is not CAT(0).

If ℓ and m intersect, then $\mathcal{X}$ is a cone over a double cover Σ of $\mathbb{S}^2$ branching at two pairs (x,y) and (v,w) of antipodal points. Suppose $|x-v|_{\mathbb{S}^2} = \ell < \frac{\pi}{2}$. Note that the inverse image of $[xv]_{\mathbb{S}^2}$ is a closed geodesic of length $4\cdot\ell < 2\cdot\pi$. Therefore, by the generalized Hadamard–Cartan theorem, Σ is not CAT(1). Hence $\mathcal{X}$ is not CAT(0) by Theorem 11.7.

9.71. Let us do the second part first. Assume A has nonempty interior. Note that the space $\tilde{\mathcal{U}}$ is simply connected and locally isometric to the doubling $\mathcal{W}$

of $\mathcal{U}$ in A; that is, any point in $\tilde{\mathcal{U}}$ has a neighborhood that is isometric to a neighborhood of a point in $\mathcal{W}$.

By the Reshetnyak gluing theorem (see 9.39), $\mathcal{W}$ is CAT(0). Therefore $\tilde{\mathcal{U}}$ is locally CAT(0); it remains to apply the Hadamard–Cartan theorem (see 9.65).

Let us come back to the general case. The above argument can be applied to a closed ε-neighborhood of A. After that we need to pass to a limit as $\varepsilon \to 0$.

The first part of the problem follows since a geodesic is a convex set.

9.72. By the globalization theorem there is a nontrivial homotopy class in the set of all closed curves of length $< 2 \cdot \varpi^\kappa$. Consider a shortest noncontractible closed curve γ in $\mathcal{K}$; note that such a curve exists. Indeed, let L be the infimum of lengths of all noncontractible closed curves in $\mathcal{X}$. Compactness and local contractibility of $\mathcal{X}$ imply that any two sufficiently close closed curves in $\mathcal{X}$ are homotopic. Then choosing a sequence of unit speed noncontractible curves whose lengths converge to L, an Arzelá–Ascoli type of argument shows that these curves subconverge to a noncontractible curve of length L.

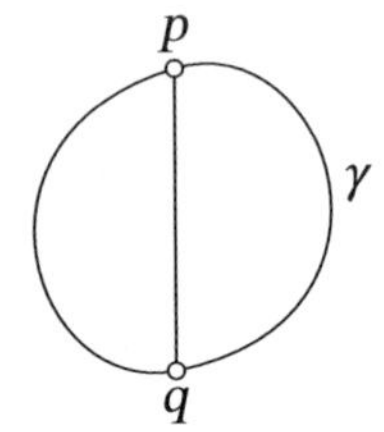

Assume that γ is not a geodesic circle, that is, there are two points p and q on γ such that the distance $|p-q|$ is shorter than the lengths of the arcs, say α_1 and α_2, of γ from p to q. Consider the products, say γ_1 and γ_2, of $[qp]$ with α_1 and α_2. Then $\operatorname{length}\gamma_1, \operatorname{length}\gamma_2 < \operatorname{length}\gamma$.

Therefore, showing that γ_1 or γ_2 is noncontractible in the set of curves of length $< 2 \cdot \varpi^\kappa$, will lead to a contradiction. The case $\kappa \leqslant 0$ is trivial. For the case $\kappa > 0$, argue as in 9.67.

Remark. The statement of the exercise fails if the requirement that $\mathcal{X}$ be compact is replaced by the assumption that it is proper. For example, the surface of revolution of the graph of $y = e^x$ around the x-axis is locally CAT(0) but has no closed geodesics.

9.75. Let $p \mapsto \bar{p}$ denote the closest-point projection to K. We need to show that $|\bar{p} - \bar{q}| \leqslant |p - q|$ for any $p, q \in \mathcal{U}$.

Assume $p \neq \bar{p} \neq \bar{q} \neq q$. Note that in this case $\measuredangle[\bar{p}\,{}^{p}_{\bar{q}}] \geqslant \frac{\pi}{2}$ and $\measuredangle[\bar{q}\,{}^{q}_{\bar{p}}] \geqslant \frac{\pi}{2}$. Otherwise a point on the geodesic $[\bar{p}\bar{q}]$ would be closer to p or to q than $\bar{p}$ or $\bar{q}$, respectively. The latter is impossible since K is convex and therefore $[\bar{p}\bar{q}] \subset K$.

Applying the arm lemma (9.63), we get the statement.

The cases $p = \bar{p} \neq \bar{q} \neq q$ and $p \neq \bar{p} \neq \bar{q} = q$ can be done similarly. The rest of the cases are trivial.

9.76. A more transparent, but less elementary solution via gradient flow is given by Alexander Lytchak and the third author [**107**].

Without loss of generality, we may assume that $p \in K$.

If $\operatorname{dist}_K x \geqslant \pi$, then set $\Psi(x) = p$.

Otherwise, if $\operatorname{dist}_K x < \pi$, by the closest-point projection lemma (9.74), there is a unique point $x^* \in K$ that minimizes distance to x; that is, $|x^* - x| = \operatorname{dist}_K x$. Let us define ℓ_x, ϕ_x and ψ_x using the following identities:

$$\begin{aligned} \ell_x &= |p - x^*|, \\ \phi_x &= \tfrac{\pi}{2} - |x^* - x|, \\ \sin \psi_x &= \sin \phi_x \cdot \sin \ell_x, \quad 0 \leqslant \psi_x \leqslant \tfrac{\pi}{2}. \end{aligned}$$

Let

$$\Psi(x) = \operatorname{geod}_{[px^*]}(\psi_x).$$

Note that Ψ is a retraction to K; that is, $\Psi(x) \in K$ for any $x \in \mathcal{U}$ and $\Psi(a) = a$ for any $a \in K$.

Let us show that Ψ is short. Given $x, y \in \mathrm{B}(K, \tfrac{\pi}{2})$, let

$$\begin{aligned} x' &= \Psi(x), & y' &= \Psi(y), \\ r &= |x - y|, & r' &= |x' - y'|, \\ d &= |x^* - y^*|, & \alpha &= \tilde\measuredangle^1 \left(p\,{}^{x^*}_{y^*}\right). \end{aligned}$$

Note that

$$\cos r \leqslant \cos \phi_x \cdot \cos \phi_y - \cos d \cdot \sin \phi_x \cdot \sin \phi_y. \tag{1}$$

Indeed, if $x, y \notin K$, then $\measuredangle\left[x^*\,{}^{x}_{y^*}\right], \measuredangle\left[y^*\,{}^{y}_{x^*}\right] \geqslant \tfrac{\pi}{2}$ and inequality (1) follows from the arm lemma (9.63). If $x \in K$ and $y \notin K$, we obtain (1) by the angle comparison (see 9.14(c)) since $\measuredangle\left[y^*\,{}^{y}_{x^*}\right] \geqslant \tfrac{\pi}{2}$. In the same way, (1) is proved if $x \notin K$ and $y \in K$. Finally, if $x, y \in K$, then $\phi_x = \phi_y = \tfrac{\pi}{2}$ and $r = d$; that is, the inequality trivially holds.

Further note that

$$\cos \alpha = \frac{\cos d - \cos \ell_x \cdot \cos \ell_y}{\sin \ell_x \cdot \sin \ell_y}.$$

Applying the angle-sidelength monotonicity (9.15), we have

$$\begin{aligned} \cos r' &\geqslant \cos \psi_x \cdot \cos \psi_y - \cos \alpha \cdot \sin \psi_x \cdot \sin \psi_y = \\ &= \cos \psi_x \cdot \cos \psi_y - (\cos d - \cos \ell_x \cdot \cos \ell_y) \cdot \sin \phi_x \cdot \sin \phi_y \geqslant \\ &\geqslant \cos \psi_x \cdot \cos \psi_y - \cos d \cdot \sin \phi_x \cdot \sin \phi_y. \end{aligned}$$

Note that $\psi_x \leqslant \phi_x$ and $\psi_y \leqslant \phi_y$; in particular,

$$\cos \phi_x \cdot \cos \phi_y \leqslant \cos \psi_x \cdot \cos \psi_y.$$

Hence

$$\cos r' \geqslant \cos r;$$

that is, the restriction $\Psi|_{B(K,\frac{\pi}{2})}$ is short. Clearly Ψ is continuous. Since the complement of $B(K, \frac{\pi}{2})$ is mapped to p, Ψ is short; that is,

$$r' \leqslant r \tag{2}$$

for any $x, y \in \mathcal{U}$.

If we have equality in (2) then

$$\cos \ell_x \cdot \cos \ell_y \cdot \sin \phi_x \cdot \sin \phi_y = 0.$$

If $K \subset B(p, \frac{\pi}{2})$, then $\ell_x, \ell_y < \frac{\pi}{2}$, which implies that $x \in K$ or $y \in K$. Without loss of generality, we may assume that $x \in K$.

It remains to show that if $y \notin K$, then the inequality (2) is strict. If $\operatorname{dist}_K y \geqslant \geqslant \frac{\pi}{2}$, then (2) holds since the left-hand side is $< \frac{\pi}{2}$ while the right-hand side is $\geqslant \frac{\pi}{2}$. If $\operatorname{dist}_K y < \frac{\pi}{2}$, then $\phi_y > 0$. Clearly $\psi_y < \phi_y$, hence inequality (2) is strict.

Below you will find a geometric way to think about the given construction; it is close to the construction in the proof of Kirszbraun's theorem (10.14).

Geometric interpretation of the map Ψ. Let $\mathring{\mathcal{U}} = \operatorname{Cone} \mathcal{U}$, and denote by $\mathring{K}$ the subcone of $\mathring{\mathcal{U}}$ spanned by K. The space $\mathcal{U}$ can be naturally identified with the unit sphere in $\mathring{\mathcal{U}}$, that is, the set

$$\left\{ z \in \mathring{\mathcal{U}} : |z| = 1 \right\}.$$

According to Theorem 11.7, $\mathring{\mathcal{U}}$ is CAT(0). Note that $\mathring{K}$ forms a convex closed subset of $\mathring{\mathcal{U}}$. According to Lemma 9.74, for any point x there is a unique point $\hat{x} \in \mathring{K}$ that minimizes the distance to x, that is, $|\hat{x} - x| = \operatorname{dist}_K x$. (If $|\hat{x}| \neq 0$, then in the notation above we have $x^* = \frac{1}{|\hat{x}|} \cdot \hat{x}$.)

Consider the half-line $t \mapsto t \cdot p$ in $\mathring{\mathcal{U}}$. By comparison, for given $s \in \mathring{\mathcal{U}}$ the geodesics $\operatorname{geod}_{[s\, t \cdot p]}$ converge as $t \to \infty$ to a half-line, say $\alpha_s : [0, \infty) \to \mathring{\mathcal{U}}$.

Note that if $|x| = 1$, then $|\hat{x}| \leqslant 1$. By assumption, for any $a \in K$ the function $t \mapsto |\alpha_a(t)|$ is monotonically increasing. Therefore there is a unique value $t_x \geqslant 0$ such that $|\alpha_{\hat{x}}(t_x)| = 1$. Define $\Psi : \mathcal{U} \to K$ by

$$\Psi(x) = \alpha_{\hat{x}}(t_x).$$

9.81. Prove that the angle comparison (see 9.14(c)) holds.

9.82. Mimic the proof of 9.65, the Hadamard–Cartan theorem.

10.5. Note that it is sufficient to show that any finite set of points $x^1, \dots, x^n \in \mathcal{X}$ lies in an isometric copy of a Euclidean polyhedron.

Observe that $\mathcal{X}$ is CBB(0) and CAT(0) at the same time. Show that there is a unique point p that minimizes the sum $|p - x^1| + \dots + |p - x^n|$. Note that the vectors $v^i = \log[px^i]$ lie in a linear subspace of T_p. Moreover if K is the convex hull of v_i, then the origin of T_p lies in the interior of K relative to its affine hull. Finally observe that the exponential map is defined on all of K and is distance-preserving. The statement follows since the exponential map sends $v^i \mapsto x^i$ for each i.

10.6. The answers are $s \leqslant \sqrt{3}$ and $s \leqslant 2$, respectively.

Let us start with the CAT(0) case. The upper bound $s \leqslant \sqrt{3}$ follows from (2+2)-point comparison. The Euclidean space works as an example if s is smaller than the large diagonal of the double pyramid with unit side (that is, if $s \leqslant 2\cdot\sqrt{2/3}$). Otherwise it can be embedded into a product of the real line with a two-dimensional cone.

For the CBB(0) case, the needed space can be constructed by doubling a polyhedron $K \subset \mathbb{E}^3$ in its boundary. The obtained space is CBB(0) by Theorem 12.5; the same follows from Perelman's doubling theorem [**127**]. We assume that the points correspond to vertices of a regular tetrahedron with three vertices on the boundary of K and one in its interior; this point corresponds to a pair of points in the doubling at distance s from each other.

Remark. The CAT(0) case also follows from [**105, 157**].

10.7. Choose a quadruple of points p, q, r, s. Suppose that it admits a distance-preserving embedding into some $\mathbb{M}^2(\mathrm{K})$ for some $\mathrm{K} \geqslant \kappa$. Then

$$\tilde\measuredangle^{\mathrm{K}}(p\,{}^q_r) + \tilde\measuredangle^{\mathrm{K}}(p\,{}^r_s) + \tilde\measuredangle^{\mathrm{K}}(p\,{}^a_q) \leqslant 2\cdot\pi.$$

Applying monotonicity of the function $\kappa \mapsto \tilde\measuredangle^{\kappa}(p\,{}^q_r)$ (see 1.1(d)) shows that

$$\tilde\measuredangle^{\kappa}(p\,{}^q_r) + \tilde\measuredangle^{\kappa}(p\,{}^r_s) + \tilde\measuredangle^{\kappa}(p\,{}^s_q) \leqslant 2\cdot\pi.$$

Since the quadruple p, q, r, s is arbitrary, the if part follows.

Now let us prove the only-if part. Denote by σ the exact upper bound on values $\mathrm{K} \geqslant \kappa$ such that all model triangles with the vertices p, q, r, s are defined.

Recall that $\tilde\measuredangle^{\mathrm{K}+}(p\,{}^q_r)$ denotes extended angle (Definition 8.49). Observe that if

$$\tilde\measuredangle^{\mathrm{K}+}(p\,{}^q_r) + \tilde\measuredangle^{\mathrm{K}+}(p\,{}^r_s) + \tilde\measuredangle^{\mathrm{K}+}(p\,{}^s_q) = 2\cdot\pi \tag{3}$$

for some $\sigma \geqslant \mathrm{K} \geqslant \kappa$, then the quadruple admits a distance-preserving embedding into $\mathbb{M}^2(\mathrm{K})$.

Observe that the left-hand side of (3) is continuous in K. Since $\mathcal{L}$ is CBB(κ), for $\mathrm{K} = \kappa$ the left-hand side cannot exceed $2\cdot\pi$. Therefore it remains smaller than $2\cdot\pi$ for all $\sigma \geqslant K \geqslant \kappa$; moreover the same holds for all permutations of the labels p, q, r, s.

Note that we can assume the perimeter of the triple q, r, s is $2\cdot\varpi^\sigma$, and use this and the overlap lemma (see 10.2) to arrive at a contradiction.

According to our definition, the real line is CBB(κ) for any $\kappa \in \mathbb{R}$, but it does not satisfy the property for $\kappa > 0$. The condition $\kappa \leqslant 0$ was used just once to ensure that the κ-model triangles with the vertices p, q, r, s are defined. One can assume instead that perimeters of all triangles in $\mathcal{L}$ are at most $2\cdot\varpi^\kappa$. This condition holds for all complete length CBB(κ) spaces of dimension at least 2; see 8.44.

10.9. Let $\tilde p, \tilde x_1, \dots, \tilde x_n$ be the array in $\mathbb{E}^n$ provided by the $(1+n)$-point comparison (see 10.8). We may assume that $\tilde p$ is the origin of $\mathbb{E}^n$.

Consider an $n{\times}n$-matrix $\tilde M$ with components

$$\tilde m_{i,j} = \tfrac12\cdot(|\tilde x_i - \tilde p|^2 + |\tilde x_j - \tilde p|^2 - |\tilde x_i - \tilde x_j|^2).$$

Note that $\tilde m_{i,j} = \langle \tilde x_i, \tilde x_j\rangle$. It follows that $\tilde M = A\cdot A^\top$ for an $n{\times}n$-matrix A that defines a linear transformation sending the standard basis to the array $\tilde x_1, \dots, \tilde x_n$. Therefore

$$\boldsymbol{s}\cdot\tilde M\cdot\boldsymbol{s}^\top = |A^\top\cdot\boldsymbol{s}^\top|^2 \geqslant 0$$

for any vector $\boldsymbol{s}$. Further show and use that

$$\boldsymbol{s}\cdot M\cdot\boldsymbol{s}^\top \geqslant \boldsymbol{s}\cdot\tilde M\cdot\boldsymbol{s}^\top$$

for any vector $\boldsymbol{s} = (s_1, \dots, s_n)$ with nonnegative components.

10.10. Apply the (5+1)-point comparison (see 10.8).

10.11. It is sufficient to construct a metric on the set of points $\{p, x^1, x^2, x^3, x^4\}$ that does not satisfy (1+4)-point comparison but does satisfy all (1+3)-point comparisons. To do this, set x^i to be the vertices of a regular tetrahedron in $\mathbb{E}^3$. Suppose p is its center and reduce the distances $|p - x^i|$ slightly.

Remark. There are examples of 6-point metric spaces that satisfy all (1+5)-point comparisons, but do not admit embedding into a complete length CBB(0) space [**103**].

10.12. By the $(1+n)$-point comparison (see 10.8), there is a point array $\tilde p, \tilde a^0, \dots, \tilde a^m \in \mathbb{M}^{m+1}(\kappa)$ such that

$$|\tilde p - \tilde a^i| = |p - a^i| \quad\text{and}\quad |\tilde a^i - \tilde a^j| \geqslant |a^i - a^j|$$

for all i and j.

For each i, set $\tilde\xi^i = \uparrow_{[\tilde p \tilde a^i]} \in \mathbb{S}^m = \Sigma_{\tilde p}(\mathbb{M}^{m+1}(\kappa))$. Note that

$$|\tilde\xi^i - \tilde\xi^j|_{\mathbb{S}^m} \geqslant \tilde\measuredangle^\kappa\left(p\,{}^{a^i}_{a^j}\right) > \tfrac{\pi}{2}.$$

Consider two matrices S and $\tilde{S}$ with components $s_{i,j} = \langle\tilde{\xi}^i, \xi^j\rangle$ and $\tilde{s}_{i,j} = \cos[\tilde{\measuredangle}^\kappa(p\,^{a^i}_{a^j})]$. By construction, $S \geqslant 0$; that is $\boldsymbol{v}\cdot S\cdot\boldsymbol{v}^\top \geqslant 0$ for any vector $\boldsymbol{v}$.

Observe that it is sufficient to show that $\tilde{S} \geqslant 0$. The latter follows since $s_{i,j} \leqslant \tilde{s}_{i,j} \leqslant 0$ if $i \neq j$ and $s_{i,j} = \tilde{s}_{i,j} = 1$ if $i = j$.

10.18. Set $\tilde{Q} = \operatorname{Conv}\{\tilde{x}^0, \tilde{x}^1, \dots, \tilde{x}^k\}$. By Kirszbraun's theorem, the map $\tilde{x}^i \mapsto x^i$ can be extended to a short map $F\colon \tilde{Q} \to \mathcal{L}$; it remains to show that the map F is distance-preserving.

Consider the logarithm map $G\colon x \mapsto \log[x_0 x]$; note that G is short. Observe that the composition $G\circ F$ is distance-preserving. Therefore F is distance-preserving; in particular we can take $Q = F(\tilde{Q})$.

10.19. Consider vectors $v^i = \log[x^0 x^i] \in \mathrm{T}_{x^0}$. Show that all the v^i lie in a linear subspace of T_{x^0} and that $x^i \mapsto v^i$ is distance-preserving. It follows that we can identify the convex hull K of the v^i with the convex hull of the $\tilde{x}^i$.

Note that the gradient exponential map $\operatorname{gexp}_{x_0}$ maps v^i to x^i. By assumption,

$$|v^i - v^j| = |x^i - x^j| \tag{4}$$

for all i and j. By 16.36, $\operatorname{gexp}_{x_0}$ is a short map. By (4), $\operatorname{gexp}_{x_0}$ cannot be strictly short at a pair of points in K. That is, $\operatorname{gexp}_{x_0}$ is distance-preserving on K.

10.20. Apply 10.17 for each of the following maps:

- $f_0\colon \tilde{x} \mapsto x$, $\tilde{p}^1 \mapsto p^1$, $\tilde{q}^1 \mapsto q^1$;
- $f_i\colon \tilde{p}^i \mapsto p^i$, $\tilde{p}^{i+1} \mapsto p^{i+1}$, $\tilde{q}^i \mapsto q^i$, $\tilde{q}^{i+1} \mapsto q^{i+1}$ for $1 \leqslant i < n$;
- $f_n\colon \tilde{y} \mapsto y$, $\tilde{p}^n \mapsto p^n$, $\tilde{q}^n \mapsto q^n$.

Denote by F_i the short extension of f_i. Observe and use that $F_{i-1}(\tilde{z}_i) = F_i(\tilde{z}_i)$ for each i.

10.23. Consider the space $\mathcal{Y}^\mathcal{X}$ of all maps $\mathcal{X} \to \mathcal{Y}$ equipped with the product topology.

Denote by $\mathfrak{S}_F$ the set of maps $h \in \mathcal{Y}^\mathcal{X}$ such that the restriction $h|_F$ is short and agrees with f in $F \cap A$. Note that the sets $\mathfrak{S}_F \subset \mathcal{Y}^\mathcal{X}$ are closed and any finite intersection of these sets is nonempty.

According to Tikhonov's theorem, $\mathcal{Y}^\mathcal{X}$ is compact. By the finite intersection property, the intersection $\bigcap_F \mathfrak{S}_F$ for all finite sets $F \subset X$ is nonempty. Hence the statement follows.

Source: This statement appears in [**128**]; it is an analogous of the finite+one lemma (see 10.16).

10.24. The Kuratowsky embedding is a distance-preserving map of $\mathcal{X}$ into the space of bounded functions $\mathcal{X}$ equipped with the metric induced by the sup-norm (see 2.G). It remains to show that the latter space is injective.

The second part of the exercise is a classical result of John Isbell [**86**] which was rediscovered several times after him; for more on the subject see lecture notes of the third author [**134**].

11.4. It is sufficient to show that the natural map $\mathcal{B} \times_g \mathcal{F} \to \mathcal{B} \times_f \mathcal{F}$ is short. The latter follows from the fiber-independence theorem (see 11.3).

11.5. Show and use that any geodesic path in $\mathrm{Cone}^\kappa \mathcal{F}$ projects to a reparametrized geodesic in $\mathcal{F}$ of length less than π.

11.9. By Theorem 11.6(a), the space $\mathcal{U}$, $\mathcal{V}$, or $\mathcal{U} \star \mathcal{V}$ is CBB(1) if and only if $\mathrm{Cone}\,\mathcal{U}$, $\mathrm{Cone}\,\mathcal{V}$, or $\mathrm{Cone}(\mathcal{U} \star \mathcal{V}) = \mathrm{Cone}\,\mathcal{U} \times \mathrm{Cone}\,\mathcal{V}$ is CBB(0), respectively.

By Theorem 11.7(a), the space $\mathcal{U}$, $\mathcal{V}$, or $\mathcal{U} \star \mathcal{V}$ is CAT(1) if and only if $\mathrm{Cone}\,\mathcal{U}$, $\mathrm{Cone}\,\mathcal{V}$, or $\mathrm{Cone}(\mathcal{U} \star \mathcal{V}) = \mathrm{Cone}\,\mathcal{U} \times \mathrm{Cone}\,\mathcal{V}$ is CAT(0), respectively.

It remains to show that the product of two spaces is CBB(0) or CAT(0) if and only if each space is CBB(0) or CAT(0), respectively.

12.3. Apply Reshetnyak gluing theorem (see 9.39) several times.

12.4. Assume $\mathcal{P}$ is not CAT(0). Then by Theorem 12.2, a link Σ of some simplex contains a closed local geodesic α with length $4 \cdot \ell < 2 \cdot \pi$. We can assume that Σ has minimal possible dimension; then by Theorem 12.2, Σ is locally CAT(1).

Divide α into two equal arcs α_1 and α_2.

Assume α_1 and α_2 are length-minimizing, and parametrize them by $[-\ell, \ell]$. Fix a small $\delta > 0$ and consider the two curves in $\mathrm{Cone}\,\Sigma$ given in polar coordinates by

$$\gamma_i(t) = \left(\alpha_i\left(\arctan \tfrac{t}{\delta}\right), \sqrt{\delta^2 + t^2}\right).$$

Show that the curves γ_1 and γ_2 are geodesics in $\mathrm{Cone}\,\Sigma$ having common endpoints.

Observe that a small neighborhood of the tip of $\mathrm{Cone}\,\Sigma$ admits a distance-preserving embedding into $\mathcal{P}$. Hence we can construct two geodesics γ_1 and γ_2 in $\mathcal{P}$ with common endpoints.

It remains to consider the case where α_1 (and therefore α_2) is not length-minimizing.

Pass to a maximal length-minimizing arc $\bar\alpha_1$ of α_1. Since Σ is locally CAT(1), by the no-conjugate-point theorem (see 9.46) there is another geodesic $\bar\alpha_2$ in Σ_p that shares endpoints with $\bar\alpha_1$. It remains to repeat the above construction for the pair $\bar\alpha_1$, $\bar\alpha_2$.

Remark. By 9.8 the converse holds as well. This problem was suggested by Dmitri Burago.

12.6. Apply Theorem 13.1, Exercise 12.5, and Theorem 12.2.

12.11. Observe and use that (1) in the barycentric subdivision every vertex corresponds to a simplex of the original triangulation, and (2) a simplex of the subdivision corresponds to a decreasing sequence of simplexes in the original triangulation.

Remark. The second statement, *any finite simplicial complex is homeomorphic to a compact length* CAT(1) *space*, is due to Valerii Berestovskii [**23**].

12.13. Use induction on the dimension to prove that if in a spherical simplex $\triangle$ every edge is at least $\frac{\pi}{2}$, then all dihedral angles of $\triangle$ are at least $\frac{\pi}{2}$.

The rest of the proof goes along the same lines as the proof of the flag condition (see 12.10). The only difference is that a geodesic may spend time *at least* π on each visit to Star_v.

Remark. It is not sufficient to assume only that all the dihedral angles of the simplexes are at least $\frac{\pi}{2}$. Indeed, the two-dimensional sphere with the interior of a small rhombus removed is a spherical polyhedral space glued from four triangles with angles at least $\frac{\pi}{2}$. On the other hand, the boundary of the rhombus is a closed local geodesic in this space and has length less than $2\cdot\pi$. Therefore the space cannot be CAT(1).

12.14. Observe that if we glue two copies of spaces along A_i, then the copies of A_j for some $j \neq i$ form a convex subset in the glued space. Use this and the Reshetnyak gluing theorem (see 9.39) n times, once for each label of the edges.

12.15. The space $\mathcal{T}_n$ has a natural cone structure whose vertex is the completely degenerate tree—all its edges have zero length.

Note that the space Σ over which the cone is taken comes naturally with a triangulation by right-angled spherical simplexes. Each simplex corresponds to the combinatorics of a possibly degenerate tree.

Note that the link of any simplex of this triangulation satisfies the no-triangle condition. Indeed, fix a simplex $\triangle$ of the complex; suppose it is described by a possibly degenerate topological tree t. A triangle in the link of $\triangle$ can be described by three ways to resolve a degeneracy of t by adding one edge, where (1) any pair of these resolutions can be done simultaneously, but (2) all three cannot be done simultaneously. Direct inspection shows that this is impossible.

By Proposition 12.8, our complex is flag. It remains to apply the flag condition (see 12.10) and Theorem 11.6(a).

12.16. Apply the flag condition (see 12.10) and Theorem 11.7(a).

13.8. Consider a cube in the ℓ^2-space defined by $|x_i| \leqslant 1$.

13.13 and 13.14. Apply the strong angle lemmas (Lemmas 9.35 and 8.43).

13.19. Apply Exercise 13.13.

13.21. Apply Theorem 8.11.

13.27. Since α is Lipschitz, so is $f \circ \alpha$. By the standard Rademacher theorem, the derivative $(f \circ \alpha)'$ is defined almost everywhere. In particular,

$$(\boldsymbol{d}_{\alpha(t)} f)(\alpha^+(t)) + (\boldsymbol{d}_{\alpha(t)} f)(\alpha^-(t)) \overset{\text{a.e.}}{=\!=} 0.$$

Further, by the extended Rademacher theorem (more precisely, its one-dimensional case; see Proposition 13.9), we have

$$\alpha^+(t) + \alpha^-(t) \overset{\text{a.e.}}{=\!=} 0.$$

In particular,

$$\langle \nabla_{\alpha(t)} f, \alpha^+(t) \rangle + \langle \nabla_{\alpha(t)} f, \alpha^-(t) \rangle \overset{\text{a.e.}}{=\!=} 0.$$

Finally, by the definition of gradient, we have

$$\langle \nabla_{\alpha(t)} f, \alpha^\pm(t) \rangle \geqslant (\boldsymbol{d}_{\alpha(t)} f)(\alpha^\pm(t)).$$

Hence the result follows.

13.35. Let us pass to the ultrapower $\mathcal{L}^\omega \supset \mathcal{L}$. Argue as in Exercise 13.14 to show that there is a geodesic $[pb]_{\mathcal{L}^\omega}$ such that

$$\boldsymbol{d}_p \operatorname{dist}_b(\uparrow_{[pa]}) = -\langle \uparrow_{[pb]}, \uparrow_{[pa]} \rangle.$$

It follows that

$$\begin{aligned}(\boldsymbol{d}_p \operatorname{dist}_a)(\nabla_p \operatorname{dist}_b) &\leqslant -\langle \uparrow_{[pa]}, \nabla_p \operatorname{dist}_b \rangle \leqslant \\ &\leqslant -\boldsymbol{d}_p \operatorname{dist}_b(\uparrow_{[pa]}) = \\ &= \langle \uparrow_{[pb]}, \uparrow_{[pa]} \rangle = \\ &= \cos \measuredangle [p\,{}^a_b]_{\mathcal{L}^\omega} \leqslant \\ &\leqslant \cos \tilde{\measuredangle}^\kappa (p\,{}^a_b).\end{aligned}$$

14.8. Choose an orbit $\{p_1, \ldots, p_n\}$; observe that its barycenter with equal masses is a fixed point.

14.18. Let k be the maximal integer such that any k subsets have a common points. By assumption $k \geqslant m+1$.

Suppose $k < n$. We can assume that $n = k+1$ and $K_1 \cap \dots \cap K_n = \varnothing$. Choose a point

$$p_i \in \bigcap_{j \neq i} K_j$$

for each i. Argue as in 14.12 to show that the simplex with vertices $p_1, \dots, p_n$ is nondegenerate and arrive at a contradiction.

15.17. Suppose that $\mathcal{L}$ is infinite dimensional. Denote by $\Omega_m \subset \mathcal{L}$ the set of all points p with $\operatorname{rank}_p \geqslant m$. Evidently $\Omega_1 \supset \Omega_2 \supset \cdots$, and Ω_m is open for each m.

By Theorem 15.6(C), each Ω_m is dense in $\mathcal{L}$. Hence there is a G-delta dense set of points $p \in \mathcal{L}$ such that $\operatorname{rank}_p = \infty$. It follows that Σ_p is not compact.

Source: It was suggested by Alexander Lytchak.

16.1. Choose a finite sequence $t_0 < \dots < t_n$. Denote by Φ_{t_i} the composition of the closest-point projections to $K_{t_0}, \dots, K_{t_i}$. Pass to a limit of the Φ_{t_i} as the sequence becomes denser in the parameter interval. Show and use that the limit ϕ_t does not depend on the choice of the sequences.

16.8. Let $\ell(t) = |\alpha(t) - \alpha(t_3)|$. Note that

$$\ell'(t) \leqslant -\langle \nabla_{\alpha(t)} f, \uparrow_{[\alpha(t)\alpha(t_3)]} \rangle.$$

Observe that the function $t \mapsto f \circ \alpha(t)$ is nondecreasing; in particular, $f(\alpha(t_1)) \leqslant f(\alpha(t_2)) \leqslant f(\alpha(t_3))$. Therefore

$$\langle \nabla_{\alpha(t)} f, \uparrow_{[\alpha(t)\alpha(t_3)]} \rangle \geqslant \boldsymbol{d}_{\alpha(t)} f(\uparrow_{[\alpha(t)\alpha(t_3)]}) \geqslant 0$$

for any $t \in [t_1, t_2]$. Therefore $\ell' \leqslant 0$ for any $t \in [t_1, t_2]$. Hence the statement.

16.10. Without loss of generality, we may assume that $(f \circ \alpha)'(t) > 0$ for any t. Let $\hat{\alpha}$ be the arclength reparametrization of α. Note that

$$(f \circ \hat{\alpha})'(s) \geqslant |\nabla_{\hat{\alpha}(s)} f|$$

almost everywhere. Therefore, by Theorem 16.3, $\hat{\alpha}$ is a gradient-like curve. It remains to apply Lemma 16.9.

16.11. Use Exercise 16.10 to prove the only-if part.

To prove the if part, set $h(z) = \frac{1}{2} \cdot |x - z|^2$. If α is an f-gradient curve, then

$$\begin{aligned}(h \circ \alpha)^+ &\geqslant |\alpha(t) - x| \cdot \langle \uparrow_{[\alpha(t)x]}, \nabla_{\alpha(t)} f \rangle \geqslant \\ &\geqslant |\alpha(t) - x| \cdot \boldsymbol{d}_{\alpha(t)} f(\uparrow_{[\alpha(t)x]}) \geqslant \\ &\geqslant f(x) - f \circ \alpha(t).\end{aligned}$$

It remains to integrate the inequality and observe that $f \circ \alpha$ is nondecreasing.

16.35. Consider $(x,\varkappa)$- and $(z,\varkappa)$-radial curves that start at y, and observe that they form a geodesic from x to z. (Compare to Exercise 10.19.)

16.38. Set $q = p + v$ and $q' = \mathrm{gexp}_p v$. By radial comparison, $|q' - x| \leqslant |q - x|$ for any $x \in \mathcal{L}$. If $q \in \mathcal{L}$, this implies that $q = q'$. Otherwise note that q' lies on the boundary line of $\mathcal{L}$, and $\mathrm{proj}(q)$ is the only point on this line that satisfies the inequality.

16.39. By the angle comparison, $|\nabla_x \mathrm{dist}_p| \geqslant -\cos\tilde{\measuredangle}^\varkappa\left(x\,{}^p_q\right)$.

Choose a $(p,\varkappa)$-radial curve α that starts at p. Observe that

$$(\mathrm{dist}_p \circ \alpha)^+(t) \geqslant -|\alpha^+(t)| \cdot \cos\tilde{\measuredangle}^\varkappa\left(\alpha(t)\,{}^p_q\right)$$

and

$$(\mathrm{dist}_q \circ \alpha)^+(t) \geqslant -|\alpha^+(t)|.$$

Therefore $t \mapsto \tilde{\measuredangle}^\varkappa\left(q\,{}^{\alpha(t)}_p\right)$ is nondecreasing, hence the result.

Bibliography

[1] Stephanie B. Alexander, I. David Berg, and Richard L. Bishop, *Geometric curvature bounds in Riemannian manifolds with boundary*, Trans. Amer. Math. Soc. **339** (1993), no. 2, 703–716, DOI 10.2307/2154294. MR1113693

[2] Stephanie B. Alexander and Richard L. Bishop, *Gauss equation and injectivity radii for subspaces in spaces of curvature bounded above*, Geom. Dedicata **117** (2006), 65–84, DOI 10.1007/s10711-005-9011-6. MR2231159

[3] Stephanie B. Alexander and Richard L. Bishop, *Comparison theorems for curves of bounded geodesic curvature in metric spaces of curvature bounded above*, Differential Geom. Appl. **6** (1996), no. 1, 67–86, DOI 10.1016/0926-2245(96)00008-3. MR1384880

[4] Stephanie B. Alexander and Richard L. Bishop, *The Fary-Milnor theorem in Hadamard manifolds*, Proc. Amer. Math. Soc. **126** (1998), no. 11, 3427–3436, DOI 10.1090/S0002-9939-98-04423-2. MR1459103

[5] Stephanie B. Alexander and Richard L. Bishop, *The Hadamard-Cartan theorem in locally convex metric spaces*, Enseign. Math. (2) **36** (1990), no. 3-4, 309–320. MR1096422

[6] Stephanie B. Alexander and Richard L. Bishop, *Warped products admitting a curvature bound*, Adv. Math. **303** (2016), 88–122, DOI 10.1016/j.aim.2016.07.005. MR3552521

[7] S. B. Alexander and R. L. Bishop, *Curvature bounds for warped products of metric spaces*, Geom. Funct. Anal. **14** (2004), no. 6, 1143–1181, DOI 10.1007/s00039-004-0487-2. MR2135163

[8] Stephanie Alexander, Vitali Kapovitch, and Anton Petrunin, *Alexandrov meets Kirszbraun*, Proceedings of the Gökova Geometry-Topology Conference 2010, Int. Press, Somerville, MA, 2011, pp. 88–109. MR2931882

[9] Stephanie Alexander, Vitali Kapovitch, and Anton Petrunin, *An optimal lower curvature bound for convex hypersurfaces in Riemannian manifolds*, Illinois J. Math. **52** (2008), no. 3, 1031–1033, `https://projecteuclid.org/euclid.ijm/1254403729`. MR2546022

[10] Stephanie Alexander, Vitali Kapovitch, and Anton Petrunin, *An invitation to Alexandrov geometry: CAT(0) spaces*, SpringerBriefs in Mathematics, Springer, Cham, 2019, DOI 10.1007/978-3-030-05312-3. MR3930625

[11] A. Alexandroff, *Existence of a convex polyhedron and of a convex surface with a given metric* (Russian, with English summary), Rec. Math. [Mat. Sbornik] N.S. **11/53** (1942), 15–65. MR0013540

[12] A. Alexandroff, *The inner geometry of an arbitrary convex surface*, C. R. (Doklady) Acad. Sci. URSS (N.S.) **32** (1941), 467–470. MR0009146

[13] A. D. Alexandroff, *On the gluing of convex surfaces*, C. R. (Doklady) Acad. Sci. URSS (N.S.) **54** (1946), 99–101. MR0019948

[14] A. D. Aleksandrov, *Vnutrennyaya Geometriya Vypuklyh Poverhnosteĭ* (Russian), OGIZ, Moscow-Leningrad, 1948. MR29518

[15] A. D. Aleksandrov and V. A. Zalgaller, *Intrinsic geometry of surfaces*, Translations of Mathematical Monographs, Vol. 15, American Mathematical Society, Providence, RI, 1967. Translated from the Russian by J. M. Danskin. MR216434

[16] A. Alexandroff, *The inner geometry of an arbitrary convex surface*, C. R. (Doklady) Acad. Sci. URSS (N.S.) **32** (1941), 467–470. MR9146

[17] A. D. Alexandrow, *Über eine Verallgemeinerung der Riemannschen Geometrie* (German), Schr. Forschungsinst. Math. **1** (1957), 33–84. MR87119

[18] A. D. Aleksandrov, *A theorem on triangles in a metric space and some of its applications* (Russian), Trudy Mat. Inst. Steklov. **38** (1951), 5–23. MR0049584

[19] Luigi Ambrosio, Nicola Gigli, and Giuseppe Savaré, *Gradient flows in metric spaces and in the space of probability measures*, 2nd ed., Lectures in Mathematics ETH Zürich, Birkhäuser Verlag, Basel, 2008. MR2401600

[20] Werner Ballmann, *Singular spaces of nonpositive curvature*, Sur les groupes hyperboliques d'après Mikhael Gromov (Bern, 1988), Progr. Math., vol. 83, Birkhäuser Boston, Boston, MA, 1990, pp. 189–201, DOI 10.1007/978-1-4684-9167-8_10. MR1086658

[21] Werner Ballmann, *Lectures on spaces of nonpositive curvature*, DMV Seminar, vol. 25, Birkhäuser Verlag, Basel, 1995. With an appendix by Misha Brin, DOI 10.1007/978-3-0348-9240-7. MR1377265

[22] W. Ballmann, *On the geometry of metric spaces*, MPIM, 2004, `https://people.mpim-bonn.mpg.de/hwbllmnn/archiv/sin40827.pdf`.

[23] V. N. Berestovskiĭ, *Borsuk's problem on metrization of a polyhedron* (Russian), Dokl. Akad. Nauk SSSR **268** (1983), no. 2, 273–277. MR0688238

[24] V. N. Berestovskij and I. G. Nikolaev, *Multidimensional generalized Riemannian spaces*, Geometry, IV, Encyclopaedia Math. Sci., vol. 70, Springer, Berlin, 1993, pp. 165–243, 245–250, DOI 10.1007/978-3-662-02897-1_2. MR1263965

[25] I. D. Berg and I. G. Nikolaev, *Quasilinearization and curvature of Aleksandrov spaces*, Geom. Dedicata **133** (2008), 195–218, DOI 10.1007/s10711-008-9243-3. MR2390077

[26] A. S. Besicovitch and P. A. P. Moran, *The measure of product and cylinder sets*, J. London Math. Soc. **20** (1945), 110–120, DOI 10.1112/jlms/s1-20.2.110. MR16448

[27] Louis J. Billera, Susan P. Holmes, and Karen Vogtmann, *Geometry of the space of phylogenetic trees*, Adv. in Appl. Math. **27** (2001), no. 4, 733–767, DOI 10.1006/aama.2001.0759. MR1867931

[28] E. Bilokopytov, *Is it possible to connect every compact set?*, MathOverflow (May 5, 2020), `https://mathoverflow.net/q/359390`.

[29] Wilhelm Blaschke, *Kreis und Kugel* (German), Chelsea Publishing Co., New York, 1949. MR76364

[30] A. Blass, *Selective ultrafilter and bijective mapping*, MathOverflow, `https://mathoverflow.net/q/324261`.

[31] B. H. Bowditch, *Notes on locally* CAT(1) *spaces*, Geometric group theory (Columbus, OH, 1992), Ohio State Univ. Math. Res. Inst. Publ., vol. 3, de Gruyter, Berlin, 1995, pp. 1–48. MR1355101

[32] Brian H. Bowditch, *Median and injective metric spaces*, Math. Proc. Cambridge Philos. Soc. **168** (2020), no. 1, 43–55, DOI 10.1017/s0305004118000555. MR4043820

[33] H. Brézis, *Opérateurs maximaux monotones et semi-groupes de contractions dans les espaces de Hilbert* (French), North-Holland Mathematics Studies, No. 5, North-Holland Publishing Co., Amsterdam-London; American Elsevier Publishing Co., Inc., New York, 1973. Notas de Matemática [Mathematical Notes], No. 50. MR348562

[34] Martin R. Bridson and André Haefliger, *Metric spaces of non-positive curvature*, Grundlehren der mathematischen Wissenschaften [Fundamental Principles of Mathematical Sciences], vol. 319, Springer-Verlag, Berlin, 1999, DOI 10.1007/978-3-662-12494-9. MR1744486

[35] Elia Bruè, Andrea Mondino, and Daniele Semola, *The metric measure boundary of spaces with Ricci curvature bounded below*, Geom. Funct. Anal. **33** (2023), no. 3, 593–636, DOI 10.1007/s00039-023-00626-x. MR4597638

[36] Dmitri Burago, *Hard balls gas and Alexandrov spaces of curvature bounded above*, Proceedings of the International Congress of Mathematicians, Vol. II (Berlin, 1998), Doc. Math. **Extra Vol. II** (1998), 289–298. MR1648079

[37] Dmitri Burago, Yuri Burago, and Sergei Ivanov, *A course in metric geometry*, Graduate Studies in Mathematics, vol. 33, American Mathematical Society, Providence, RI, 2001, DOI 10.1090/gsm/033. MR1835418

[38] D. Burago, S. Ferleger, and A. Kononenko, *A geometric approach to semi-dispersing billiards*, Ergodic Theory Dynam. Systems **18** (1998), no. 2, 303–319, DOI 10.1017/S014338579811564X. MR1619558

[39] D. Burago, S. Ferleger, and A. Kononenko, *Uniform estimates on the number of collisions in semi-dispersing billiards*, Ann. of Math. (2) **147** (1998), no. 3, 695–708, DOI 10.2307/120962. MR1637663

[40] D. Burago, S. Ferleger, and A. Kononenko, *Topological entropy of semi-dispersing billiards*, Ergodic Theory Dynam. Systems **18** (1998), no. 4, 791–805, DOI 10.1017/S0143385798108246. MR1645377

[41] D. Burago, S. Ferleger, and A. Kononenko, *Unfoldings and global bounds on the number of collisions for generalized semi-dispersing billiards*, Asian J. Math. **2** (1998), no. 1, 141–152, DOI 10.4310/AJM.1998.v2.n1.a4. MR1656555

[42] D. Burago, S. Ferleger, and A. Kononenko, *A geometric approach to semi-dispersing billiards*, Hard ball systems and the Lorentz gas, Encyclopaedia Math. Sci., vol. 101, Springer, Berlin, 2000, pp. 9–27, DOI 10.1007/978-3-662-04062-1_2. MR1805324

[43] D. Burago, S. Ferleger, B. Kleiner, and A. Kononenko, *Gluing copies of a 3-dimensional polyhedron to obtain a closed nonpositively curved pseudomanifold*, Proc. Amer. Math. Soc. **129** (2001), no. 5, 1493–1498, DOI 10.1090/S0002-9939-01-05554-X. MR1707510

[44] Yu. Burago, M. Gromov, and G. Perel′man, *A. D. Aleksandrov spaces with curvatures bounded below* (Russian, with Russian summary), Uspekhi Mat. Nauk **47** (1992), no. 2(284), 3–51, 222, DOI 10.1070/RM1992v047n02ABEH000877; English transl., Russian Math. Surveys **47** (1992), no. 2, 1–58. MR1185284

[45] Herbert Busemann, *Spaces with non-positive curvature*, Acta Math. **80** (1948), 259–310, DOI 10.1007/BF02393651. MR29531

[46] S. V. Bujalo, *Shortest paths on convex hypersurfaces of a Riemannian space* (Russian, with English summary), Zap. Naučn. Sem. Leningrad. Otdel. Mat. Inst. Steklov. (LOMI) **66** (1976), 114–132, 207. Studies in topology, II. MR643664

[47] É. Cartan, *Leçons sur la Géométrie des Espaces de Riemann* (French), Gauthier-Villars, Paris, 1946. 2d ed. MR20842

[48] Nicola Cavallucci, *The L^2-completion of the space of Riemannian metrics is* CAT(0)*: a shorter proof*, Proc. Amer. Math. Soc. **151** (2023), no. 7, 3183–3187, DOI 10.1090/proc/16360. MR4579388

[49] Ruth Charney and Michael Davis, *Singular metrics of nonpositive curvature on branched covers of Riemannian manifolds*, Amer. J. Math. **115** (1993), no. 5, 929–1009, DOI 10.2307/2375063. MR1246182

[50] Ruth M. Charney and Michael W. Davis, *Strict hyperbolization*, Topology **34** (1995), no. 2, 329–350, DOI 10.1016/0040-9383(94)00027-I. MR1318879

[51] Jeff Cheeger and Tobias H. Colding, *Lower bounds on Ricci curvature and the almost rigidity of warped products*, Ann. of Math. (2) **144** (1996), no. 1, 189–237, DOI 10.2307/2118589. MR1405949

[52] Jeff Cheeger, *Comparison and Finiteness Theorems for Riemannian Manifolds*, ProQuest LLC, Ann Arbor, MI, 1967. Thesis (Ph.D.)–Princeton University, `https://gateway.proquest.com/openurl?url_ver=Z39.88-2004&rft_val_fmt=info:ofi/fmt:kev:mtx:dissertation&res_dat=xri:pqdiss&rft_dat=xri:pqdiss:6802465`. MR2616706

[53] Jeff Cheeger and Detlef Gromoll, *The splitting theorem for manifolds of nonnegative Ricci curvature*, J. Differential Geometry **6** (1971/72), 119–128, `https://projecteuclid.org/euclid.jdg/1214430220`. MR303460

[54] Brian Clarke, *Geodesics, distance, and the* CAT(0) *property for the manifold of Riemannian metrics*, Math. Z. **273** (2013), no. 1-2, 55–93, DOI 10.1007/s00209-012-0996-x. MR3010152

[55] S. Cohn-Vossen. *Totalkrümmung und geodätische Linien auf einfachzusammenhängenden offenen vollständigen Flächenstücken*, German. *Mat. Sb.* **1** (1936), no. 2, 43.

[56] Aris Daniilidis, Olivier Ley, and Stéphane Sabourau, *Asymptotic behaviour of self-contracted planar curves and gradient orbits of convex functions* (English, with English and French summaries), J. Math. Pures Appl. (9) **94** (2010), no. 2, 183–199, DOI 10.1016/j.matpur.2010.03.007. MR2665417

[57] Tamás Darvas, *The Mabuchi completion of the space of Kähler potentials*, Amer. J. Math. **139** (2017), no. 5, 1275–1313, DOI 10.1353/ajm.2017.0032. MR3702499

[58] Michael W. Davis, *Exotic aspherical manifolds*, Topology of high-dimensional manifolds, No. 1, 2 (Trieste, 2001), ICTP Lect. Notes, vol. 9, Abdus Salam Int. Cent. Theoret. Phys., Trieste, 2002, pp. 371–404. MR1937019

[59] Michael W. Davis, *Groups generated by reflections and aspherical manifolds not covered by Euclidean space*, Ann. of Math. (2) **117** (1983), no. 2, 293–324, DOI 10.2307/2007079. MR690848

[60] Michael W. Davis, *Exotic aspherical manifolds*, Topology of high-dimensional manifolds, No. 1, 2 (Trieste, 2001), ICTP Lect. Notes, vol. 9, Abdus Salam Int. Cent. Theoret. Phys., Trieste, 2002, pp. 371–404. MR1937019

[61] Anna Dyubina and Iosif Polterovich, *Explicit constructions of universal* $\mathbb{R}$*-trees and asymptotic geometry of hyperbolic spaces*, Bull. London Math. Soc. **33** (2001), no. 6, 727–734, DOI 10.1112/S002460930100844X. MR1853785

[62] David A. Edwards, *The structure of superspace*, Studies in topology (Proc. Conf., Univ. North Carolina, Charlotte, N.C., 1974; dedicated to Math. Sect. Polish Acad. Sci.), Academic Press, New York-London, 1975, pp. 121–133. MR401069

[63] Ryszard Engelking, *Dimension theory*, North-Holland Mathematical Library, vol. 19, North-Holland Publishing Co., Amsterdam-Oxford-New York; PWN—Polish Scientific Publishers, Warsaw, 1978. Translated from the Polish and revised by the author. MR482697

[64] J.-H. Eschenburg, *The splitting theorem for space-times with strong energy condition*, J. Differential Geom. **27** (1988), no. 3, 477–491, `https://projecteuclid.org/euclid.jdg/1214442005`. MR940115

[65] Lucas C. F. Ferreira and Julio C. Valencia-Guevara, *Gradient flows of time-dependent functionals in metric spaces and applications to PDEs*, Monatsh. Math. **185** (2018), no. 2, 231–268, DOI 10.1007/s00605-017-1037-y. MR3748335

[66] Zdeněk Frolík, *Concerning topological convergence of sets* (English, with Russian summary), Czechoslovak Math. J. **10(85)** (1960), 168–180. MR116303

[67] Jian Ge and Nan Li, *Gluing of multiple Alexandrov spaces*, Adv. Math. **399** (2022), Paper No. 108248, 26, DOI 10.1016/j.aim.2022.108248. MR4383015

[68] Nicola Gigli, *An overview of the proof of the splitting theorem in spaces with non-negative Ricci curvature*, Anal. Geom. Metr. Spaces **2** (2014), no. 1, 169–213, DOI 10.2478/agms-2014-0006. MR3210895

[69] Nicola Gigli, *The splitting theorem in non-smooth context*, Preprint, `arXiv:1302.5555`, 2013.

[70] R. E. Greene and H. Wu, *On the subharmonicity and plurisubharmonicity of geodesically convex functions*, Indiana Univ. Math. J. **22** (1972/73), 641–653, DOI 10.1512/iumj.1973.22.22052. MR422686

[71] Misha Gromov, *Metric structures for Riemannian and non-Riemannian spaces*, Reprint of the 2001 English edition, Modern Birkhäuser Classics, Birkhäuser Boston, Inc., Boston, MA, 2007. Based on the 1981 French original; With appendices by M. Katz, P. Pansu and S. Semmes; Translated from the French by Sean Michael Bates. MR2307192

[72] Mikhael Gromov, *Groups of polynomial growth and expanding maps*, Inst. Hautes Études Sci. Publ. Math. **53** (1981), 53–73, `http://www.numdam.org/item?id=PMIHES_1981__53__53_0`. MR623534

[73] M. Gromov, *Almost flat manifolds*, J. Differential Geometry **13** (1978), no. 2, 231–241, `https://projecteuclid.org/euclid.jdg/1214434488`. MR540942

[74] M. Gromov, *Asymptotic invariants of infinite groups*, Geometric group theory, Vol. 2 (Sussex, 1991), London Math. Soc. Lecture Note Ser., vol. 182, Cambridge Univ. Press, Cambridge, 1993, pp. 1–295. MR1253544

[75] Michael Gromov, *Curvature, diameter and Betti numbers*, Comment. Math. Helv. **56** (1981), no. 2, 179–195, DOI 10.1007/BF02566208. MR630949

[76] M. Gromov, *Hyperbolic groups*, Essays in group theory, Math. Sci. Res. Inst. Publ., vol. 8, Springer, New York, 1987, pp. 75–263, DOI 10.1007/978-1-4613-9586-7_3. MR919829

[77] Karsten Grove and Peter Petersen V, *Bounding homotopy types by geometry*, Ann. of Math. (2) **128** (1988), no. 1, 195–206, DOI 10.2307/1971439. MR951512

[78] Karsten Grove and Burkhard Wilking, *A knot characterization and 1-connected nonnegatively curved 4-manifolds with circle symmetry*, Geom. Topol. **18** (2014), no. 5, 3091–3110, DOI 10.2140/gt.2014.18.3091. MR3285230

[79] J. Hadamard, *Sur la forme des lignes géodésiques à l'infini et sur les géodésiques des surfaces réglées du second ordre* (French), Bull. Soc. Math. France **26** (1898), 195–216, `https://www.numdam.org/item?id=BSMF_1898__26__195_0`. MR1504321

[80] Stephanie Halbeisen, *On tangent cones of Alexandrov spaces with curvature bounded below*, Manuscripta Math. **103** (2000), no. 2, 169–182, DOI 10.1007/s002290070018. MR1796313

[81] Felix Hausdorff, *Grundzüge der Mengenlehre* (German), Chelsea Publishing Co., New York, 1949. MR31025

[82] Juha Heinonen, Pekka Koskela, Nageswari Shanmugalingam, and Jeremy T. Tyson, *Sobolev spaces on metric measure spaces: An approach based on upper gradients*, New Mathematical Monographs, vol. 27, Cambridge University Press, Cambridge, 2015, DOI 10.1017/CBO9781316135914. MR3363168

[83] Wu-Yi Hsiang and Bruce Kleiner, *On the topology of positively curved 4-manifolds with symmetry*, J. Differential Geom. **29** (1989), no. 3, 615–621, `https://projecteuclid.org/euclid.jdg/1214443064`. MR992332

[84] Thakyin Hu and W. A. Kirk, *Local contractions in metric spaces*, Proc. Amer. Math. Soc. **68** (1978), no. 1, 121–124, DOI 10.2307/2040922. MR464180

[85] Witold Hurewicz and Henry Wallman, *Dimension Theory*, Princeton Mathematical Series, vol. 4, Princeton University Press, Princeton, NJ, 1941. MR6493

[86] J. R. Isbell, *Six theorems about injective metric spaces*, Comment. Math. Helv. **39** (1964), 65–76, DOI 10.1007/BF02566944. MR182949

[87] S. Ivanov, *On Helly's theorem in geodesic spaces*, Electron. Res. Announc. Math. Sci. **21** (2014), 109–112, DOI 10.3934/era.2014.21.109. MR3260133

[88] S. Ivanov, *To what extent is convexity a local property?*, MathOverflow (June 9, 2012), `https://mathoverflow.net/q/22091`.

[89] Jürgen Jost, *Nonlinear Dirichlet forms*, New directions in Dirichlet forms, AMS/IP Stud. Adv. Math., vol. 8, Amer. Math. Soc., Providence, RI, 1998, pp. 1–47, DOI 10.1090/amsip/008/01. MR1652278

[90] Jürgen Jost, *Nonpositive curvature: geometric and analytic aspects*, Lectures in Mathematics ETH Zürich, Birkhäuser Verlag, Basel, 1997, DOI 10.1007/978-3-0348-8918-6. MR1451625

[91] C. Jun, *Pursuit-evasion and time-dependent gradient flow in singular spaces*, Ph.D. thesis, University of Illinois at Urbana-Champaign, 2012, `http://gateway.proquest.com/openurl?url_ver=Z39.88-2004&rft_val_fmt=info:ofi/fmt:kev:mtx:dissertation&res_dat=xri:pqm&rft_dat=xri:pqdiss:3570501`.

[92] C. Jun, *Time-dependent gradient curves on CAT(0) spaces*, Preprint, `arXiv:1801.04583`, 2018.

[93] M. B. Karmanova, *Metricheskaya teorema Rademakhera i formula ploshchadi dlya lipshitsevykh otobrazheniĭ so znacheniyami v metricheskom prostranstve*, Russian. Vestn. NGU. Ser. matem., mekh., inform. **6** (2006), no. 4, 50–69.

[94] M. Katětov, *On universal metric spaces*, General topology and its relations to modern analysis and algebra, VI (Prague, 1986), Res. Exp. Math., vol. 16, Heldermann, Berlin, 1988, pp. 323–330. MR952617

[95] Bernd Kirchheim, *Rectifiable metric spaces: local structure and regularity of the Hausdorff measure*, Proc. Amer. Math. Soc. **121** (1994), no. 1, 113–123, DOI 10.2307/2160371. MR1189747

[96] Bruce Kleiner, *The local structure of length spaces with curvature bounded above*, Math. Z. **231** (1999), no. 3, 409–456, DOI 10.1007/PL00004738. MR1704987

[97] Bruce Kleiner and Bernhard Leeb, *Rigidity of quasi-isometries for symmetric spaces and Euclidean buildings*, Inst. Hautes Études Sci. Publ. Math. **86** (1997), 115–197 (1998), `https://www.numdam.org/item?id=PMIHES_1997__86__115_0`. MR1608566

[98] N. N. Kosovskiĭ, *Gluing of Riemannian manifolds of curvature* $\geq \kappa$ (Russian, with Russian summary), Algebra i Analiz **14** (2002), no. 3, 140–157; English transl., St. Petersburg Math. J. **14** (2003), no. 3, 467–478. MR1921991

[99] U. Lang and V. Schroeder, *Kirszbraun's theorem and metric spaces of bounded curvature*, Geom. Funct. Anal. **7** (1997), no. 3, 535–560, DOI 10.1007/s000390050018. MR1466337

[100] Urs Lang and Viktor Schroeder, *On Toponogov's comparison theorem for Alexandrov spaces*, Enseign. Math. **59** (2013), no. 3-4, 325–336, DOI 10.4171/LEM/59-3-6. MR3189039

[101] Nina Lebedeva and Anton Petrunin, *Curvature bounded below: a definition a la Berg-Nikolaev*, Electron. Res. Announc. Math. Sci. **17** (2010), 122–124, DOI 10.3934/era.2010.17.122. MR2735031

[102] Nina Lebedeva and Anton Petrunin, *Local characterization of polyhedral spaces*, Geom. Dedicata **179** (2015), 161–168, DOI 10.1007/s10711-015-0072-x. MR3424662

[103] Nina Lebedeva, Anton Petrunin, and Vladimir Zolotov, *Bipolar comparison*, Geom. Funct. Anal. **29** (2019), no. 1, 258–282, DOI 10.1007/s00039-019-00481-9. MR3925110

[104] N. Lebedeva and A. Petrunin, *5-point Toponogov theorem*, Preprint, `arXiv:2202.13049`, 2022.

[105] Nina Lebedeva and Anton Petrunin, *5-point* CAT(0) *spaces after Tetsu Toyoda*, Anal. Geom. Metr. Spaces **9** (2021), no. 1, 160–166, DOI 10.1515/agms-2020-0126. MR4298102

[106] J. Liberman, *Geodesic lines on convex surfaces*, C. R. (Doklady) Acad. Sci. URSS (N.S.) **32** (1941), 310–313. MR0010994

[107] Alexander Lytchak and Anton Petrunin, *Short retractions of CAT(1) spaces*, Proc. Amer. Math. Soc. **149** (2021), no. 3, 1247–1257, DOI 10.1090/proc/15268. MR4211878

[108] Alexander Lytchak and Stephan Stadler, *Conformal deformations of* CAT(0) *spaces*, Math. Ann. **373** (2019), no. 1-2, 155–163, DOI 10.1007/s00208-018-1703-y. MR3968869

[109] A. Lytchak and S. Stadler, *Curvature bounds of subsets in dimension two*, `https://arxiv.org/abs/2105.00726`, 2021.

[110] A. Lytchak, *Differentiation in metric spaces*, Algebra i Analiz **16** (2004), no. 6, 128–161, DOI 10.1090/S1061-0022-05-00888-5; English transl., St. Petersburg Math. J. **16** (2005), no. 6, 1017–1041. MR2117451

[111] A. Lytchak, *Open map theorem for metric spaces*, Algebra i Analiz **17** (2005), no. 3, 139–159, DOI 10.1090/S1061-0022-06-00916-2; English transl., St. Petersburg Math. J. **17** (2006), no. 3, 477–491. MR2167848

[112] Alexander Lytchak, *On the geometry of subsets of positive reach*, Manuscripta Math. **115** (2004), no. 2, 199–205, DOI 10.1007/s00229-004-0491-8. MR2098470

[113] Hans von Mangoldt, *Ueber diejenigen Punkte auf positiv gekrümmten Flächen, welche die Eigenschaft haben, dass die von ihnen ausgehenden geodätischen Linien nie aufhören, kürzeste Linien zu sein* (German), J. Reine Angew. Math. **91** (1881), 23–53, DOI 10.1515/crll.1881.91.23. MR1579879

[114] G. A. Margulis and G. D. Mostow, *Some remarks on the definition of tangent cones in a Carnot-Carathéodory space*, J. Anal. Math. **80** (2000), 299–317, DOI 10.1007/BF02791539. MR1771529

[115] Uwe F. Mayer, *Gradient flows on nonpositively curved metric spaces and harmonic maps*, Comm. Anal. Geom. **6** (1998), no. 2, 199–253, DOI 10.4310/CAG.1998.v6.n2.a1. MR1651416

[116] Karl Menger, *Untersuchungen über allgemeine Metrik* (German), Math. Ann. **100** (1928), no. 1, 75–163, DOI 10.1007/BF01448840. MR1512479

[117] Alexander Mielke, Riccarda Rossi, and Giuseppe Savaré, *Nonsmooth analysis of doubly nonlinear evolution equations*, Calc. Var. Partial Differential Equations **46** (2013), no. 1-2, 253–310, DOI 10.1007/s00526-011-0482-z. MR3016509

[118] A. D. Milka, *Metric structure of a certain class of spaces that contain straight lines* (Russian), Ukrain. Geometr. Sb. **4** (1967), 43–48. MR0256327

[119] Nicolas Monod, *Superrigidity for irreducible lattices and geometric splitting*, J. Amer. Math. Soc. **19** (2006), no. 4, 781–814, DOI 10.1090/S0894-0347-06-00525-X. MR2219304

[120] Jun-iti Nagata, *Modern dimension theory*, Revised edition, Sigma Series in Pure Mathematics, vol. 2, Heldermann Verlag, Berlin, 1983. MR715431

[121] F. Nazarov, *Intrinsic metric with no geodesics* (February 18, 2010), `https://mathoverflow.net/q/15720`.

[122] Shin-ichi Ohta, *Gradient flows on Wasserstein spaces over compact Alexandrov spaces*, Amer. J. Math. **131** (2009), no. 2, 475–516, DOI 10.1353/ajm.0.0048. MR2503990

[123] D. Panov and A. Petrunin, *Ramification conjecture and Hirzebruch's property of line arrangements*, Compos. Math. **152** (2016), no. 12, 2443–2460, DOI 10.1112/S0010437X16007648. MR3594282

[124] Athanase Papadopoulos, *Metric spaces, convexity and non-positive curvature*, 2nd ed., IRMA Lectures in Mathematics and Theoretical Physics, vol. 6, European Mathematical Society (EMS), Zürich, 2014, DOI 10.4171/132. MR3156529

[125] G. Perelman and A. Petrunin, *Quasigeodesics and Gradient curves in Alexandrov spaces*, `https://www.math.psu.edu/petrunin/`.

[126] G. Ya. Perel′man, *Elements of Morse theory on Aleksandrov spaces* (Russian, with Russian summary), Algebra i Analiz **5** (1993), no. 1, 232–241; English transl., St. Petersburg Math. J. **5** (1994), no. 1, 205–213. MR1220498

[127] G. Perelman. *Alexandrov spaces with curvature bounded from below II.* Preprint LOMI, 1991. 35pp. `http://www.math.psu.edu/petrunin/`.

[128] Anton Petrunin and Stephan Stadler, *Metric-minimizing surfaces revisited*, Geom. Topol. **23** (2019), no. 6, 3111–3139, DOI 10.2140/gt.2019.23.3111. MR4039186

[129] Anton Petrunin, *A globalization for non-complete but geodesic spaces*, Math. Ann. **366** (2016), no. 1-2, 387–393, DOI 10.1007/s00208-015-1295-8. MR3552243

[130] Anton Petrunin, *Alexandrov meets Lott-Villani-Sturm*, Münster J. Math. **4** (2011), 53–64. MR2869253

[131] Anton Petrunin, *Quasigeodesics in multidimensional Alexandrov spaces*, ProQuest LLC, Ann Arbor, MI, 1995. Thesis (Ph.D.)–University of Illinois at Urbana-Champaign, `https://gateway.proquest.com/openurl?url_ver=Z39.88-2004&rft_val_fmt=info:ofi/fmt:kev:mtx:dissertation&res_dat=xri:pqdiss&rft_dat=xri:pqdiss:9543694`. MR2693118

[132] Anton Petrunin, *Semiconcave functions in Alexandrov's geometry*, Surveys in differential geometry. Vol. XI, Surv. Differ. Geom., vol. 11, Int. Press, Somerville, MA, 2007, pp. 137–201, DOI 10.4310/SDG.2006.v11.n1.a6. MR2408266

[133] Anton Petrunin, *Applications of quasigeodesics and gradient curves*, Comparison geometry (Berkeley, CA, 1993), Math. Sci. Res. Inst. Publ., vol. 30, Cambridge Univ. Press, Cambridge, 1997, pp. 203–219. MR1452875

[134] A. Petrunin, *Pure metric geometry: introductory lectures*, Preprint, `arXiv:2007.09846`, 2020.

[135] Robert R. Phelps, *Convex functions, monotone operators and differentiability*, 2nd ed., Lecture Notes in Mathematics, vol. 1364, Springer-Verlag, Berlin, 1993. MR1238715

[136] P. Pizzetti, *Paragone fra due triangoli a lati uguali*. Atti della Reale Accademia dei Lincei, Rendiconti (5). Classe di Scienze Fisiche, Matematiche e Naturali, **16** (1907), no. 1, 6–11.

[137] C. Plaut, *Toponogov's theorem for metric spaces II*, MPIM Preprints (1990), `https://archive.mpim-bonn.mpg.de/id/eprint/4029`.

[138] Conrad Plaut, *Spaces of Wald curvature bounded below*, J. Geom. Anal. **6** (1996), no. 1, 113–134, DOI 10.1007/BF02921569. MR1402389

[139] Conrad Plaut, *Metric spaces of curvature $\geq k$*, Handbook of geometric topology, North-Holland, Amsterdam, 2002, pp. 819–898. MR1886682

[140] L. Pontrjagin, *Sur une hypothèse fondamentale de la théorie de la dimension*, French. C. R. Acad. Sci., Paris, **190** (1930), 1105–1107.

[141] Yu. G. Reshetnyak, *Two-dimensional manifolds of bounded curvature* (Russian), Geometry, 4 (Russian), Itogi Nauki i Tekhniki, Akad. Nauk SSSR, Vsesoyuz. Inst. Nauchn. i Tekhn. Inform., Moscow, 1989, pp. 7–189, 273–277, 279. MR1099202

[142] Ju. G. Rešetnjak, *Non-expansive maps in a space of curvature no greater than K* (Russian), Sibirsk. Mat. Ž. **9** (1968), 918–927. MR244922

[143] Willi Rinow, *Die innere Geometrie der metrischen Räume*, Die Grundlehren der mathematischen Wissenschaften, Band 105, Springer-Verlag, Berlin-Göttingen-Heidelberg, 1961. MR123969

[144] Walter Rudin, *Homogeneity problems in the theory of Čech compactifications*, Duke Math. J. **23** (1956), 409–419, `https://projecteuclid.org/euclid.dmj/1077466953`. MR80902

[145] I. Kh. Sabitov, *Around the proof of the Legendre-Cauchy lemma on convex polygons* (Russian, with Russian summary), Sibirsk. Mat. Zh. **45** (2004), no. 4, 892–919, DOI 10.1023/B:SIMJ.0000035837.80962.0a; English transl., Siberian Math. J. **45** (2004), no. 4, 740–762. MR2091654

[146] Takashi Sato, *An alternative proof of Berg and Nikolaev's characterization of* CAT(0)*-spaces via quadrilateral inequality*, Arch. Math. (Basel) **93** (2009), no. 5, 487–490, DOI 10.1007/s00013-009-0057-9. MR2563595

[147] Thomas Foertsch and Viktor Schroeder, *Minkowski versus Euclidean rank for products of metric spaces*, Adv. Geom. **2** (2002), no. 2, 123–131, DOI 10.1515/advg.2002.002. MR1895343

[148] Giuseppe Savaré, *Gradient flows and diffusion semigroups in metric spaces under lower curvature bounds* (English, with English and French summaries), C. R. Math. Acad. Sci. Paris **345** (2007), no. 3, 151–154, DOI 10.1016/j.crma.2007.06.018. MR2344814

[149] V. A. Šarafutdinov, *The Pogorelov-Klingenberg theorem for manifolds that are homeomorphic to* $\mathbf{R}^n$ (Russian), Sibirsk. Mat. Ž. **18** (1977), no. 4, 915–925, 958. MR0487896

[150] Katsuhiro Shiohama, *An introduction to the geometry of Alexandrov spaces*, Lecture Notes Series, vol. 8, Seoul National University, Research Institute of Mathematics, Global Analysis Research Center, Seoul, 1993. MR1320267

[151] R. M. Solovay, *Maps preserving measures*, `https://math.berkeley.edu/~solovay/Preprints/Rudin_Keisler.pdf`, 2011.

[152] Stephan Stadler, *The structure of minimal surfaces in CAT(0) spaces*, J. Eur. Math. Soc. (JEMS) **23** (2021), no. 11, 3521–3554, DOI 10.4171/jems/1075. MR4310811

[153] K. T. Sturm, *Metric spaces of lower bounded curvature*, Exposition. Math. **17** (1999), no. 1, 35–47. MR1687468

[154] K.-T. Sturm, *The space of spaces: curvature bounds and gradient flows on the space of metric measure spaces*, Preprint, `arXiv:1208.0434`, 2012.

[155] Karl-Theodor Sturm, *Probability measures on metric spaces of nonpositive curvature*, Heat kernels and analysis on manifolds, graphs, and metric spaces (Paris, 2002), Contemp. Math., vol. 338, Amer. Math. Soc., Providence, RI, 2003, pp. 357–390, DOI 10.1090/conm/338/06080. MR2039961

[156] V. A. Toponogov, *Riemann spaces with curvature bounded below* (Russian), Uspehi Mat. Nauk **14** (1959), no. 1(85), 87–130. MR0103510

[157] Tetsu Toyoda, *An intrinsic characterization of five points in a* CAT(0) *space*, Anal. Geom. Metr. Spaces **8** (2020), no. 1, 114–165, DOI 10.1515/agms-2020-0111. MR4141370

[158] Cédric Villani, *Optimal transport: Old and new*, Grundlehren der mathematischen Wissenschaften [Fundamental Principles of Mathematical Sciences], vol. 338, Springer-Verlag, Berlin, 2009, DOI 10.1007/978-3-540-71050-9. MR2459454

[159] A. Wald, *Begründung eiiner Koordinatenlosen Differentialgeometrie der Flächen*, German. Ergebnisse eines mathematischen Kolloquium, **6** (1935), 24–46.

[160] R. A. Wijsman, *Convergence of sequences of convex sets, cones and functions. II*, Trans. Amer. Math. Soc. **123** (1966), 32–45, DOI 10.2307/1994611. MR196599

[161] Sumio Yamada, *On the geometry of Weil-Petersson completion of Teichmüller spaces*, Math. Res. Lett. **11** (2004), no. 2-3, 327–344, DOI 10.4310/MRL.2004.v11.n3.a5. MR2067477

[162] Takao Yamaguchi, *Collapsing and pinching under a lower curvature bound*, Ann. of Math. (2) **133** (1991), no. 2, 317–357, DOI 10.2307/2944340. MR1097241

Index

Selected Published Titles in This Series

242 **Jayadev S. Athreya and Howard Masur,** Translation Surfaces, 2024
238 **Julio González-Díaz, Ignacio García-Jurado, and M. Gloria Fiestras-Janeiro,** An Introductory Course on Mathematical Game Theory and Applications, Second Edition, 2023
237 **Michael Levitin, Dan Mangoubi, and Iosif Polterovich,** Topics in Spectral Geometry, 2023
236 **Stephanie Alexander, Vitali Kapovitch, and Anton Petrunin,** Alexandrov Geometry, 2024
235 **Bennett Chow,** Ricci Solitons in Low Dimensions, 2023
234 **Andrea Ferretti,** Homological Methods in Commutative Algebra, 2023
233 **Andrea Ferretti,** Commutative Algebra, 2023
232 **Harry Dym,** Linear Algebra in Action, Third Edition, 2023
231 **Luís Barreira and Yakov Pesin,** Introduction to Smooth Ergodic Theory, Second Edition, 2023
230 **Barbara Kaltenbacher and William Rundell,** Inverse Problems for Fractional Partial Differential Equations, 2023
229 **Giovanni Leoni,** A First Course in Fractional Sobolev Spaces, 2023
228 **Henk Bruin,** Topological and Ergodic Theory of Symbolic Dynamics, 2022
227 **William M. Goldman,** Geometric Structures on Manifolds, 2022
226 **Milivoje Lukić,** A First Course in Spectral Theory, 2022
225 **Jacob Bedrossian and Vlad Vicol,** The Mathematical Analysis of the Incompressible Euler and Navier-Stokes Equations, 2022
224 **Ben Krause,** Discrete Analogues in Harmonic Analysis, 2022
223 **Volodymyr Nekrashevych,** Groups and Topological Dynamics, 2022
222 **Michael Artin,** Algebraic Geometry, 2022
221 **David Damanik and Jake Fillman,** One-Dimensional Ergodic Schrödinger Operators, 2022
220 **Isaac Goldbring,** Ultrafilters Throughout Mathematics, 2022
219 **Michael Joswig,** Essentials of Tropical Combinatorics, 2021
218 **Riccardo Benedetti,** Lectures on Differential Topology, 2021
217 **Marius Crainic, Rui Loja Fernandes, and Ioan Mărcuţ,** Lectures on Poisson Geometry, 2021
216 **Brian Osserman,** A Concise Introduction to Algebraic Varieties, 2021
215 **Tai-Ping Liu,** Shock Waves, 2021
214 **Ioannis Karatzas and Constantinos Kardaras,** Portfolio Theory and Arbitrage, 2021
213 **Hung Vinh Tran,** Hamilton–Jacobi Equations, 2021
212 **Marcelo Viana and José M. Espinar,** Differential Equations, 2021
211 **Mateusz Michałek and Bernd Sturmfels,** Invitation to Nonlinear Algebra, 2021
210 **Bruce E. Sagan,** Combinatorics: The Art of Counting, 2020
209 **Jessica S. Purcell,** Hyperbolic Knot Theory, 2020
208 **Vicente Muñoz, Ángel González-Prieto, and Juan Ángel Rojo,** Geometry and Topology of Manifolds, 2020
207 **Dmitry N. Kozlov,** Organized Collapse: An Introduction to Discrete Morse Theory, 2020
206 **Ben Andrews, Bennett Chow, Christine Guenther, and Mat Langford,** Extrinsic Geometric Flows, 2020
205 **Mikhail Shubin,** Invitation to Partial Differential Equations, 2020

For a complete list of titles in this series, visit the
AMS Bookstore at **www.ams.org/bookstore/gsmseries/**.